Direct Characterization of Fineparticles

CHEMICAL ANALYSIS

A SERIES OF MONOGRAPHS ON ANALYTICAL CHEMISTRY AND ITS APPLICATIONS

VOLUME 61

A WILEY-INTERSCIENCE PUBLICATION

JOHN WILEY & SONS

New York / Chichester / Brisbane / Toronto

Direct Characterization of Fineparticles

BRIAN H. KAYE

Institute for Fineparticles Research
Laurentian University
Sudbury, Ontario, Canada

A WILEY-INTERSCIENCE PUBLICATION

JOHN WILEY & SONS

New York / Chichester / Brisbane / Toronto

Library of Congress Cataloging in Publication Data:

Kaye, Brian H. (Brian Howard), 1932–
Direct characterization of fineparticles.

(Chemical analysis, ISSN 0069-2883; v. 61)
"A Wiley-Interscience publication."
Includes index.
1. Particle size determination. 2. Aerosols—Analysis. I. Title. II. Series.

TA418.8.K39 620′.43 81-1734
ISBN 0-471-46150-4 AACR2

Printed in the United States of America

10 9 8 7 6 5 4 3 2 1

Dedicated to the memory of my mother
Jessie Harrison Kaye

PREFACE

The writer started his first draft of a chapter for a book on fineparticle characterization 17 years ago. The pressure of work and the rapid development of new technology always seemed to overtake the drafts that were already written. Over that 17-year period the subject of fineparticle science has expanded rapidly, and when a sabbatical year made it possible to get down to a serious attempt to complete the project, it was decided that the original project should be split into three different books. First of all, in writing a description of the methodology for characterizing fineparticles, there was the constant question of how much detail to give for laboratory methods that the students or the beginning practitioners of fineparticle characterization in industry would carry out for themselves. It was decided that the inclusion of experimental detail in the main text of the book would make it inordinately long; therefore, it was decided to supplement textbook material with a laboratory handbook that would also contain problems designed to give the student familiarity with the data manipulations carried out in fineparticle characterization. Further, the subject matter dealing with characterization technology has grown so rapidly that it was decided to split the textbook into two volumes entitled respectively *Direct Characterization of Fineparticles* and *Indirect Characterization of Fineparticles*. It is hoped that the latter volume will be written in the not-too-distant future.

When describing work carried out by the writer, if the work has been published in referred scientific journals, the work is referred to as "Kaye has shown." On the other hand, if the writer is giving an opinion that has not been subjected to peer review, the statement "in the writer's opinion . . ." appears.

Over the years, the writer has been fortunate in having worked with many of the people who have made significant contributions to the subject of fineparticle characterization. His own knowledge of the subject has been developed through conversations and/or cooperative projects with M. J. Groves, Reg Davies, Meryl Jackson, Terry Allen, Brian Scarlett, and Dick Karuhn and in recent years with colleagues at Laurentian University: Dolph Rubin, Florian Robillard, S.J., Tom Patitsas, Ian Robb, and Tony Naylor. In particular, Ian Robb and Tony Naylor have read many parts of the manuscript and given helpful comments. Thanks are due to all of these people and to Miss Cynthia Taylor, Mrs. June Talbot, Miss Judy Mittermayer, and Mrs. B. Degroot, who typed the manuscript.

In the early years of work in the field of fineparticle characterization, it was a pleasure to meet personally such pioneers as Lincoln T. Work of the United States, Dr. Hans Rumpf of Germany, and Professor Harold Heywood of Great Britain—all of whom encouraged the writer to continue working in such a complex field as fineparticles. It has also been a pleasure to meet with such active workers as Dr. Henry Hausner, Dr. Milton Kerker, Dr. Walter McCrone, Dr. Clyde Orr, Dr. Roger Loveland, and Dr. Søren Berg of Denmark.

I hope this book will be useful to students and industrialists.

BRIAN H. KAYE

Sudbury, Ontario, Canada
June 1981

ACKNOWLEDGMENTS

We wish to thank the following people for permission to publish the material in various figures:

Joy Manufacturing Company	2.12
A. Lieberman, Royco Instruments	3.6
C. J. Stairmand	5.8
	5.10
	7.2
	7.4
	7.5
V. Timbrell	7.8
Dr. L. M. Fingerson	7.9
Dr. Werner Stöber	7.17
	7.19
Cassella Co.	7.15
	7.16
Warman Proprietary	7.24
Coulter Electronics, Great Britain	8.3
R. W. DeBlois	8.5
W. J. Yanta	8.13
Graticules Limited, Tonbridge, England	9.3
E. T. White	10.1
	10.2
	10.3
R. Davies	10.8
A. I. Medalia	10.10
Keith Beddow	10.11
R. Ehrlich	10.16
C. J. McCarthy	10.24

B.H.K.

CONTENTS

GLOSSARY OF NEW TERMS

A major research project proceeding at the Institute for Fineparticles Research, Laurentian University, is concerned with the design of an efficient computer-based organization and retrieval system for the literature of fineparticle science and technology. In the short term, the project aims to set up a keyword retrieval system of titles and abstracts. In the longer term, a more ambitious project aims to produce automatic translation of titles and abstracts by computer word processing technology. As a result of the perceived needs of automatic literature processing algorithms, the writer has evolved a specialist terminology that could greatly speed literature dissemination and retrieval in fineparticle science and technology. This terminology is introduced and used in the text of this book. The salient features of this new vocabulary are summarized in this glossary.

Centridisk. Several characterization procedures depend on the sedimentation of fineparticles in a viscous fluid under centrifugal force. These instruments have been referred to either by proprietary tradenames or by the awkward phrase "centrifugal disk sedimentometer." The term "centridisk" has been coined as a keyword and as a generic term to describe sedimentation procedures in which the fineparticles are studied in a disk centrifuge. A term denoting how the movement of the fineparticles being studied can then be added to describe, for example, a photocentridisk as being a centrifugal disk sedimentometer in which the movement of the fineparticles is monitored by using a light beam.

Centrifugal Disk Sedimentometer. See "Centridisk."

Climet Counter. The Climet counter is a photozone counter in which the size of the fineparticle is deduced by the light scattered by the fineparticle out of a beam of light.

Coulter Counter. This is a proprietary form of resistazone counter.

Direct Methods of Fineparticle Characterization. This classification of methodology for characterizing fineparticles includes methods in which the measurements are made directly on the individual fineparticles. Thus sedimentation studies are direct methods for characterizing fineparticles whereas permeability methods in which the surface of the powder is deduced from the fluid drag experienced by a fluid moving through a compact of the powder is an indirect method for characterizing fineparticles.

Eriometry. Thomas Young, who established the wave nature of light by his pioneer experiments, originally studied the fineness of wool by looking at the diffraction patterns created by a random array of fibres. From the Greek word for wool, *erios*, he coined the term "eriometry" to describe the measurement of the fineness of a system from its diffraction pattern. This term is still widely used in the fiber industry. Recently the meaning of the term has been expanded to describe procedures in which the fineparticles are characterized by studying the diffraction pattern created by a random array of those fineparticles. Since the size of the fineparticles has to be deduced in a complicated manner from the diffraction pattern, eriometry methods are classified as indirect methods of fineparticle characterization.

Fineparticle. This is a preferred term used to describe all finely divided substances. It is used in preference to "particle," which is too well established in the field of atomic physics to be useful as a term in powder science information retrieval systems.

HIAC Counter. This is a proprietary photozone counter in which the size of the fineparticle is deduced from the energy removed by the fineparticle from a forward-traveling white-light beam.

Iconometrics. This is the science of abstracting measurements from images using computers.

Indirect Methods of Fineparticle Characterization. See "Direct Methods of Fineparticle Characterization."

Joyce–Loebel Centrifuge. See "Centridisk."

Magiscan. This is an automated iconometric system marketed by the Joyce–Loebel Corporation. See "Iconometrics."

Micromesh Sieves. This term was introduced to describe new, very small aperture sieves, smaller than those created by weaving with wire. A better term for such sieves is "electroformed sieves" or "photoetched sieves," depending on the appropriate manufacturing technique.

Morphology. This term is used to describe any shape characteristic of a fineparticle. In general, it is used to distinguish between the "topology" of a fineparticle (the basic structure of the fineparticle with respect to included holes) and the "topography" of a fineparticle. Topography can best be appreciated by considering an outline of a profile as if it were an island. The basic "in's and out's" of the island corresponding to estuaries and bays would be topographical features. The "texture" of the fineparticle is the detailed structure of elements of the surface of the profile.

Omnicon. This is a tradename for a computerized system for evaluating fineparticle images. It is manufactured and distributed by the Bausch and Lomb Corporation. The name derives from two Greek words meaning "all images," based on the idea that it is a universal analytical system. See "Iconometrics."

Particle. This term is often used to describe a fineparticle, but its use in this context is very confusing to physicists to whom a particle is a structural component of the atom. The preferred term is "fineparticle."

Particle Size Analysis. This is the term previously used to describe general procedures for measuring the size of fineparticles. However, the growing interest in industrial hygiene and air pollution where a fineparticle is subjected to chemical analysis makes it desirable to reserve the term "fineparticle analysis" for studies of the chemical nature of a fineparticle and to use the term "fineparticle characterization" to describe methods for studying the shape and the size of a fineparticle.

Particulate. There has been growing use in the published literature of this term to describe fineparticles. However, indiscriminate use of this term as a noun and as an adjective can be confusing to people whose native language is not English, and the term deserves to be allowed to sink into oblivion.

Photosed. This is a suggested keyword for the description of all photosedimentometers.

Photosedimentometers. This term refers to sedimentation procedures in which the movement of fineparticles is monitored with a light beam. See "photosed."

Photozone Counters. These are stream counters in which the size of a fineparticle is deduced from the changes it causes in the light signal leaving the interrogation zone.

Quantimet. This is a proprietary automated iconometric system manufactured and distributed by the Cambridge Instrument Company. See "Iconometrics."

Resistazone Counter. This is the generic name used to describe stream counters in which the interrogation zone is monitored for changes in electrical resistance due to the presence of a fineparticle.

Royco Stream Counter. The Royco Corporation manufactures several different types of stream counter. Those that work with fineparticles in the gas stream are referred to as aerosol stream counters. They are basically photozone counters, but different models use different optical geometry in assessing the optical signal generated by the fineparticle. See "Photozone."

Sieve Analysis. This term was previously used to describe the characterization of powders with sieves. Unfortunately, however, the name would indicate that the sieve and not the powder is being analyzed. This term has been dropped in favor of the phrase "sieve fractionation studies."

Stream Counter. In stream counters the fineparticles to be characterized are presented to an interrogation zone as a one-dimensional stream. The size of the fineparticle is deduced from the physical changes that it causes in the interrogation zone. The different types of stream counter are named for the physical signal used to monitor the properties of the zone. See "Resistazone," "Photozone," and "Sonizone."

Texture of a Fineparticle. See "Morphology."

Topography of a Fineparticle. See "Morphology."

Topological Structure of a Fineparticle. See "Morphology."

CHAPTER

1

THE BASIC CONCEPTS INVOLVED IN THE DIRECT CHARACTERIZATION OF FINEPARTICLES

1.1 INTRODUCTION

Some examples of important fineparticle systems are cosmetic powders, powdered drugs, smokes, mists, paints, cement, powdered fuel for rocket systems, dusts, food powders, and powders used in the metallurgical and chemical engineering industries [1]. When describing the behavior of fineparticle systems, a knowledge of the dimensions and surface characteristics of individual fineparticles is essential. The vocabulary used to describe fineparticles has evolved concurrently in the many industries. As a result of the way in which it developed, the terminology employed lacks consistency. In an interdisciplinary subject like fineparticle science, the development of a systematic terminology is important not only for the benefit of students, but also for the efficient organization of computer-based automated information retrieval systems. As a result of research into the design of an information system for fineparticle science, the writer would like to see the adoption of a new systematic terminology that would facilitate information retrieval [2]. In this suggested terminology, the term "fineparticle" is used in preference to "particulate" or "particle" because any use of "particle" as a keyword in an automated retrieval system results in a flood of information from atomic physics publications on anything from quarks to neutrons.

Classical physics is concerned with macroscopic concepts and forces such as weight, elasticity, viscosity, and inertia. Colloidal science is concerned with systems in which surface forces are the dominant physical phenomena. Fineparticle science is concerned with systems intermediate between colloidal systems and the systems of classical physics. Sharp boundaries cannot be drawn between the three subjects, but in general, fineparticles have dimensions of the order of 0.1–2000 μm. In fineparticle science, systems that can be described only in statistical terms are often dealt with, and anyone wishing to understand fineparticle systems requires a working knowledge of surface physics, chemistry, classical physics, and mathematical statistics.

For anything other than a fineparticle system composed of smooth dense spheres, the various methods for characterizing fineparticle systems can generate widely differing estimates of the size of the fineparticles. This apparent lack of agreement between characterization data arises from the fact that different physical aspects of the structure of a fineparticle are explored in the various characterization methods that utilize different physical principles. For this reason, all size data reported for a fineparticle system should clearly indicate the method used in arriving at the data.

When describing an individual fineparticle, the term "topography" is used to describe obvious gross structural features of the fineparticle, whereas morphology refers to the overall shape of the fineparticle. The surface texture refers to any small-scale structure or features of the surface that are much smaller than the topographical features of the fineparticle. In general, the morphology of a fineparticle is important in the description of the dynamics of the fineparticle, the ability of a fineparticle to lodge in the lung, or the covering power of a pigment in a paint. The surface texture is important in the absorption of gas by the fineparticle and in the reaction rate of a fineparticle with chemicals. Topographical features are important when considering the packing and flow behavior of a powder. It should be noted that a crystallographer talking about the morphology of a fineparticle may be referring to the internal crystal structure of the bulk material of the fineparticle. However, in such a situation the meaning to be given to the term "morphology" will be obvious from the context. The description of the surface state of a fineparticle is concerned with the possible presence of a chemisorbed layer of molecules on the surface tightly held to that surface, and with the possible presence of a dynamically absorbed layer of molecules arriving and leaving the surface of the fineparticle. Many fineparticles have oxide films. Because of the way in which they are made, some fineparticles may have a few surface layers of atoms that have been disturbed from their normal crystalline structure for a depth of a few atom diameters into the fineparticle. For example, fineparticles of crystalline silica, prepared in a ball mill, often have a glassy amorphous layer a few atoms thick on their surfaces because of the locally high temperatures created by friction between fineparticles tumbling over each other in the mill. Methods for characterizing the surfaces of fineparticles are not described in this book; however, as the surface state of a fineparticle can be a factor influencing the reliability of the data generated in a characterization experiment, it is essential to have a working knowledge of the various terms given previously.

1.2 DIRECT–INDIRECT CHARACTERIZATION OF A FINEPARTICLE

In theory, one can specify the location in space of a fineparticle surface by means of polar coordinates using a convenient point to establish the coordinate system for the mathematical equation describing the fineparticle surface [3]. In practice, the resultant equations are too difficult to handle, except for such simple symmetrical geometric shapes as spheres, cubes, ellipsoids, and laminae. Usually, the technologist has to be satisfied with a more limited description of a fineparticle's dimensions and structure based either on the geometric dimensions and structure or the functional behavior of the fineparticle. In general, size estimates of a fineparticle that correspond to some physical dimension of a fineparticle are defined as geometric diameter or descriptors. In some special circumstances, some fineparticles are large enough to enable the scientist to take direct magnitude measurements of their dimensions. However, usually one has to create an image of the fineparticle with some kind of imaging system before taking direct dimensional measurements. Procedures for measuring geometric diameters are described in Chapter 9. Measurements of the dimensions of a fineparticle constitute the most direct method for evaluating a fineparticle. The term "direct method" as used in this book is also intended to cover the measurement of size descriptors deduced by direct observation of the physical behavior of the fineparticle. Thus methods for evaluating operational diameters such as Stokes diameters, which are deduced from a study of the terminal velocity of fineparticles in a viscous fluid, fall within the scope of this book. A fineparticle placed in the path of a laser generates a diffraction pattern from which characteristic parameters for describing the shape and the size of the fineparticle can be deduced. However, the theory is quite complex and does not involve direct physical measurements on the fineparticle itself. Therefore, size characterization procedures based on diffraction studies are not included in this book. Other indirect characterization procedures not included in this book are permeability procedures, in which surface area magnitudes are deduced from a study of fluid flow through a plug of powder, mercury intrusion methods, gas adsorption studies, x-ray line-broadening studies, and stereological procedures.

1.3 CLASSIFICATION OF DIRECT CHARACTERIZATION METHODS FOR MEASURING FINEPARTICLE SIZE

In sieve fractionation studies of powder, the dimensions of the fineparticles are evaluated by splitting the powder into undersize and oversize fractions with the use of sieves. These methods are fully described in Chapter 4.

Methods that evaluate the Stokes diameter of a fineparticle from experimental measurements made on its settling behavior in a passive fluid are known as *sedimentation methods*. The term "passive fluid" denotes the fact that the fineparticles are moving through a fluid that, within its container walls, is essentially at rest except for the disturbances created by the moving fineparticles. For convenience in this book, sedimentation methods are split into two groups. Those in which the fineparticles are subjected to gravitational forces are described in Chapter 5, and those using centrifugal force are described in Chapter 6. When the fineparticles are moving through a fluid that itself is undergoing gross flow, the techniques are described as *elutriation methods*. These are discussed in Chapter 7. In stream methods for characterizing fineparticles, described in Chapter 8, the fineparticles are moved through an interrogation zone and inspected by means of some physical probe. Methods for evaluating a fineparticle based on direct measurement of fineparticle dimensions are described in Chapter 9.

In the early 1970s, as industry started to move to highly automated systems, there began to be increasing demands for so-called on-line methods for characterizing fineparticle systems to provide rapid feedback information for controlling fineparticle systems. Any method for characterizing the fineparticle system can be made into an on-line method, provided one is prepared to invest enough money in the conversion process. Possible modifications that facilitate the modification of standard procedures for use as on-line procedures are discussed throughout the text [4].

Sampling a fineparticle system to obtain a representative set of fineparticles to be used in a characterization procedure is often a major problem. Sampling procedures for aerosol systems (smokes, clouds, mists, etc.) are described in Chapter 2, and those for use with powders are described in Chapter 3.

1.4 THE OPERATIONAL INTEGRITY OF A FINEPARTICLE

When selecting a sampling and characterization procedure for studying fineparticles, one must ensure that the treatment of the fineparticles en route to the characterization study does not change their structure.

For example, many smokes contain assemblies of fineparticles that have been made into strong agglomerates. Resuspension of filtered smoke fineparticles using ultrasonic agitation could result in the breakdown of the agglomerates into the constituent units that are often much smaller than the agglomerate. This could cause problems in relating size information to the behavior of the smoke since the fallout kinetics of the smokes would be dominated by the kinetics of agglomerates. Thus the large agglomerates would settle very quickly, whereas the very small individual fineparticles created by the dispersion technique would remain suspended in air currents for very long periods if they existed in the smoke.

The words "agglomerate" and "aggregate" are used in a very confusing manner in fineparticle science. The situation is made worse by the fact that standard definitions for terminology in fineparticle science have managed to oppositely define the meaning of "aggregate" and "agglomerate" in American and British English.

In this book a *fineparticle* is defined as a single unit of material that can be clearly discerned, either directly or indirectly, and that maintains its integrity throughout the operational functioning of the fineparticle system being studied. Thus, in a smoke, the agglomerated individual spheres constitute an effective fineparticle that maintains its operational integrity throughout the dispersion and settling of the smoke. It follows that any sampling technology should not treat the system being sampled with forces of such severity that the operational integrity of the fineparticle is degraded. In this book the fineparticle that has been formed by the obvious combination of small particles into a group that maintains its identity throughout operational systems is called *agglomerate*, whereas an association of fineparticles that is only a loose confederation likely to disintegrate during the operational life of the fineparticle is termed an *aggregate*. Dispersion techniques used to prepare a fineparticle system should be comparable to those encountered by the fineparticle during its operational existence. Thus, if a loosely aggregated pharmaceutical powder is being sampled, and if this powder is to be sampled in the making of a medication in such a manner that the aggregates are broken down, then the preparatory technique for sedimentation study should use a dispersion technology of severity comparable to that used when making the medication.

When preparing a suspension for a sedimentation study, the scientist can err in two directions. For example, if a pigment is being characterized by sedimentation procedures, and if only very light dispersion technology is used on an aggregated assembly of pigment fineparticles to prepare the suspension, then very large effective fineparticles will be present in the suspension. However, in the making of a paint, the pigment is subjected to severe dispersion techniques so that the loosely aggregated system of the sedimentation study will bear no resemblance to the entities present in the paint. Thus, for example, the writer has seen sedimentation studies of titanium dioxide, a white pigment, of 1–7-μm Stokes diameters when the dispersed pigment in the paint probably had no fineparticles larger than 1 μm. No general rules can be given for selecting a dispersion technique; however, it should be appreciated that when viewed from the fineparticle's perspective, some apparently mild dispersion techniques are very severe. For example, for preparation of fineparticles for viewing under the microscope, a common procedure is as follows. First a small portion of the powder to be characterized is made into a thick paste by using a drop of viscous oil. Then the operator is instructed "to place a drop of the paste onto a glass slide and smear it out gently with a glass rod." The movement of the glass rod over the slide may look gentle to the outside observer; however, to the fineparticle trapped in the oil, the glass rod looks like a steam roller and the severe shear conditions immediately between the rod and the glass plate can destroy the operational integrity of many fineparticles.

Again, a widely used procedure for dispersing fineparticles in a liquid prior to a sedimentation study is to agitate the suspension by placing it in an ultrasonic cleaning bath or alternatively to insert directly into the suspension a probe for generating ultrasonic waves. To the external observer, the ultrasonic agitation of a suspension appears to be a relatively mild procedure. However, the mechanism by which ultrasonic treatment of a suspension aids dispersion of the fineparticles is the creation of cavitation zones within the suspension by the ultrasonic waves. The shear forces and impact forces within a cavitating zone can be very severe, resulting in the complete degradation of a fineparticle system. Before employing any preparatory technique for a fineparticle system, the forces experienced by the individual fineparticles in the system should be assessed to evaluate the effect of that preparatory system on the integrity of the fineparticle.

1.5 DATA TRANSFORMATION

Many previous books on fineparticle science have contained chapters devoted to data processing or size distribution functions. One of the reasons why previous writers on fineparticle science included such a chapter on data processing is probably because the statistical nature of the procedures used in transforming the data is often unfamiliar to scientists trained in the classical sciences. In the opinion of the writer, the scientist is either familiar with statistical reasoning or needs far more help than can be given in an individual chapter of a book on fineparticle characterization. The fineparticle scientist should avoid unnecessary data transformations. Often a set of raw data is converted to a distribution function for no other purpose than to display the data in a familiar form. Whenever possible, experimental data should be used directly in the decision-making process because data transformation usually involves loss of information in the grouping of data or in the digitization of a continuous curve.

1.6 CORRELATION BETWEEN SIZE CHARACTERIZATION DATA OBTAINED BY DIFFERENT CHARACTERIZATION PROCEDURES

Many so-called problems of correlating information for a given fineparticle population obtained by different procedures for characterizing nonspherical fineparticles arise from an invalid anticipation of agreement between data that measure different attributes of the fineparticle population. This invalid anticipation arises from an unjustified extrapolation to sets of irregular fineparticles of the fact that for smooth, dense, hard spheres, size characterization data from all characterization procedures are directly compatible. Thus the Stokes diameter of a dense, smooth, glass sphere has the same magnitude as its physical diameter when measured by examining it through a microscope. The fact that the characterization data for glass beads from different methods can be made to agree with each other has led to the simplistic anticipation by many workers new to fineparticle science that any difference between sets of data from different characterization methods for any population of fineparticles can be forced to agree by expending more and more effort on eliminating what is incorrectly viewed as "experimental error." To illustrate the semantic problems of correlating data, consider a typical problem that could arise in environmental science, the characterization of agglomerates of spherical fineparticles being emitted from a furnace. Assume that the various types of fineparticle to be evaluated are similar to those shown in Figure 1.1. Let us assume that the fineparticles are to be characterized by three procedures: microscope examination, measurement of the Stokes diameter, and the use of a resistazone stream counter. In a resistazone stream counter such as the Coulter counter® equipment, the change in the electrical properties of a cylinder of electrolyte, with and without the fineparticle under study, is correlated to the magnitude of the particle (see Chapter 8). To help explain the problems of correlating data from the three methods, we use the concepts of set theory. The first way in which we could group the agglomerates is according to the number of subunits they contain. The straight lines in Figure 1.1 separate the agglomerates into sets according to this descriptor. If we now attempt to organize the set of agglomerates that have the same projected length as viewed under the microscope, then the agglomerates such as those indicated by the

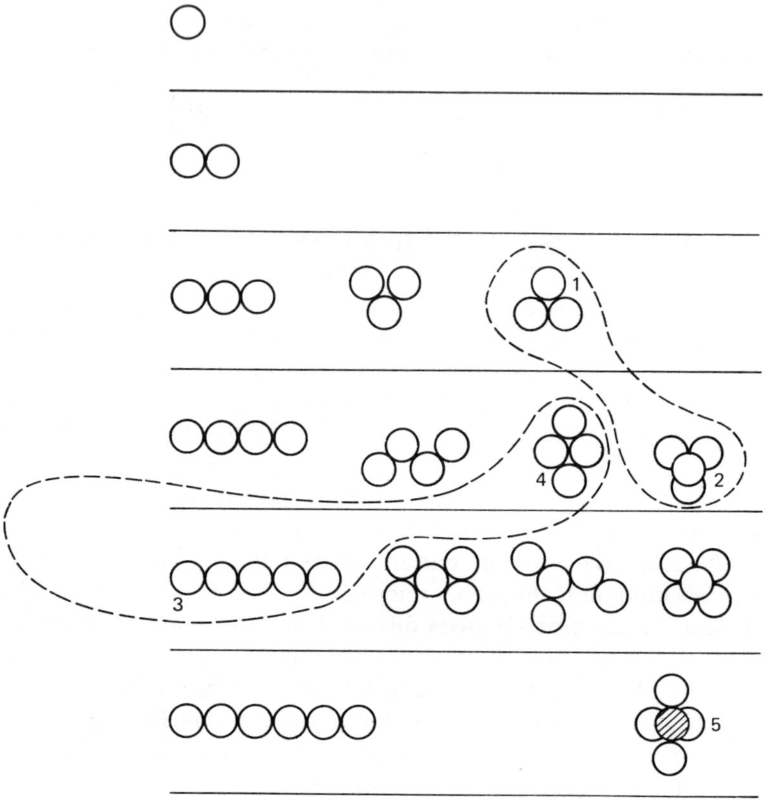

Fig. 1.1 A one-dimensional descriptor classification of nonspherical fineparticles can group together, according to one descriptor, fine particles that differ greatly when described by an alternate descriptor. Descriptor "projected diameter" would describe 1 and 2 as identical. Descriptor "settling speed" would link 4 and 3 as identical. Note that one sphere in cluster 5 is hidden at the bottom of the cluster.

linkage in Figure 1.1 would be identical. On the other hand, if we now attempt to draw up the classification of fineparticles according to their Stokes diameter (i.e., measured settling velocity), we note that the same fineparticles can have different diameters depending on the initial orientation of the fineparticle when placed in the viscous fluid and that apparently very different fineparticles can be identical with respect to the Stokes diameter parameter (see Figure 1.2).

Again, with the resistazone equipment, we find that the recorded fineparticle diameters of agglomerates entering the interrogation zone of the instrument will depend on the orientation of the agglomerates and their position with respect to the axis of the cylindrical interrogation zone

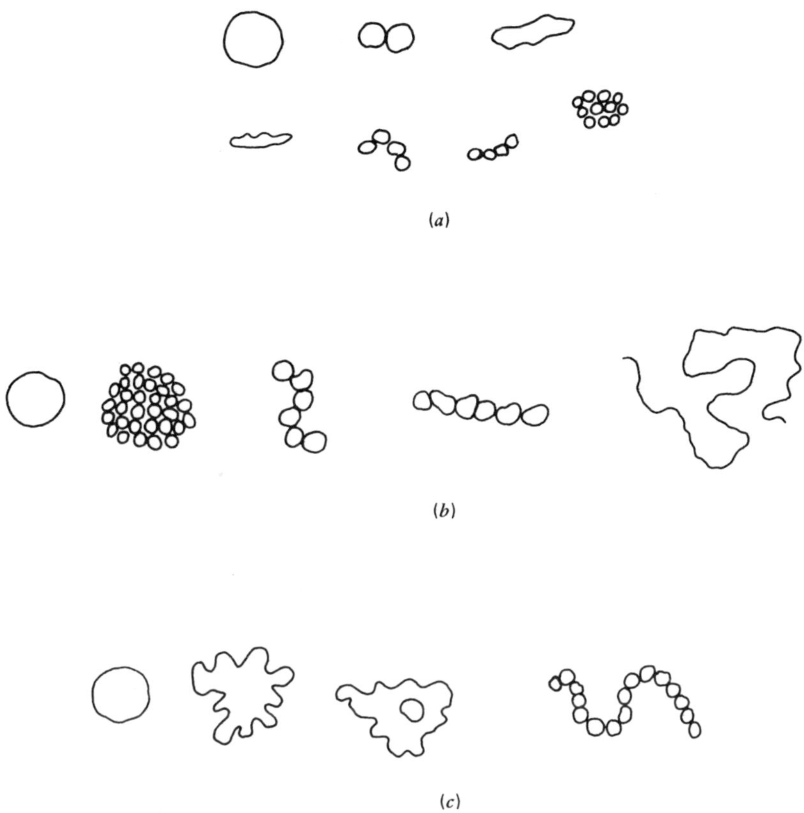

Fig. 1.2 Fineparticles that look different can belong to the same set when described by a one-dimensional descriptor: (*a*) set of fineparticles that have the same descriptor parameter "projected length"; (*b*) possible members of the set of fineparticles of identical Stokes diameter; (*c*) possible members of the set of "all fineparticles of the same volume."

used in such instruments. Because of these factors, agglomerates of identical structure may end up in different size descriptor classes in the course of any measurement procedure. With resistazone counters there may be also problems arising from a failure to respect the integrity of the agglomerate. In instruments such as the Coulter counter, and the Celloscope® (see Chapter 6), relatively severe shear conditions exist at the entrance of the interrogation zone that could cause the breakdown of the agglomerates of Figure 1.1, with the result that all the fineparticles would appear to be the same size. This type of degradation of the agglomerates could also occur in the primary sampling procedure. Thus a common method used to sample smokes and airborne dusts in an industrial environment is the

Greenberg–Smith impinger (see Chapter 3). In this technique all the fineparticles are impinged onto a surface under a liquid. Before a suspension of the fineparticles collected in this way is characterized by any of the three methods, it may be subjected to ultrasonic agitation "to disperse the fineparticles." This treatment may result in the breakup of all the agglomerates.

In the summary, characterization data for nonspherical fineparticles based on one-dimensional or two-dimensional descriptors cannot be expected to be coincident when those descriptors are based on different physical measurements. Even if all problems associated with inaccuracy, imprecision, reality discrepancies, inappropriate preparatory techniques, and sampling uncertainty are avoided, there will still be differences between sets of data for irregular fineparticles obtained using characterization procedures based on different physical principles. These differences do not represent "error"; rather, they constitute information on the shape of the fineparticles (see Chapter 9).

1.7 STANDARD FINEPARTICLES FOR INSTRUMENT CALIBRATION

Many of the published discussions of the physical bases of the various procedures used to characterize fineparticles describe the appropriate theory in terms of dense smooth spheres moving through the characterization equipment. These idealized discussions often seem to indicate that the characterization data generated by the instrument can be used to deduce fineparticle magnitude parameters from physical theory without recourse to calibration procedure. In practice, however, the working scientist rarely has access to all the relevant physical properties of the fineparticles to be characterized, and it is usually necessary to calibrate the equipment with standard fineparticles [5]. Standard fineparticles are available from two commercial companies operating fineparticle reference material banks [6, 7]. Latex spheres are widely used as standards. These were originally available from the Dow Chemical Company but are now available from Dow Diagnostics and some of the companies involved in selling fineparticle characterization instrumentation [5–9]. Pollens are sometimes used as ideal reference materials; however, since they are of biological origin, their exact point of origin, their age, the storage conditions to which they have been subjected, and their swelling behavior when placed in a fluid can affect their external size. Because of these factors, size data quoted from pollens should be interpreted with care in any given experimental situation. Recently Wilson has described reference quartz powders characterized by many methods [10].

NOTES AND REFERENCES

1. "Powders in Industry," symposium, September 1960; proceedings published as SCI Monograph No. 14, Society of Chemical Industry, 14 Belgrave Square, London, England SW1.
2. B. H. Kaye and J. C. Mittermayer, "Automated Retrieval of Literature on Fineparticle Science and Technology," Powder Technology Conference, Chicago, March 10–12, 1977; published in the conference proceedings.
3. H. P. Schwartz and K. C. Shane, "Measurement of Particle Shape by Fourier Analysis," *Sedimentology*, 13 (1969), 213–231.
4. B. H. Kaye, "The Data Processing Revolution and the Characterization of Fineparticles," Plenary Lecture, 1st European Symposium on Particle Characterization, Nuremberg, West Germany, September 16–19, 1975, published in the conference proceedings.
5. B. H. Kaye, "Standard Fineparticles—Who Needs Them?," Chemtech, January, 1980.
6. Duke Standards Company, 445 Sherman Avenue, Palo Alto, California 94306.
7. Particle Information Service, 600 South Spring Road, Los Altos, California 94022.
8. Dow Diagnostics, The Dow Chemical Company, P. O. Box 68511G, Indianapolis, Indiana 46268.
9. D. F. Alliet, "A Study of Available Particle Size Standards for Calibrating Electrical Sensing Zone Methods," *Powder Technol.*, 13 (1976), 3–7.
10. R. Wilson, "Some Recently Certified Particle Size Reference Materials," Partikel Technologie, Second European Symposium on Particle Characterization, Nuremberg, West Germany, September 24–26, 1979, pp. 59–77.

CHAPTER

2

SAMPLING A POWDER OR A SUSPENSION

2.1 TERMINOLOGY

A fineparticle system being studied is described as the parent population. A representative sample is identical in constitution to the parent population. A biased sample has a constitution that differs from that of the parent population. A random sample is one taken from a supply of powder without detailed study as to whether such a sample is representative or biased. When a sample of powder is taken from a moving stream of powder, either on a conveyor belt or in a cascading stream within a process, the resultant sample is often described as a "snatch sample." Snatch samples can be either representative or biased, depending on how they are taken from the powder stream. The volumes of powder and the range of fineparticle sizes involved in any particular sampling problems can vary enormously.

It is useful to distinguish between three different sampling procedures. There are total processing procedures, selected region sampling, and "blend and snatch" sampling. Total processing procedures are usually more efficient than other sampling strategies but can be expensive and time consuming. When dealing with very large volumes of powder, selected region snatch sampling is often used for economy reasons. In such procedures snatch samples are taken at various locations and/or time intervals from the bulk supply in such a way that they are representative of the total bulk of powder. These are then combined to form a gross sample that is then subdivided by efficient bulk processing equipment to obtain a representative subsample.

In blend and snatch sampling procedures, the powder to be sampled is processed in a powder blender so that any subvolume of the powder is as representative as any other subvolume within the blender. Any snatch sample from such a homogenized bulk supply of powder is a representative sample. Because of the poor or unknown performance of traditional powder blending equipment, this approach has not been widely used in past sampling studies. However, new developments in blending equipment are making it possible to consider the adoption of the blend and snatch approach to the sampling of a powder system.

2.2 SAMPLING LARGE QUANTITIES OF POWDER

Taking samples of powder in a plant producing tens of thousands of tons of powdered substance a day is a highly complex task and involves special installations costing hundreds of thousands of dollars. Several companies build and install specialist sampling devices for such large-scale operations. Literature describing these sampling systems is available from these companies [1, 2]. Specific industries concerned with the sampling of large tonnage products such as cement and powdered coal have set up standard procedures under the auspices of such organizations as the American Society for Test Materials and the British Standards Institute. Details of such procedures can be found in the literature of such organizations.

A specific sampling program within a large plant is usually evolved in dialog between production company engineers and consultant engineers from the manufacturers of sampling equipment. A full description of the various heavy-duty "run of mine" or plant sampling devices is beyond the scope of these discussions, but we discuss briefly some of the major design problems in such sampling situations.

When large tonnages of powder are being moved by a conveyor belt, samples are often snatched either at the end of the conveyor belt or along its length, as illustrated in Figure 2.1. When sampling in this way, it is important to take into account various powder segregation mechanisms that may have been operative in the powder system being studied. Whenever a free-flowing powder is poured into a heap, segregation by size occurs as a result of the larger fineparticles rolling down the outside of the heap and collecting around the periphery. As the powder is fed onto the growing heap, it alternately builds up a cone with sharp-angled sides that slumps and dumps fines around the edge. Traces of several collapsed transient cones in the final heap create layers of segregated fineparticles. This type of dynamic segregation in a growing cone has been studied extensively by J. C. Williams and co-workers (see [3]). Fineparticles in a heap standing on a vibrating surface are subject to two segregation mechanisms—percolation and "walking." Percolation is the movement of fines down through the structure of the heap. "Walking" occurs when a relatively large fineparticle, initially in a horizontal position, tilts under vibration so that smaller fineparticles near its shoulder pour underneath it and consolidate the new position. A later random elevation of the other end of the larger fineparticle results in the elevation of the fineparticle relative to its original position. Over a period of time, the net effect of many vibrations and powder movement within the heap is to "walk" the larger fineparticles to the surface of the heap (the same process results in

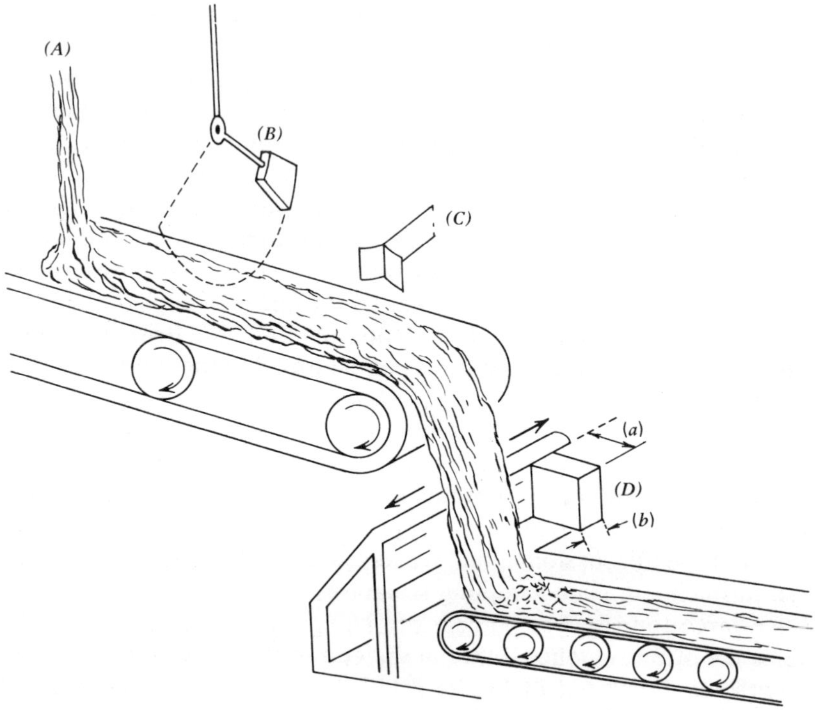

Fig. 2.1 Many sampling devices can be used to take samples from moving conveyor belts and powder streams. Symbols: *A*, powder stream; *B*, pendulum or swing arm sampler; *C*, plough sampler pushed across the belt; *D*, cutter sampler.

large flints or boulders working their way to the top of a field of soil). Thus an aged heap will have the larger fineparticles concentrated on its surface with fines percolated to the bottom.

Powder on a conveyor belt is often poured onto it in a cone-like deposit. Therefore, efficient sampling from a conveyor belt should allow for the segregation effects that can occur in a cone as discussed earlier. Thus for efficient operation, the pendulum sampler shown in Figure 2.1 must sweep right across the belt and should be in touch with the conveyor belt right across the belt. If it only bites into the top of a fresh heap, it will probably generate a biased sample. It is difficult to adjust the angle of swing of a pendulum sampler to take a good, deep, clean bite. For this reason, and also because it is cheaper to build and install, a plough sampler as illustrated in Figure 2.1 is often used. The V-shaped mouth of the plough should be big enough to hold a "bite" of powder as it ploughs across. It should be built so that overflow turbulence is not created at the entrance of the plough. The plough discharges its sample at the opposite side of the belt into a hopper (not shown in Figure 2.1).

The cutter sampler shown in Figure 2.1 consists of a collecting box, called the *cutter*, that moves back and forth through a powder stream leaving a conveyor belt. Because of their higher momentum, larger fineparticles are projected farther from the end of the belt than smaller ones. Because of this within-stream segregation, the dimension *a* of the cutter box must be greater than the depth of the powder stream. Operational experience with cutter samples has shown that the width of the mouth of the cutter, shown as *b* in Figure 2.1, should be equal to or greater than 20:1 with respect to the longest dimension of the largest fineparticle present in the stream of powder. The volume of the cutter should be at least one "stream bite"; otherwise, overflow from a full cutter traversing the whole stream will bias the sample toward the smaller fineparticle content of the stream.

When sampling in the processing industry, one should never delay sampling until the powder is in a storage bin of conventional design. (The qualifying term "of conventional design" is added because some industries are moving to turbulent fluidized bins for storage of large quantities of fineparticles. Such storage bins, called "homogenization storage bins," constitute huge continuous blenders, and snatch sampling from such bins constitute representative samples.) Segregation in a traditional storage bin can arise from the alternate buildup and collapse of transient cones in the bin. Also, the air forced out of the bin by the incoming powder can flush out the smaller fineparticles from the collapsing transient cones. This latter effect operating over the whole depth of the bin can lead to the final settled layer of powder at the top of the hopper being very enriched in smaller fineparticles. Subsequent vibration of the bin could lead to the "walking" of coarser fineparticles to the top of the bin.

Many of the design problems of conventional snatch sampling operating in large-scale powder operations could be overcome by the use of passive blenders placed at appropriate positions in the powder flow system. In a passive blender, also known in industry as a *motionless mixer* because it has no moving parts, the aim is to randomize the position of a fineparticle leaving the blender so that its probability of exit is uniform across the exit throat of the blender and independent of the fineparticle's size and point of entry into the blender. To illustrate the concepts embodied in a passive blender, consider the system shown in Figure 2.2. This system is currently undergoing development at Laurentian University. The fineparticles are made to undergo a random walk through the system by means of the randomizing rods. The rods are placed in such a way that the fineparticles never get a chance to build up momentum in any given direction,

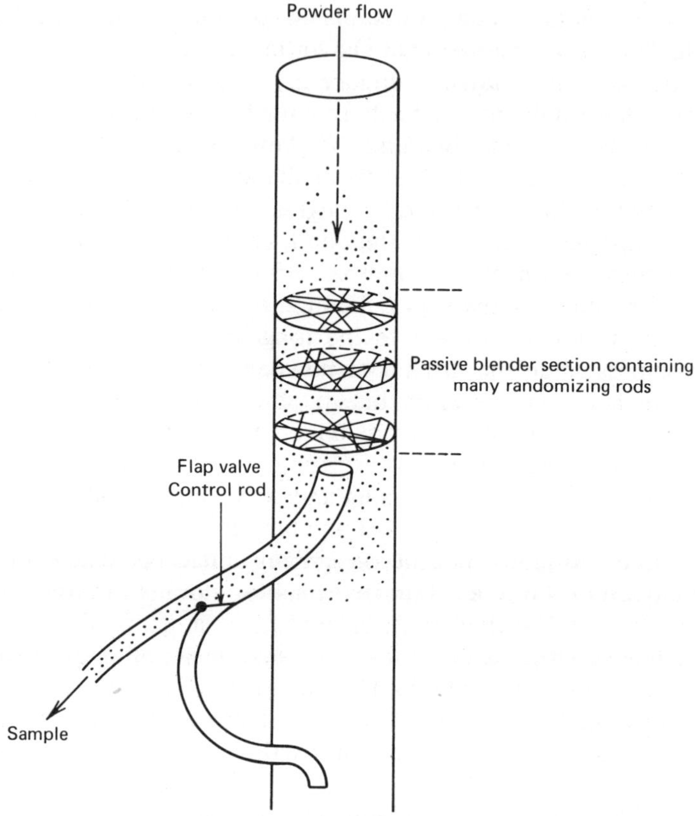

Fig. 2.2 A passive blender can be used to sample a stream of powder.

and their exit positions relative to their entry positions are randomized. A sampling pipe can be installed immediately after the passive blender as shown in Figure 2.2. Powder flowing down this sampling pipe can either be diverted to a holding bin or fed back into the processing stream by operating the flap valve shown. Other passive blenders available commercially could be used in the powder stream above the sampling pipe [4–6].

The simplest sampling device that can be used to extract powder from the depths of a heap, a spear sampler, is a piece of hollow metal pipe with the end sharpened to aid penetration into the heap. A spear sampler is thrust upward into the body of a heap to avoid losing powder from the inside of the pipe as it is withdrawn from the heap. It is difficult to isolate individual portions of the powder sample along the length of the spear, and this can be a disadvantage in some sampling situations. A pneumatic sampling probe is another sampling device made by attaching a vacuum pump with filter and holding chamber to one end of a hollow metal tube. This can be inserted vertically into a heap and is useful for sampling from cargo holds and railway wagons. Sampling with a pneumatic probe can, however, result in bias in the sample toward the smaller fineparticles in the heap. In any given situation, investigations into possible bias using several suction velocities should be carried out and, if necessary, correction factors applied to the sampling performance of the equipment.

The structure of a sampling thief used for taking samples from a bulk quantity of powder is shown in Figure 2.3. Initially the inner tube is twisted so that the two sets of entry ports do not match. With the tubes in this position, the thief is thrust into the powder system to be sampled. Then the inner tube is rotated to line up the two sets of entry ports so that powder flows into the sampling chambers. When they are full, the inner tube is given a further twist to close the openings and the thief is withdrawn. A thief sampler is fairly easy to use with relatively coarse, free-flowing powder. However, when sampling fine powders it is not always easy to rotate the inner tube because fineparticles become trapped between the walls of the tube, binding the tubes together. The use of excessive force to push the thief into the powder can crush fragile powders, resulting in a size reduction of fineparticles along the path of the thief. When sampling cohesive powders, the passage of the thief can compact the powder so that it does not flow into the openings in the tubes.

A simple scoop or shovel may be used to scrape samples from the

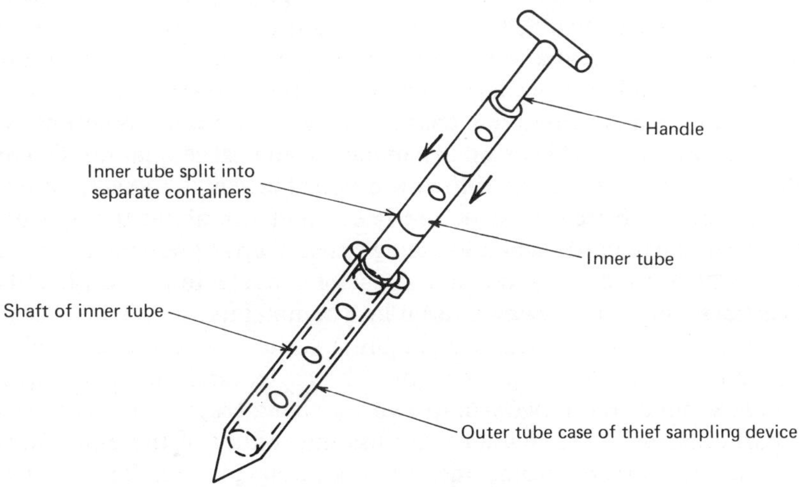

Fig. 2.3 A sampling thief for sampling powder from a heap.

surface of a heap of powder. When sampling this way, potential segregation effects at the mouth of the scoop due to diversion of fineparticles by the sides of the scoop should be avoided. Also, the scoop should not be overfilled because there will be a danger of the preferential loss from the scoop of larger fineparticles rolling down the heap of powder on the scoop as it is moved to place the powder sample in a container. If the fineparticle constitution of a heap of powder varies at random, then any sequential set of samples scooped from the heap constitutes a subset of representative samples. On the other hand, if it is known that any systematic segregation mechanisms have operated during the construction of a heap, sampling should be done in a systematic manner at the various locations where variations are anticipated. Thus if the heap was constructed from the top, samples should be taken at the base, the top, and, if possible, the center of the base of the heap.

2.3 TOTAL PROCESSING SAMPLING PROCEDURES

A widely used total sampling procedure is the so-called cone-and-quarter sampling technique, where the original quantity of powder is poured onto a polished surface in the shape of a cone. This cone is then split up into four equal parts by means of a thin sheet of metal. Theoretically, the chance that any individual fineparticle will be found in any of the quarters is the same and all four quarters will have the same fineparticle constitution. Therefore, any quarter is a representative sample. A sample of any required size can be taken from any original quantity of the powder by taking a quarter from the prime cone and using it to create a second cone that, in turn, is quartered, and so on. The procedure was originally devised for sampling from large quantities (i.e., several tons) of ores or crushed coal that are relatively coarse powders. It is also sometimes used in the laboratory to achieve small samples from a large quantity delivered to the laboratory. It is often difficult to pour the powder exactly onto the top of the cone so that its structure is symmetrical about the axis of the cone. Also, particularly when handling small heaps of powder in the laboratory, say on a glass slide, it is possible that a few important large fineparticles might roll away and be lost completely.

In the chute riffler illustrated in Figure 2.4, the powder to be subdivided is poured into the V-shaped trough. Chutes leading out of this trough alternately direct the powder into two receiving trays. The chances that any particular grain of powder will pass into either of the two receiving trays are equal. Therefore, ideally, on passing a portion of powder through the riffler, two subsamples of identical constitution should be

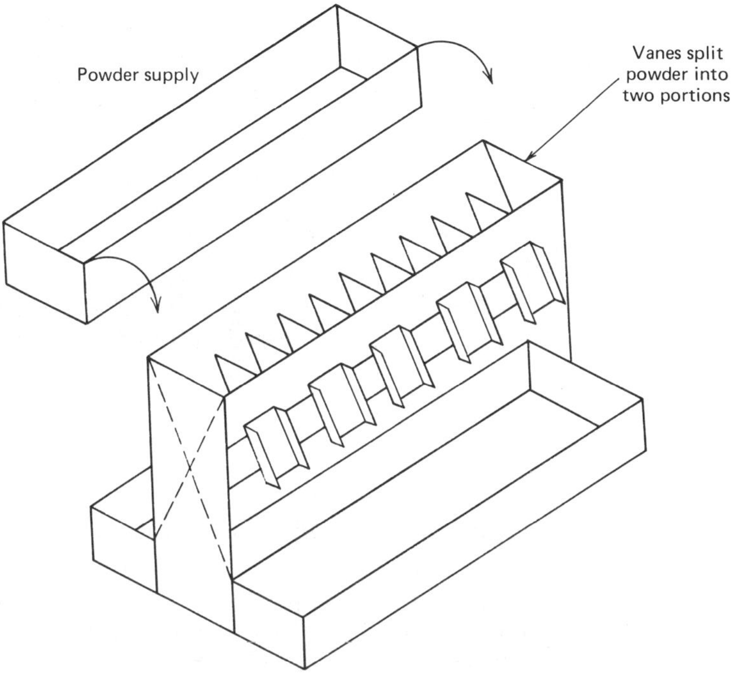

Fig. 2.4 A chute riffler for powder sampling.

obtained. In practice, samples obtained by multiple passage through a chute riffler may differ in quantity and/or fineparticle constitution. Quantity variation comes from uneven flow across the feed trough of the device. Inequality of constitution comes from a failure of the randomizing mechanism, that is, interception of the powder stream by adjacent oppositely directed chutes, to cope with inhomogeneity of the powder feed. The feature of a chute riffler controlling its ability to sample a powder supply efficiently is the width of its chutes. This fact has not always been appreciated by technologists working in this area of fineparticle technology. Thus Batel, in a review of the performance of the chute riffler, stated that there was little to be gained by increasing the number of chutes beyond 10 [9]. The number of chutes controls the processing capacity of the device; the essential randomization mechanism is the ability to direct adjacent fineparticles in opposite directions, and this is governed by the width of adjacent chutes. This aspect of chute riffler performance is illustrated clearly by the data obtained by Kaye using the chute riffler shown in Figure 2.5. The design of the system shown enables the feed rate to the chutes to be controlled more easily than in the standard type of chute

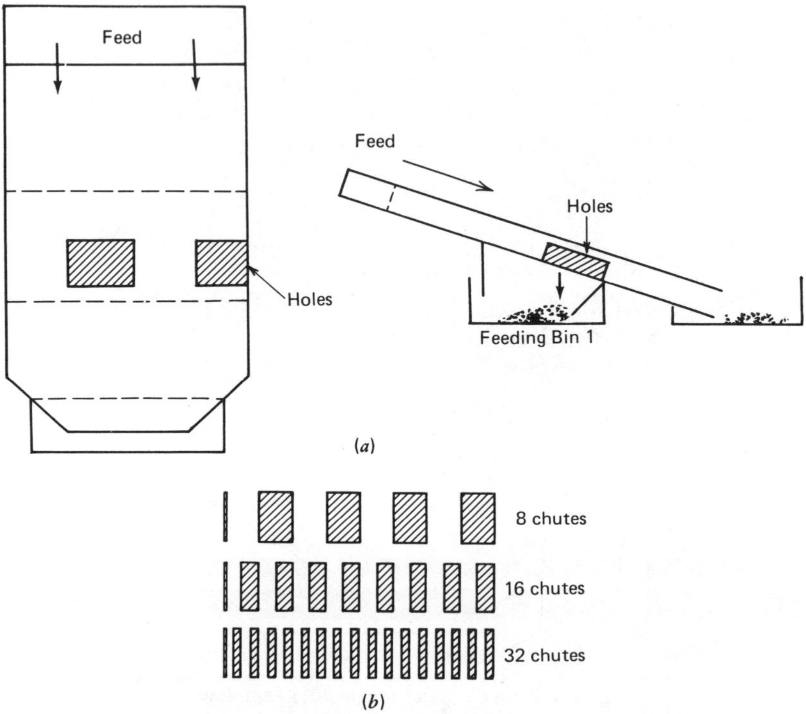

Fig. 2.5 Schematic representation of the chute riffler used in sampling efficiency experiment [7]: (*a*) a chute riffler using an inclined plane (half the powder falls through holes fed by chutes; the other half passes down to the bin at the base); (*b*) chute pattern of rifflers used in the sampling efficiency experiment.

riffler. Four rifflers of the same overall width having 4, 8, 16, and 32 chutes, respectively, were constructed. The efficiency of these rifflers was tested with a glass bead powder made from two fractions. The fraction of larger beads had a median of 1320 μm, with diameters ranging 820–1700 μm. The smaller-diameter beads ranged 230–320 μm, with a median diameter of 280 μm. These fractions could be easily separated on a 25-mesh sieve. Then 200 g each of the large and small beads were placed in a 12-ounce glass bottle and mixed. The powder was then poured through the first chute riffler. Sample subdivision was repeated to achieve a set of samples, each of which contained approximately $\frac{1}{32}$nd of the original powder. The oversize percentage of each sample was measured by using a 25-mesh sieve. Sampling efficiency can be quantified by using the variable $S = A - B$, where A is the measured oversize percentage of sample and B is the oversize percentage of parent powder. The cumulative percentage occurrence of the variable S for the set of subsamples was calculated. The experiment was repeated for each riffler having 4, 8, 16, and 32 chutes, respectively. The data from this set of experiments are summarized in Figure 2.6. The graph paper used in this diagram is Gaussian probability paper. Since this type of graph paper will not be familiar to all readers, its structure is described briefly.

If the frequency of occurrence of a physical variable plotted against magnitude of the variable produces a bell-type curve of the form shown in Figure 2.7, the data are said to conform to the Gaussian probability distribution. The terms "bell curve," "arithmetic probability function," and "normal probability function" are often used as synonyms for this type of distribution. It is common practice to plot the magnitude of the variable describable by the Gaussian probability function as deviations from the mean of the variable. The cumulative transformed version of the Gaussian

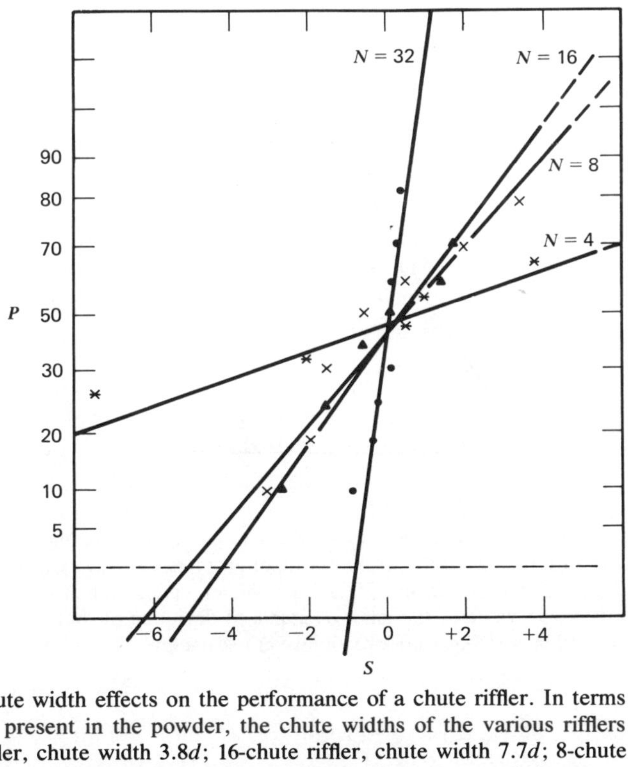

Fig. 2.6 Chute width effects on the performance of a chute riffler. In terms of maximum fineparticles present in the powder, the chute widths of the various rifflers tested were: 32-chute riffler, chute width 3.8d; 16-chute riffler, chute width 7.7d; 8-chute riffler, chute width 15.4d; 4-chute riffler, chute width 30.7d. Symbols: P, cumulative percentage of samples with values of S less than or equal to stated magnitude; S, measured oversize residue of sample—average oversize residue for all samples; N, number of chutes.

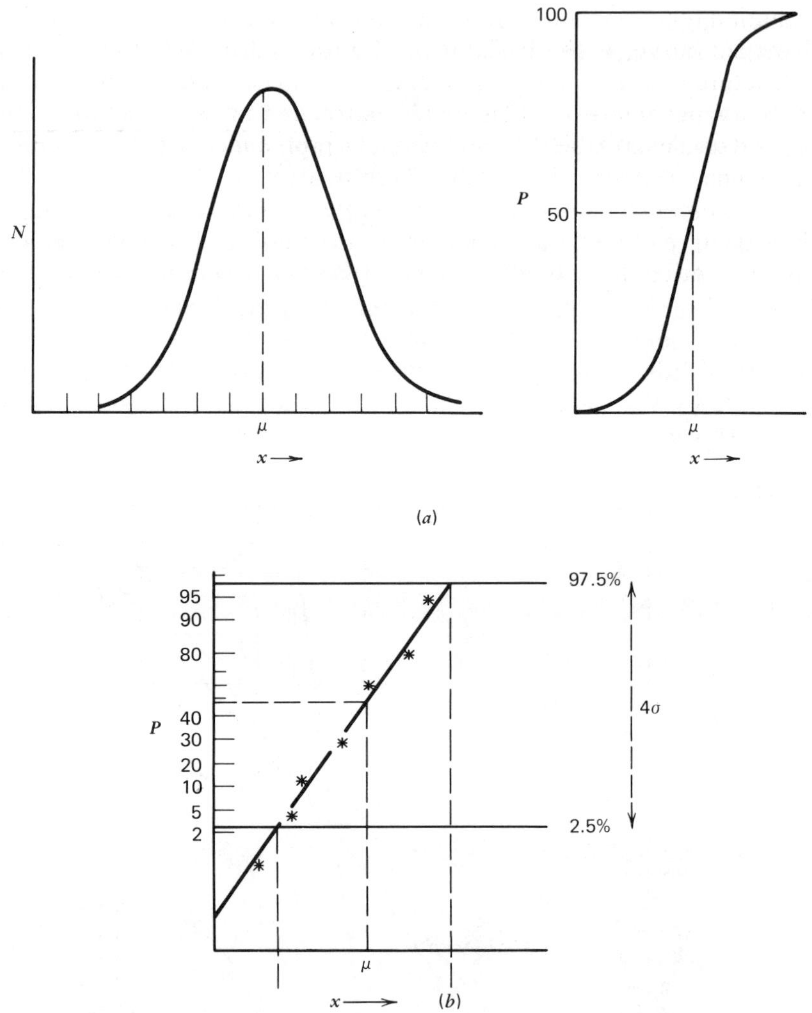

Fig. 2.7 Gaussian probability graph paper is structured so that the cumulative occurrence of data describable by the Gaussian distribution function plots as a straight line: x, variable quantity; N, number of measurements; P, cumulative percentage of measurements equal to or less than x; μ, mean value of the data; σ, standard deviation of the data; (a) Gaussian probability distribution; (b) Gaussian probability graph paper.

distribution function, shown in Figure 2.7*b*, which is less familiar than the bell-shaped curve, is the basis of the Gaussian probability graph paper, the structure of which is shown in Figure 2.8. The scales of the Gaussian probability paper are mathematically stretched so that the cumulative form of data describable by the Gaussian probability function plots as a straight line. Mathematically, this is equivalent to structuring the "cumulative percentage of occurrence" scale of Figure 2.7*b*, so that the S-shaped curve of that graph becomes a straight line. When dealing with a variable describable by the Gaussian distribution function, the cumulative

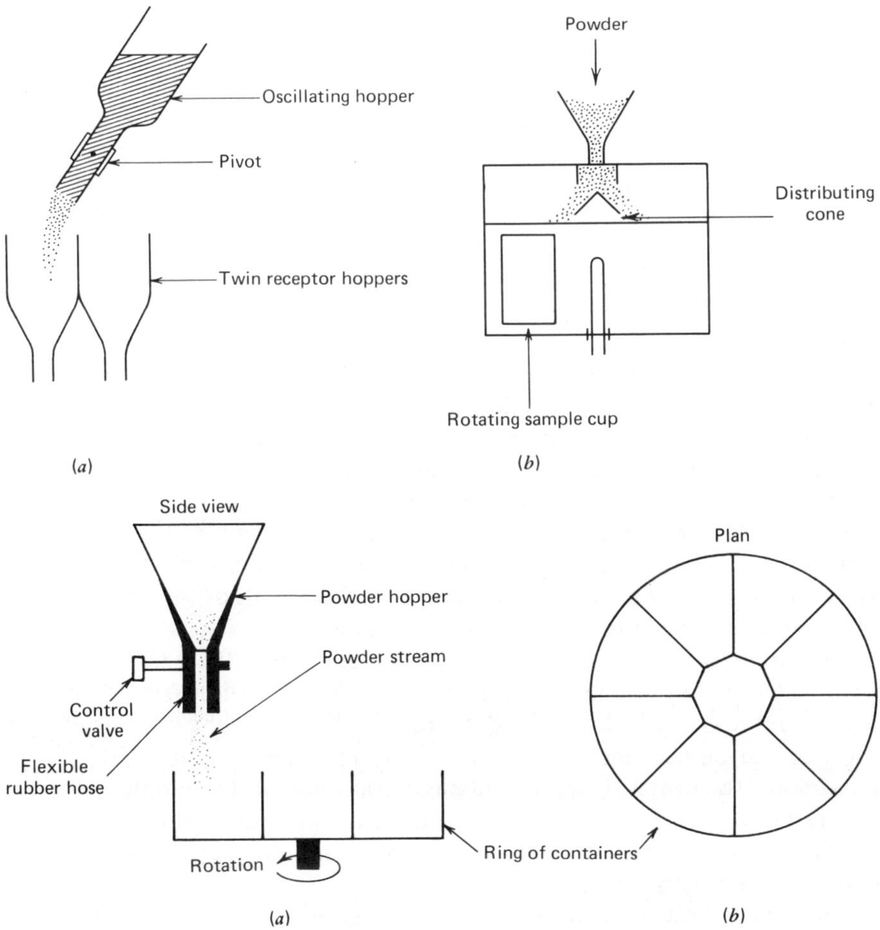

Fig. 2.8 Efficient powder sampling devices [8]: (*a*) oscillating sampler; (*b*) cone and cup sampler; (*c*) spinning riffler.

form of the raw data can be plotted on Gaussian probability paper and the best straight line drawn through the data. As shown in Figure 2.7*c*, the slope of the line is related to the standard deviation of the data. Thus when data are plotted as in Figure 2.7*c*, the greater the slope of the data line, the less the variability of the data. This kind of graph paper is also called "normal graph paper" and "arithmetic probability paper." Sometimes the logarithm of a physical variable can be described by the Gaussian probability function. In such situations it is common practice to refer to the data as being "lognormally distributed." A version of Gaussian probability graph paper employing a logarithmic scale of the physical variable axis is known as "lognormal graph paper."

When the data obtained by Kaye using the various chute rifflers are plotted on Gaussian probability paper, the series of straight-line relationships shown in Figure 2.6 are obtained. It can be seen that the smaller the chute width, the steeper the line. Thus a smaller chute width corresponds to a higher sampling efficiency. There is a big increase in efficiency as the chute width approaches the dimensions of the maximum fineparticle being sampled. From the data in Figure 2.6, it can be concluded that when building a chute riffler, the chute width should be as narrow as it is consistent with the flow of powder through the equipment and the number of chutes regulated only by cost and convenience considerations.

Two powder sampling devices described in a British Standards Institute publication are illustrated in Figure 2.8 [8]. The oscillating hopper system shown in Figure 2.8*a* allows for inhomogeneity in the feed alternating rapidly between the reception hoppers. To function efficiently, the time of powder flow through the system during a sampling operation divided by the time of one oscillation should be a large number. In the system shown in Figure 2.8*b* the powder is fed onto a cone that distributes the powder. A sample container rotates around the periphery of the distributing cone. If W is the width of the sampler cup and L is the length of the periphery of the cone, the sample reduction achieved for a single pass is W/L. Experimental studies have demonstrated that the spinning riffler, shown in Figure 2.8*c*, is a very efficient sampling device [9, 11]. It consists of a ring of containers rotating under a powder stream. If T is the time of powder flow through the device and t the time of rotation, then T/t must be a large number (of the order of 100) for efficient operation of the sampling device. A set of data illustrating the relative efficiency of various sampling procedures and demonstrating in particular the high efficiency of the spinning riffler are summarized in Figure 2.9.

There is a limit as to how fast a spinning riffler can be operated. At higher speeds the rotation of the sampling cups creates a fan action that can blow fines in the powder stream away from the sampling cups. A good

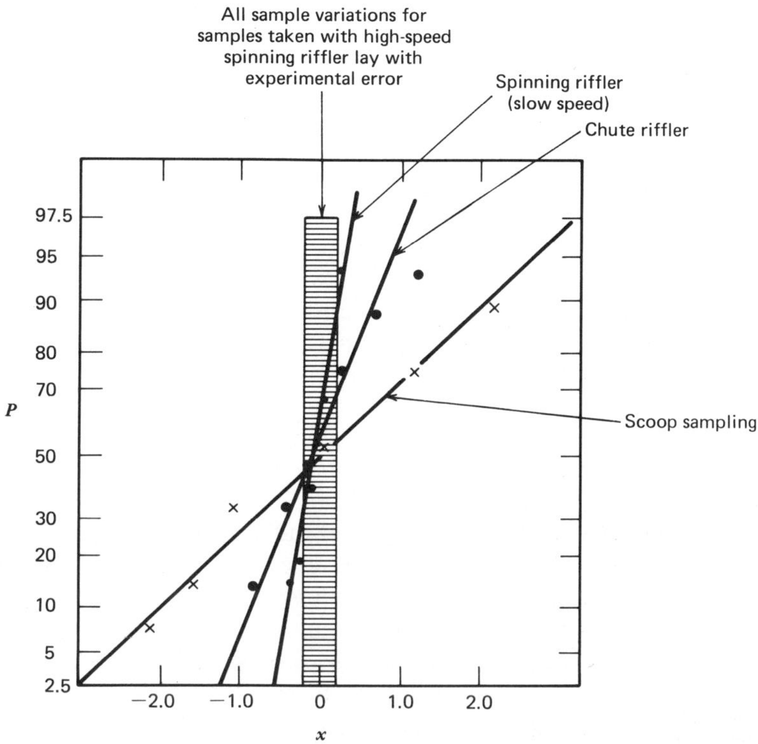

Fig. 2.9 Comparison of the sampling efficiency of several sampling procedures illustrates the efficiency of the spinning riffler sampler [11]. Each linear relationship represents data from 16 samples of a bronze powder. Symbols: x, oversize percentage of the sample—percentage oversize for parent powder (shaded vertical band represents limits of experimental error); P, cumulative percentage of samples with values of less than or equal to the stated value (Gaussian probability scale).

temporary spinning riffler can be fabricated from an old phonograph turntable and disposable coffee cups.

It is difficult to adapt the spinning riffler to liquid sampling because of the expense of making special waterproof cups that fit into the ring of containers. Some manufacturers of spinning rifflers have overcome this problem by fabricating a solid ring of funnels that rotates under the feed hopper. Test tubes or jars are fitted to the base of each funnel. A sampling device that is just as efficient as a spinning riffler and can be readily adapted for sampling liquids is the so-called Rakpak sampler, shown in Figure 2.10 [10]. When sampling liquids, rectangular plastic containers can be used and the junctions between them made waterproof by the simple box-to-box seal illustrated in Figure 2.10. Since the Rakpak works

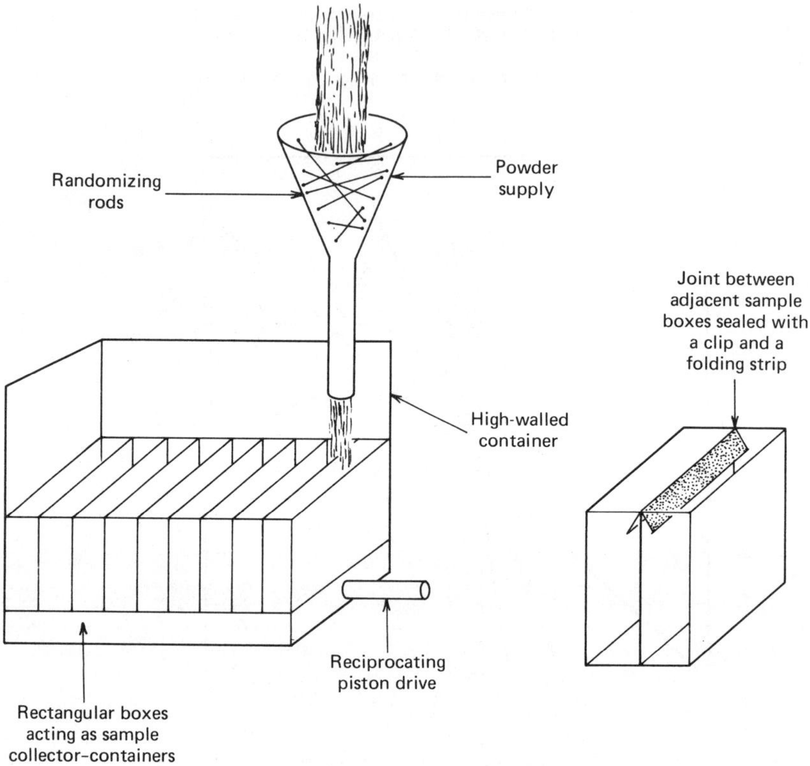

Fig. 2.10 The "Rakpak"® sampling device.

on the same sampling principle as a spinning riffler with the rotary motion replaced by a reciprocating action, its performance is directly comparable to that of the spinning riffler with the added advantage that the problem due to the fan action of the rotating cups is eliminated. When discussing the performance of a sampling device such as a Rakpak, confusion is sometimes caused by a failure to differentiate between equal and representative subsamples. A set of samples obtained by sampling devices such as the Rakpak can be unequal in volume but all equally representative of the original powder supply provided T/t, where T is the time of the powder flow and t, the time for one complete oscillation of the sampling rack, is large.

Sometimes when several kilograms of powder have to be processed to obtain a representative sample, a device known as a *sampling table* is used. The operation of a sampling table is illustrated in Figure 2.11. The powder is fed from the hopper AB down the inclined plane. The powder

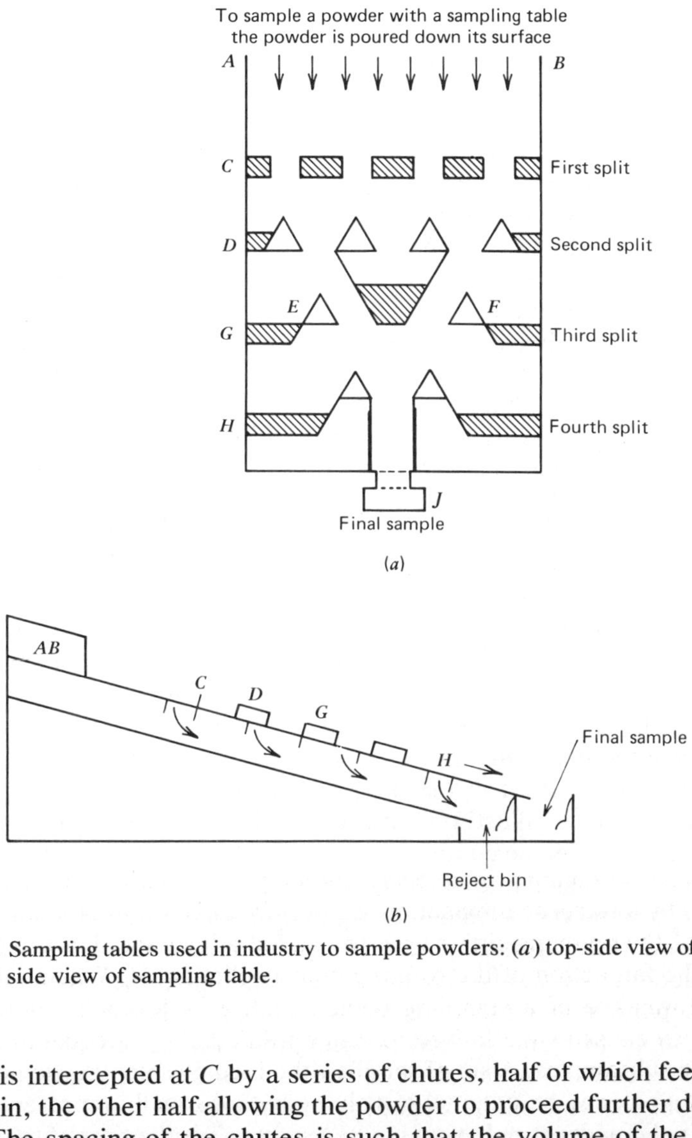

Fig. 2.11 Sampling tables used in industry to sample powders: (*a*) top-side view of sampling table; (*b*) side view of sampling table.

stream is intercepted at *C* by a series of chutes, half of which feed to the reject bin, the other half allowing the powder to proceed further down the table. The spacing of the chutes is such that the volume of the powder passing through them should equal half the powder fed to the equipment. The flowing powder is now intercepted by four prisms that again halve the available quantity of powder sending one portion to the reject bin. It is the intent of the design that the powder flowing from one set of prisms to another is blended in the throats formed by the prisms at *E* and *F*. Further subdivision of the powder is achieved at stages *G* and *H*. Theoretically, a representative $\frac{1}{16}$ of the initial powder sample should emerge from chute *J*. In practice, the sampling device does not always function efficiently.

Kaye investigated the efficiency of a sampling table by splitting the feed trough *AB* of the system shown in Figure 2.11 into eight components [7]. One compartment was filled with red glass beads and the others with white beads. The location of the compartment filled with red beads was changed from experimental run to experimental run. The rate of flow of powder down the table was varied by altering the angle of the table. When the powder was fed through the equipment, the flow pattern of the red beads could be observed directly. In a series of tests the equipment failed to sample adequately from the feed segregated in this way. Visual observation of the flow pattern showed that one of the major causes of inefficiency was the fact that fineparticles were not blended in the throats formed by the prisms. It was also observed that the blending achieved in the throats was less with denser streams of powder. The efficiency of any sampling table used in a sampling procedure should always be evaluated carefully.

When the amount of powder to be sampled is a relatively small quantity in the bottom of a bottle, many scientists would feel justified in shaking the bottle to mix the powder and then snatching a sample with a scoop. In practice, it is not easy to "shake a bottle" to achieve mixing, and it has been shown that some people can introduce segregation in a powder sample by the way in which they shake the bottle [9]. Whenever possible, when handling small quantities of powder, one should build miniature sampling devices.

The Riflo blender, illustrated in Figure 2.12, can be used as a sampling device [12]. This device contains an interlaced set of chute rifflers that act as passive blending elements. The powder being blended is picked up by the flights around the periphery of the rotating outer cylinder and poured onto the interlaced rifflers placed along the axis of the cylinder. By repeated action, the flights and the riffler randomize the position of any given fineparticle along the axis of the blending cylinder. The equipment could be used to blend several samples from a large tonnage of powder with subsequent withdrawal of a smaller portion for characterization study from the end of the cylinder. Alternately, it could be used as a sample homogenizer for the repeated withdrawal of small subsamples from a larger quantity of powder for a variety of studies. If the axis of the cylinder is tilted as illustrated in Figure 2.12, the equipment can be used as a continuous blender with a portion of the output being diverted for characterization studies. By altering the angle of slope, the rate of progress through the blender, that is, the homogenization capacity of the blender, can be varied.

Fig. 2.12 The Riflo continuous blender can be used as a sampling device [12].

It is relatively easy to obtain representative samples from large quantities of powder down to quantities of the order of 1 by using procedures outlined in the foregoing sections.

Further sampling down to 0.01 g, a quantity often required in characterization studies, is the most difficult step in a total sampling procedure. If the characterization technique requires that the powder be made into a suspension, it is convenient to make a suspension of samples of the order of 1 g with subsequent sampling of the suspension by using a modified Rakpak or spinning riffler sampling device (e.g., see the sampler developed by Burt et al. [13]). Alternatively, the 1-g quantity can be made up into a paste with a small amount of dispersing liquid. Pastes are relatively easy to sample efficiently since most differential fineparticle movements within the past are suppressed by the high internal viscosity of the paste. However, for the same reasons, pastes of fineparticles should be handled carefully since apparently harmless stirring or spreading actions applied externally to the paste can create localized regions of severe high shear within the paste that could alter the fineparticle attributes of the original powder. Paste preparation and sampling is particularly useful if the fineparticles are to be dispersed on a slide for examination through a microscope.

If dry fineparticles are placed on a microscope slide and sampled by a "cone-and quartering" procedure (recommended in some publications), it is easy to lose fines to air currents, and larger fineparticles can roll off the slide.

2.4 SUSPENSION SAMPLING

When a suspension is stored in a static container, the fineparticles tend to segregate within the fluid under the influence of gravity. Therefore, it is essential to agitate the suspension immediately prior to the insertion of the sampling probe into the suspension. If the sampling probe draws fineparticle-laden fluid out of the suspension by suction, the probe should be aligned with its mouth facing in the direction opposite to the gravitational forces acting on the fineparticles. If this precaution is omitted, gravity forces will generate a bias toward the finer constituents of the suspension in the withdrawn sample.

When the suspension sample has to be taken from a bulk supply moving through a pipe, the potential segregation effects that can arise in the vicinity of a bed or a constriction in the pipe should always be considered. For example, an elbow bend in a pipe can act as a quarter turn of a centrifuge throwing the fineparticles to the outer surface of the bend in the pipe. In fact, such fractionation action of a bend in a pipe is exploited in the Milltronics online characterization equipment discussed in Chapter 8. Fractionation due to centrifugal forces at a bend can be severe in the case of aerosol systems (these problems are discussed in Chapter 3); however, in a relatively viscous fluid such as water, rods can be used to create randomizing turbulence if it is necessary to sample immediately after a turn in the pipe. In general, the creation of general turbulence just prior to a sampling point is a useful sampling strategy. Commercial equipment for sampling from a liquid stream such as Isolock samplers create turbulent conditions in the stream before and during the sampling action (see Figure 2.13) [14].

2.5 SAMPLING PERSPECTIVE

In any particular sampling situation the variation in the measured size parameter arising from the sampling procedure should always be isolated from the uncertainty due to experimental error. All too often, excessive effort goes into attempts to improve experimental precision in the characterization procedure when data variability is generated by unsatisfactory sampling technique. For example, if repetitive characterization

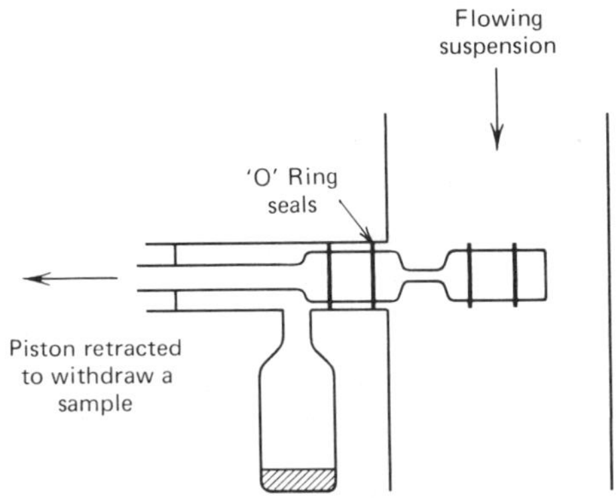

Fig. 2.13 Isolock samplers® are used to sample suspensions flowing in a pipe [14].

measurement of the same sample of powder yields a precision of 0.1% on a measured parameter and sampling variation gives an uncertainty of 2.0% in the measured parameter, there is a mismatch in characterization versus sampling effort, particularly if the high degree of precision in the analytical procedure is being achieved at relatively high cost, in time and equipment. In such a situation, if it is not possible to improve on sampling efficiency, it is worth considering switching to a faster, cheaper, and less precise method of characterization to characterize more samples to allow for sampling uncertainty rather than to persist in collecting precise data on unrepresentative samples.

NOTES AND REFERENCES

1. Sampling equipment literature is available from Gustafson, 6340 LBJ Freeway, Suite 180, Dallas, Texas 75240.
2. Sampling equipment literature is available from Gilson Screen Company, P. O. Box 99, Malinta, Ohio 43535.
3. T. Allen, *Particle Size Measurement*, 2nd ed., Chapman and Hall, London, 1976.
4. Passive blender literature is available from the Kenics Corporation, 1 Southside Road, Danvers, Massachusetts 01923.
5. Passive blender literature is available from Koch Engineering Company Inc., New York, New York.
6. Passive blender literature is available from Charles Ross and Son, 710 Old Willets Path, Hauppage, New York 11787.
7. B. H. Kaye, "Efficient Sample Reduction of Powders by Means of a Riffler Sampler," *Soc. Chem. Ind. Monogr.* 18 (1964), 159–163.

8. British Standards Methods for the Determination of Particle Size Powders. Part 1, "Subdivision of Gross Sample Down to 0.2 ml," BS3406, Part 1, 1961.
9. B. H. Kaye, "An Investigation into the Relative Efficiency of Different Sampling Procedures," *Powder Met.* [9] (1962), 213–234.
10. Information on the Rakpak and the spinning riffler systems is available from Microscal Ltd., 20 Mattock Lane, Ealing, London W5 5BH.
11. Spinning riffler equipment is available from Ladal Scientific Equipment Ltd., Warlings, Warley Edge, Warley, Halifax, Yorkshire, England HX2 7RL.
12. Joy Manufacturing Company, Hard Rock Mining Division, Box 39-J, Denver, Colorado 80239.
13. M. W. G. Burt, C. A. Fewtrell, and R. A. Wharton, *Powder Technol.* 7 (6) (1973), 327–330.
14. Isolock samplers are available from Bristol Engineering Company, 204 South Bridge St., Box 696, Yorkville, Illinois 60560.

CHAPTER

3

AEROSOL SAMPLING

3.1 THE FLUID DYNAMICS OF AEROSOL CLOUDS

Relevant fluidic dynamic concepts and relationships important in aerosol sampling are reviewed in this section. Many of these relationships are also important in stream, sedimentation, and elutriation methods for characterizing fineparticles.

To describe a type of flow that often exists when a fluid moves quietly over a surface at a low rate of velocity, as illustrated schematically in Figure 3.1, one can imagine that the layers of the fluid are sliding over each other. This type of fluid motion is known as *laminar flow*. This term comes from the Latin *lamina,* meaning "thin sheets." When laminar flow is sharply diverted by an obstacle or if the velocity of the fluid increases to the point where the layers of fluid are no longer stable with respect to each other, the liquid starts to move in eddies and vortices. This type of motion is termed *turbulent*. When a fluid is moving in a laminar flow, a velocity gradient is established in the various layers of the fluid with respect to the bounding surface or layer. Immediately above the boundary surface is a layer of fluid called the *boundary layer,* which is stagnant. If fineparticles migrate down to the boundary layer of a moving fluid, they no longer participate in the motion of that fluid but cling to the surface. When the layers of fluid moving over each other can exert measurable forces on each other, that is, when the fluid has internal friction, the liquid is said to be *viscous*. In many situations the movement of gases can be considered to be friction free, and this type of flow is described as *inviscid*.

The concept of the coefficient of viscosity is used to describe the properties of a viscous fluid in a quantitative manner. Imagine that to cause motion of the top laminar fluid element in fluid undergoing laminar flow, a force F must be exerted over an area A. To keep the top layer moving with a velocity Δv when it is at a height Δy above the boundary layer, experiments have shown that F and A are related by the equation

$$\eta = \frac{F}{A} \cdot \frac{\Delta y}{\Delta v} \tag{3.1}$$

where η = the coefficient of viscosity. The coefficient of viscosity is usu-

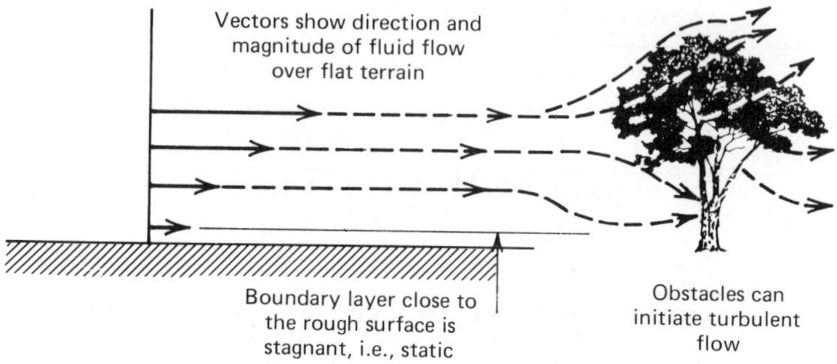

Fig. 3.1 A velocity gradient exists in a fluid flowing slowly over a surface. Symbols: → denotes flow direction.

ally measured in terms of the poise. In absolute units, 1 P equals a force of 1 dyne operating over 1 cm^2 when the velocity gradient is unity. The magnitude of various coefficients of viscosities for several substances at different temperatures are summarized in Table 3.1, where it should be noted that the viscosity of a fluid varies markedly with temperature.

When a fluid moves down a pipe under laminar flow conditions, the center of the fluid moves much faster than the fluid near the walls of the pipe. It can be shown that the velocity of a fluid moving through a pipe is given by the relationship [1]

$$u = \frac{1}{4\eta} \cdot \frac{\Delta P}{\Delta l} \cdot (r^2 - R^2) \tag{3.2}$$

where R = radius of the pipe
Δp = pressure drop over the distance Δl
u = velocity of cylindrical layer of fluid at a distance r from the center of the pipe

The fastest-moving fluid is at the axis of the pipe and moves at a velocity

$$u_{\max} = \frac{-\Delta p/\Delta l \cdot R^2}{4\eta} \tag{3.3}$$

where $u_{\max}$ denotes maximum velocity. (The minus sign appears in the formula because the velocity has positive magnitude in the sense opposite to that of the pressure gradient $\Delta p/\Delta l$.) The average velocity of fluid for laminar flow of fluid in a pipe is

$$u_{\text{av}} = \frac{-(\Delta p/\Delta l)\, R^2}{8\eta} \tag{3.4}$$

Thus the maximum velocity is twice the average. This is an important factor to consider when sucking aerosol into a sampling tube.

TABLE 3.1. Typical Values of Viscosity

Temperature, °C	Viscosity of Castor Oil, P[a]	Viscosity of Water, cP[b]	Viscosity of Air, μP[c]
0	53	1.792	171
20	9.86	1.005	181
40	2.31	0.656	190
60	0.80	0.469	200
80	0.30	0.357	209
100	0.17	0.284	218

[a]P = poise.
[b]cP = centipoise.
[c]μP = micropoise.

The total volume of fluid moving through the pipe is given by

$$Q = \frac{\pi (P_2 - P_1) R^4}{8\eta \cdot l} = A \cdot u_{\text{av}} \tag{3.5}$$

where Q is the volume of fluid flowing per second, A is the cross-sectional area of the pipe, and P_1 and P_2 respectively are the initial and final pressures at the ends of a pipe of length l. This relationship is known as *Poiseuille's law*. If the pressure gradient across the pipe is increased so that the fluid is moved faster, then at a certain flow rate the conditions of flow in the tube change over from laminar to turbulent. The velocity at which the change of flow conditions occurs is known as the *critical velocity*. Under turbulent flow conditions, the overall mass flow of fluid down through the pipe is more uniform across the cross section of the pipe, but within this mass flow there are swirling turbulent vortices. Whenever there is turbulence in an aerosol system, there is always a danger of kinetic agglomeration due to the fineparticles colliding with each other in the randomly moving fluid of the vortex. Generally, turbulence should be avoided in the sampling of aerosols unless one is deliberately trying to capture the dust fineparticles with mist droplets as in the Venturi scrubber.

Flow conditions in a fluid can be described with the aid of a dimensionless number called the *Reynolds number* [2]. (This number is named for Osborne Reynolds, 1842–1912, who made an extensive study of the flow of water in pipes.) The Reynolds number is usually written in the form

$$Re = \frac{v\ \delta \cdot \rho_L}{\eta} \tag{3.6}$$

where ρ_L denotes the density of the moving fluid and δ is some significant dimension of the system in which the fluid flow is being studied. For example, when studying the flow of fluid through a pipe, the value of the diameter of the pipe is usually substituted for δ in equation 3.6. When the movement of a sphere through a viscous fluid is being considered, the significant parameter substituted in the formula is D, the diameter of the sphere. When the fineparticle being considered is of irregular shape, δ is sometimes taken to be the square root of the projected area of the fineparticle at right angles to the direction of the fineparticle movement through the viscous fluid. Unfortunately, many elementary textbooks teach the Reynolds number only with respect to the flow of fluid through a pipe. Because of this limited treatment, it is sometimes possible for a student to gain the false impression that the Reynolds number at which the flow changes from laminar to turbulent is a fixed figure for all flow systems. In fact, the magnitude of a Reynolds number at which flow changes from laminar to turbulent depends on the system being studied, and there is no direct correlation between a critical Reynolds number at which flow changes from laminar to turbulent for fluid flowing round a following sphere and the appropriate Reynolds number for the same transition in pipe flow. Furthermore, laminar flow can persist in Reynolds number regimes where turbulent flow is normally expected if particular care is taken to remove disturbances and transient phenomena from the flow conditions and the immediate physical environment of the flow system. Strictly speaking, any quoting of a Reynolds number should carry a qualification statement describing the system for which the Reynolds number has been set up. In practice, people tend to quote the Reynolds number as if it were independent of any actual system, and this can lead to confusion when interpreting published data.

It should be noted that the Reynolds number can be written in the form

$$Re = (V \cdot \delta) \div \left(\frac{\eta}{\rho_L}\right)$$

The quantity η/ρ_L, the viscosity of a fluid divided by its density, is often referred to as the *kinematic viscosity* of the fluid. The basic unit of kinematic viscosity of a fluid is the *stokes*. Smaller units are called the *centistokes* and the *millistokes*.

When discussing the changeover of laminar to turbulent flow in a pipe, it is useful to use a factor known as the *friction factor*, defined by the relationship

$$f = \frac{\tau_R}{\rho_L} \tag{3.7}$$

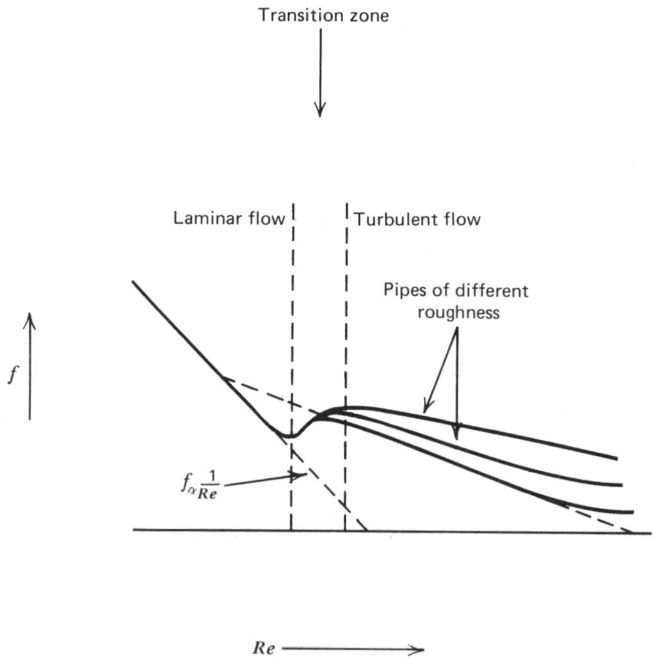

Fig. 3.2 The friction factor f, used in describing fluid flow, is related to Re, the Reynolds number, in a complex manner [4]. The Reynolds number range for transition zone for fluid flow through pipes is approximately 1000 → 3000.

where τ = a quantity known as the sheer stress at the pipe wall (see also Figure 3.2). It can be shown that for laminar flow conditions, equations 3.6 and 3.7 are related as follows:

$$f = \frac{2}{Re} \tag{3.8}$$

Therefore, a plot of $\log f$ against $\log Re$ should yield a slope of -2. Experimental data for many types of pipe are summarized in Figure 3.2. It can be seen that the anticipated relationship holds for Reynolds numbers up to, and of the order of, 2000. The changeover from laminar to turbulent flow conditions depends on complex interactions of small instabilities in fluid flow. In a situation where the equipment is vibration free, extremely steady driving pressures are operative, and the pipe is very straight and smooth, laminar flow conditions can persist at Reynolds numbers up to and in excess of 3000. On the other hand, in adverse conditions, turbulent flow can begin even earlier than Reynolds number 1500.

If aerosol is sucked through a tube at such a rate that the flow conditions are turbulent, fineparticles can be lost to the wall because of coagulation and migration from the turbulent eddies into the boundary layer onto the wall of the tube. To describe aerosol fineparticle losses in a turbulent duct, the concept of *duct half length* is used. This is the length of a duct that causes a loss of half of the fineparticles in the aerosol cloud to the walls of the duct when moving the aerosol through the duct under flow conditions specified by a given Reynolds number. A set of experimental data on fineparticle losses in a duct is summarized in Figure 3.3 and serves to illustrate the serious errors that can be induced in a sample of aerosol withdrawn from a parent population at high flow rates. Once a fineparticle comes into good contact with a clean surface, it adheres tenaciously, and it is not easy to resuspend in moving airstream simply by blowing over the surface with an air jet.

When a sphere falls through a viscous fluid, it is acted on by viscous drag forces, gravity, and inertial forces; the latter term refers to the forces

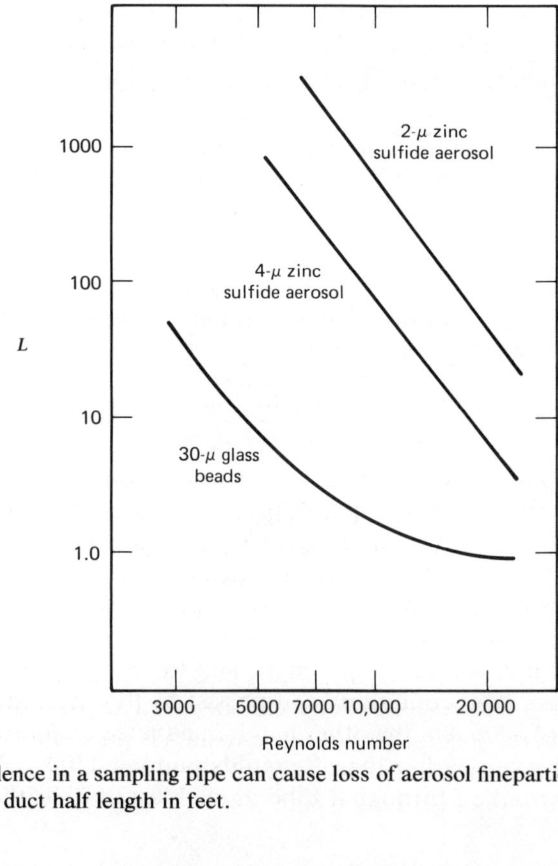

Fig. 3.3 Turbulence in a sampling pipe can cause loss of aerosol fineparticles to the wall of the duct [3]: L, duct half length in feet.

required to move the fluid from in front of the sphere to a position behind the falling sphere. It can be shown that the inertial forces can be neglected for laminar flow conditions. George Gabriel Stokes (1819–1903) solved the theoretical problem of describing the motion of a sphere falling under the influence of gravity in laminar flow conditions. The assumptions he made when deriving his solution were that the sphere (*a*) must be perfectly smooth and rigid, (*b*) should be considered to be moving in an infinite sea of fluid, and (*c*) should be considered to have reached steady terminal velocity, that is, that it had come into a situation of equilibrium under the forces of gravity and the viscous drag forces. After making these assumptions, Stokes was able to show that under laminar flow conditions the drag force is given as

$$F_D = 3\pi\, d\eta\, u \tag{3.9}$$

where u is the terminal velocity of the sphere and F_D is the drag force.

The force exerted by gravity on the falling sphere is modified by the buoyant force (sometimes referred to as the *Archimedean force* as it is an embodiment of Archimedes' principle). Therefore, the downward force on the falling sphere is given by the relationship

$$F_g = \frac{\pi}{6} \cdot (\rho_P - \rho_L) \cdot g \cdot d^3 \tag{3.10}$$

where F_g = efficient gravitational force in fluid on the sphere
ρ_P = density of the sphere
ρ_L = density of the fluid
d = diameter of the sphere

Equating these two relationships and rearranging the terms, we obtain

$$d = \sqrt{\frac{18\eta \cdot u}{(\rho_P - \rho_L)g}} \tag{3.11}$$

which is known as the *Stokes equation*. This equation is used in fineparticle science to deduce a size descriptive parameter from the measured terminal velocity of a fineparticle. This parameter is known as the *Stokes diameter* of the fineparticle and represents the equivalent sphere made of the same substance that would have the same falling speed, in the same fluid, as the fineparticle.

There have been many investigations into the range of Reynolds numbers over which Stokes equations can be used to describe the falling motion of a sphere. A summary of these results is presented graphically in Figure 3.4. In this graph use is made of the dimensionless number C_D, known as the *drag coefficient,* where

$$C_D = \frac{\text{drag force}}{\text{cross-sectional area of the sphere} \times \text{dynamic pressure on the sphere}}$$

$$= \frac{F_D}{d^2/4 \cdot \rho_L \cdot u^2/2}$$

It can be shown that for true laminar flow conditions, the drag coefficient is related to the Reynolds number of the falling sphere system by the relationship

$$C_D = \frac{24}{Re}$$

It follows from this relationship that a logarithmic plot of the drag coefficient against the logarithm of the Reynolds number of the system should be a straight-line relationship of slope −1 for systems in which the Stokes equation describes the motion of the sphere. The data summary in Figure 3.4 shows that the predicted linear relationship between C_D and Re holds up to $Re = 0.2$, and this is commonly regarded as the upper limit for systems used in fineparticle characterization studies. For systems of higher Reynolds number, the Stokes equation is no longer an adequate equation describing the motion of the fineparticle, and C_D is related to Re in a complex manner. Sometimes the behavior of aerosol fineparticles is described in terms of the magnitude of the sphere of unit density that would have the same terminal velocity as the fineparticle. This magnitude is known as the aerodynamic diameter of the aerosol fineparticle. It is

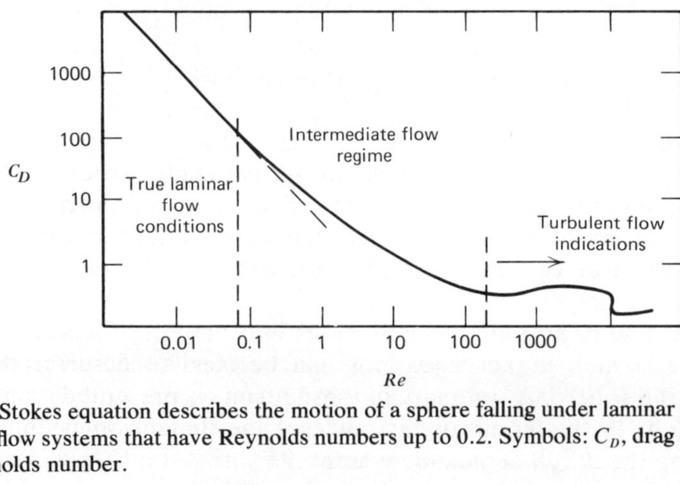

Fig. 3.4 Stokes equation describes the motion of a sphere falling under laminar flow conditions for flow systems that have Reynolds numbers up to 0.2. Symbols: C_D, drag coefficient; Re, Reynolds number.

calculated from the Stokes equation by inserting the measured terminal velocity in the equation and assuming that the term $(\rho_P - \rho_L)$ can be given the value of 1 since the density of air can be neglected with respect to the assumed unit density of the equivalent sphere.

When studying an irregularly shaped fineparticle, the appropriate dimensions in the Reynolds number and the drag coefficients relationship are the cross-sectional area of the fineparticle at right angles to the direction of movement of the fluid and the square root of this area as a linear dimension in the Reynolds number. When considering the behavior of a porous and/or irregular fineparticle, it should be realized that static fluid held by the irregularities and pores of the fineparticle alters the effective density of the falling fineparticle.

An important aspect of the falling behavior of fineparticles is that under laminar flow conditions, the orientation of an irregularly shaped fineparticle with respect to the direction of flow persists throughout the fall of the fineparticle, whereas in turbulent flow conditions forces exist that are able to change the orientation of the fineparticle. There have been many experimental studies of the falling speeds of fineparticles of various shapes at high Reynolds numbers [5].

When dealing with the dynamics of very small aerosols, the fact that the gas will not behave as a continuous medium with respect to the fineparticle has to be taken into account. In such a system the fineparticle is able to slip freely between the molecules of the gas for a certain portion of its trajectory. Thus the fineparticle experiences less drag than it would if the gas were a continuous medium. The correction for this effect is variously known as the *slip correction* or the *Cunningham correction*. Dynamics of very small aerosol sizes also involve Brownian motion, which tends to randomize the position of the fineparticle with respect to its theoretical trajectory. The detailed discussion of the effects of Brownian motion on the dynamics of very fine aerosols is beyond the scope of this text, and those interested in such phenomena should consult more advanced texts on aerosol physics [6].

When sampling a moving aerosol system, it is often necessary to know the velocity of the stream. A flowmeter widely used in aerosol science exploits Bernoulli's principle, which states that if the velocity of a moving fluid is increased, its internal pressure drops.

The design of the flowmeter is illustrated in Figure 3.5. The system shown in Figure 3.5*a* is the classical Pitot tube (after Henri Pitot, 1695–1771). The pressure registered in its manometer is the static pressure of the moving fluid. This pressure is also known as the *stagnation pressure of the fluid*. The pressure registered by the tube shown in Figure 3.5*b*, where the entry to the manometer is at the side so that the fluid is moving

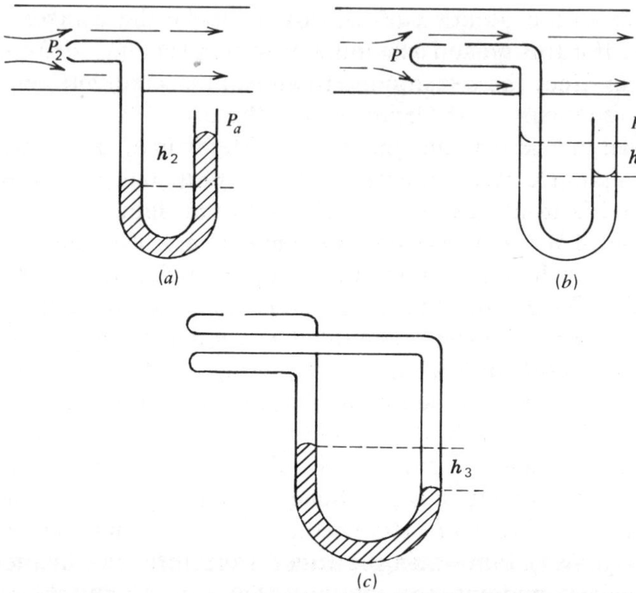

Fig. 3.5 The Prandtl tube (often misnamed the Pitot tube) is used for measuring flow velocities of moving fluids (→ denotes fluid flow lines): (*a*) Pitot tube measures static pressure of a fluid; (*b*) tube for measuring dynamic pressure in a moving fluid; (*c*) Prandtl tube for measuring velocity in a fluid.

over it rather than into it, is termed the *dynamic pressure of the fluid*. By Bernoulli's principle, the difference between the dynamic and stagnation pressures is related to the velocity of the fluid. The system shown in Figure 3.5*c* combines the two systems of Figures 3.5*a* and 3.5*b* to measure the differential pressure directly. The combination is known as the *Prandtl tube* (after Ludwig Prandtl, 1875–1953). The Prandtl tube was also known as the *Pitot static tube*, and common usage shortened this to *Pitot tube*. In fineparticle literature the term "Pitot tube" is most likely to refer to the system shown in Figure 3.5*c*, which is also known as a *pitotmeter*. The pitotmeter is used to measure aerosol flow in ducts and chimney stacks.

3.2 SAMPLING AN AEROSOL CLOUD

A sampling probe for taking a sample from an aerosol cloud is often a small pipe placed in the cloud with the sample being sucked into the pipe. It must be remembered that any bends in the pipe can cause centrifugal fractionation with deposition of the larger fineparticles on the elbows of the corners. For this reason, the number of bends in any sampling system must be minimized. Further, any narrowing of the sampling pipe will cause a change in the flow rate and hence a pressure drop (Bernoulli's principle), which in turn can cause condensation problems in an aerosol cloud.

When sampling a moving aerosol, the most efficient way to sample the aerosol cloud is to align the inlet tube with the direction of flow. If the average flow of aerosol into the sampling tube is the same as the flow velocity of the aerosol system, the sampling is said to be carried out *isokinetically* (Greek *iso*, "the same" and *kinos*, "movement"). Sampling speeds into the tube that are not at the flow velocity of the sampled system are said to be *anisokinetic*. Anisokinetic sampling can introduce bias into the population of the sampled aerosol, as illustrated by the data for unit density aerosols summarized in Figure 3.6.

As explained when discussing Poiseuille's law, the velocity profile of fluid across a sampling tube is parabolic; thus if the average sampling rate is isokinetic, the velocity of entry on the axis of the sampling tube is twice the average flow velocity into the tube. Therefore, the diameter of the sampling probe can affect the sampling efficiency. This problem has been studied extensively and possible bias tabulated [7]. The effect of sampling at various suction rates at various orientations of the sampling probe in a still aerosol cloud has also been tabulated [7].

The dimensions and the shape of the ends of a sampling probe are

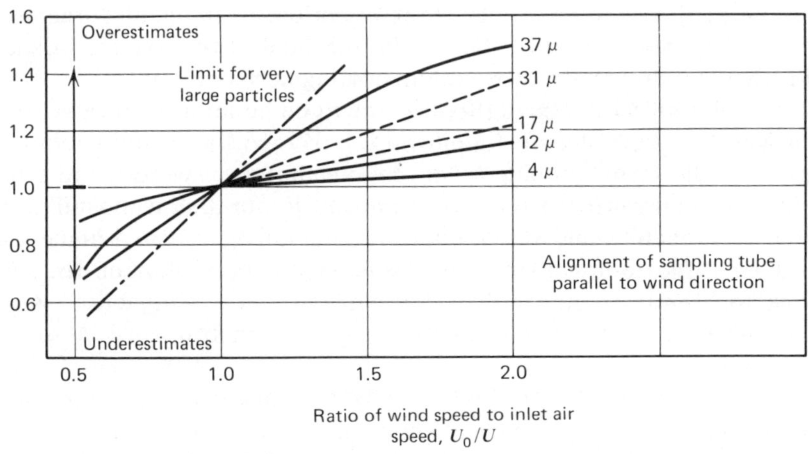

Fig. 3.6 Failure to achieve isokinetic sampling conditions can distort the measured concentration and size distribution function of an aerosol cloud. Inlet speed is the average velocity of air entering the tube [3].

important in considering their effect on a moving aerosol cloud. Thus a tapered end to the probe will have an effect on the sampling performance of the probe different from that of a blunt-ended pipe. For this reason, sampling probes intended for use in chimney stacks, where the aerosol is moving with relatively high velocity, are carefully standardized. For details of the construction of such probes, see the relevant literature of the various standards organizations [8].

In some stream methods for characterizing fineparticles, the aerosol fineparticles are examined directly in the jet of sampled aerosol emerging from a sampling tube by shining a light beam into the jet. If the jet moves directly into an area of static air, shear forces between the moving jet of air and the still air create vortices that spread the fineparticles out of the emergent jet. In an attempt to overcome this problem, the exit jet is sometimes sheathed in a jet of clean air. This extends the zone of undisturbed aerosol jet. This technique is called *fluid dynamic* or *hydrodynamic focusing* (see discussion of hydrodynamic focusing in the resistazone stream counters in Chapter 8).

In many aerosol sampling systems fineparticles are removed from a jet of aerosol by a process known as *impaction*. This type of sampling device is known as a *jet impactor*. In such an impactor the motion of the jet of aerosol is obstructed with a plate placed at right angles to the fluid flow. The presence of this plate causes the stream to split into two symmetrical curved fluid streams. From the point at the exit of the fluid stream out of the jet to the position at which horizontal flow is achieved, the fineparticles in the fluid stream can be considered to be moving in a centrifugal force field. Those fineparticles that experience sufficient force to be forced out of the ribbon of fluid while the fluid is turning are impacted onto, that is, are collected by, the deflecting surface.

In the scientific literature, there seems to be an interchangeable use of the terms "impactor" and "impinger." Early workers with aerosol systems used the term "impinger" to denote that they were using a high-velocity jet and reserved the term "impactor" for a slower fluid stream system. In recent years scientists have generally employed higher fluid speed jets, and today there is no useful distinction between the terms "impactor" and "impinger." In this book the term "impactor" is used except when for historic reasons the name of an instrument is associated with the word "impinger."

The exact theory of the collection performance of a jet impactor is complicated; however, a physical appreciation of the basic limitations and expectations of the system can be gained by using a simplified theory. Consider a jet such that the exit orifice is a rectangular slit with the width being very small compared to the length. The fluid stream coming out of the jet converges as it approaches the orifice of the jet. This convergence does not cease immediately on exit from the orifice, and the minimum cross section of the fluid stream is achieved some distance from the end of the jet. This minimum cross section is known as the *vena contracta* of the stream. Assume that for a slit system of the type defined the *vena contracta* has a width of $2W$. Assume that the linear velocity of this turning stream of fluid is u. Buoyancy forces acting on the fineparticle are neglected. Let R equal the radius of the arc along which the fluid stream is moving. From Newton's second law of motion, the radial force on the fineparticle in the turning fluid is

$$\frac{mu^2}{R} = \frac{\rho d^3}{6} \cdot \frac{u^2}{R} \cdot \pi$$

for a sphere of density ρ and diameter d.

Assume that the radial velocity v in the direction of R is small so that laminar flow conditions exist. Then from Stokes's relationship, we obtain

$$F_D = 3\pi\, d\eta\, v$$

Equating the two forces, it follows that

$$\frac{\rho d^3}{6} \cdot \frac{u^2}{R}\,\pi = 3\pi\, d\eta\, v \qquad \therefore \qquad v = \frac{\rho d^2 u^2}{18R\eta}$$

The length of the curved path followed by the aerosol stream is L, given by

$$L = \frac{2\pi R}{4}$$

Therefore, the time for which the fineparticles are subjected to the centrifugal force is

$$t = \frac{\pi R}{2} \cdot \frac{1}{u}$$

Therefore, fineparticles of diameter d move a distance s in the radial direction, where

$$s = vt$$

$$= \frac{\rho d^2 u^2 t}{18\eta R}$$

But

$$u = \frac{\pi R}{2t} \qquad \therefore \qquad s = \frac{\rho\pi\, d^2 u}{36\eta}$$

Therefore, the minimum size that is completely eliminated from the fluid stream as the jet is diverted, denoted by $d_{\min}$, is given by the relationship

$$d_{\min} = \sqrt{\frac{36\eta w}{\pi\rho u}}$$

From this relationship, it follows that the critical variables in the design of an efficient jet impactor are the fluid velocity, the width of the stream of fluid, the viscosity of the fluid, and the density of the fineparticle.

For a full discussion of the actual effect of the geometry of the orifice and a more exact derivation of the minimum diameter, see Dennis [5] and Fuchs [6].

A basic physical limitation on the fractionation that can be achieved with a jet is imposed by the fact that if the velocity of the fluid achieves that of sound in the fluid, it is no longer possible to maintain laminar flow conditions in the jet. Also, as the fluid moves out of the jet, there is a drop in the pressure that causes adiabatic expansion of the gas. This will normally cause cooling, and vapors may condense on the fineparticles. This will alter their effective mass in the fluid jet and could also cause coagulation of the fineparticles. Jet impactors should never be used to size liquid droplets because of problems associated with this adiabatic expansion, which can cause rapid changes in the size distribution of droplets in the gas stream.

In the Greenburg–Smith impinger, developed in 1922 and widely used in the mining industry, the aerosol stream is drawn through an orifice of 2.3-mm diameter at a speed of 1 ft^3/minute. The jet, immersed in about 2 cm of dust-free water, is positioned 5 mm above the base of a cylindrical container. The velocity of the jet of aerosol coming into the fluid is such that fineparticles of aerodynamic diameter of 1 μm or less are collected with 100% efficiency. A variation of the Greenburg–Smith impinger, known as the *midget impinger,* employs a jet orifice of 1 mm diameter with a flow rate through the instrument of 2.8 l. These instruments are useful for gross collection of solid fineparticles to be examined later by several methods.

The pressure cooker single-jet impactor, developed at the Stanford Research Institute, is an instrument in which the impactor jet is mounted in a 2-quart domestic pressure cooker, thus forming a convenient robust vacuum chamber. Insider the chamber a small turntable is mounted opposite the end of the jet so that fineparticles are collected at different positions during a long sampling period.

When an aerosol stream is sampled through a series of jet impactors of decreasing orifice linked in series, the system is known as a *cascade impactor*. A full discussion of these instruments is given in Chapter 7.

Small cyclones are used to collect fineparticles from an aerosol stream. The operating principles of a cyclone are illustrated in Figure 3.7. The stream of aerosol enters the top of the cyclone tangentially, is whirled downward, and escapes through an inner return flow zone. The lines of flow shown indicate general direction of flow and are not intended to show actual flow patterns. Fineparticles that are too large to follow the changes in the flow pattern impinge on the walls. The smaller fineparticles escape with the discharge stream. Fineparticles hitting the walls slide down and collect at the base. The efficiency of separation and the smallest size of fineparticles captured by a cyclone depend on the exact flow pattern established in the cyclone. Cyclones are rarely efficient at collecting fineparticles smaller than 5 μm; however, they serve as useful prefractionators to take out coarser dusts from an aerosol stream before using more expensive equipment to collect very small fineparticles.

Two of the major problems encountered in the use of jet impactors are referred to as *bounce* and *reentrainment*. Bounce occurs when fineparticles are impacted onto the collecting surface; but then, because of their kinetic energy, they are able to bounce into the airstream and are carried away. Bouncing problems can be avoided by coating the collecting surface of the impactor with grease. The grease, however, can cause problems in any subsequent study of the deposited fineparticles. Reentrainment occurs when aerosol flow over the collected fineparticles lifts them back into the flowing stream. Earlier it was stated that fineparticles lost to

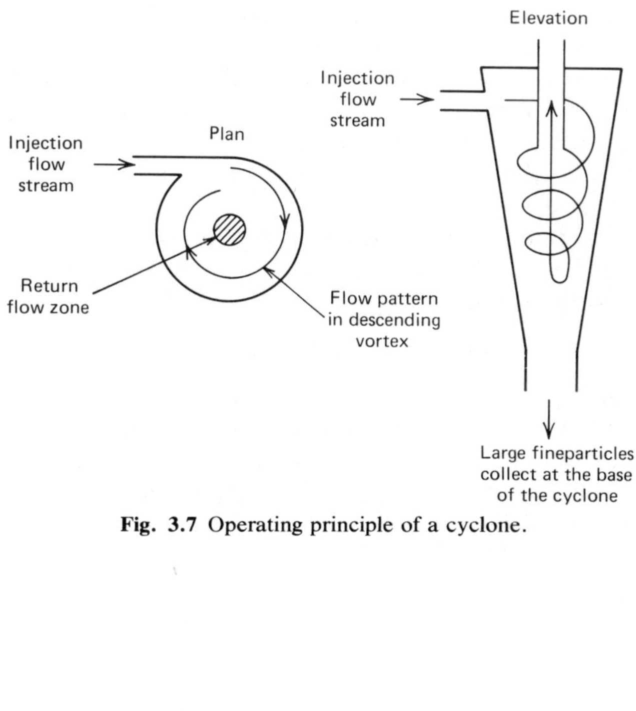

Fig. 3.7 Operating principle of a cyclone.

the boundary layer of a sampling tube are difficult to reentrain. At first sight, this seems to contradict the statement that reentrainment is a problem in impactors. However, the high-speed jet of fluid in the impactor is a much more severe dynamic environment for the deposited fineparticles than is the smooth flow of aerosol in a wide-bore sampling tube.

The virtual impactor has been developed to avoid problems due to bounce and reentrainment without using grease [10]. The basic principle of the virtual impactor is illustrated in Figure 3.8. Instead of a solid surface, the aerosol stream is diverted by a volume of static air in a reservoir. This static air constitutes a virtual surface, and hence the name of the device. The larger fineparticles unable to follow the streamlines of the aerosol stream enter the static air reservoir. Bounce cannot happen with the soft virtual surface, and reentrainment is not possible as the impacted fineparticles sink into the reservoir. Experiments demonstrated that a completely static reservoir of the type shown in Figure 3.8 suffered from turbulence at the edges of the orifice. It has been shown that better fractionation at the virtual surface is obtained if there is a very slow flow rate of gas out of the virtual impactor reservoir. This low flow rate of air out of the oversize zone is typically one-fiftieth of the airflow over the virtual surface [9, 10].

An aerosol sampling device utilizing the principles of virtual impaction is the dichotomous sampler illustrated in Figure 3.9. Loo et al. [9] found that because there are often a few fines trapped in the oversize fraction, the fractionating efficiency of a virtual sampler could be improved by using primary and secondary impactors of the same fractionating power in series.

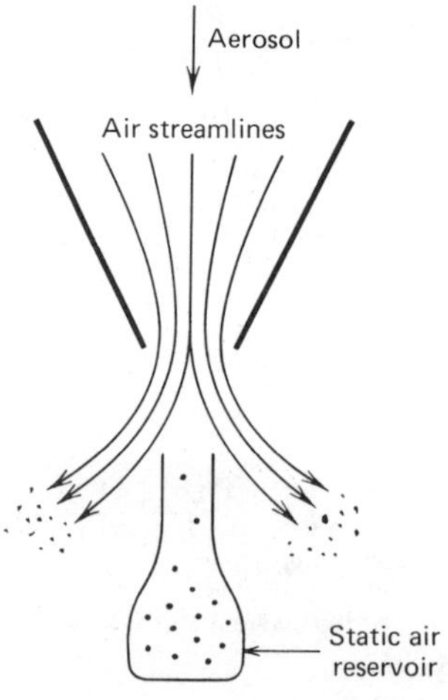

Fig. 3.8 In the virtual impactor, problems of fineparticle bounce and reentrainment are minimized by impacting the fineparticles on a virtual "air surface" [10].

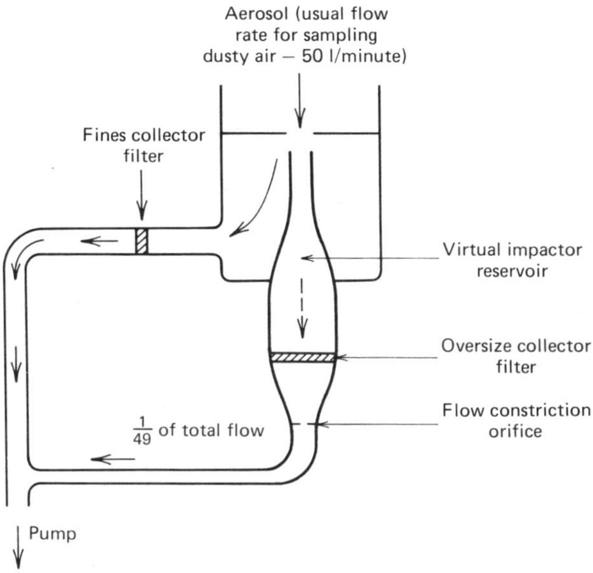

Fig. 3.9 The dichotomous sampler is a virtual impactor [10].

In studies of the characteristics of an aerosol cloud contained in a chamber, the fineparticles can sometimes be collected by letting them fall onto various types of collector plate placed at the bottom of the chamber. Strips of "two-sides sticky" tape can be placed on glass slides and used to collect solid fineparticles. In another technique a glass slide to be used to collect droplets is covered with magnesium oxide deposited from a smoke in such a manner that droplets from the chamber create craters in the magnesium oxide. From the dimensions of these craters, the size of the droplet can be evaluated. Another technique is to add dye to the droplets so that as they fall on a collecting filter paper they leave colored disks, the size of which can be related to the size of the droplet. Each filter medium tends to absorb the droplet in a different way. Therefore, there is no unique relationship between the droplet diameter and that of the disk formed on the collection material; and the spreading factor, which relates the spread-out disk diameter to that of the original droplet, has to be evaluated for any particular collection system [11].

Another technique is to add a small amount of acid to the liquid being used to generate a test cloud of aerosol. Droplets of such an aerosol when collected on litmus paper leave circles of differing color, the size of which can be related to the dimensions of the droplets.

3.3 FILTRATION OF AEROSOLS

The traditional filter paper that the student encounters in the chemistry laboratory is usually composed of a random mat of fibers of the kind used to make ordinary paper. This type of filter is described as a cellulose paper filter. Similar filters can be made from glass fibers [12]. Random fiber mat filter systems are referred to as *depth filters* because the fineparticles penetrate the depth of the array of random fibers. There is always a random chance of a relatively large tunnel existing through the structure of this type of depth filter so that some relatively large fineparticles can pass right through. This makes it necessary to quote penetration efficiencies for this type of filter. A penetration efficiency rating of a filter paper is usually a graph of the efficiency of capture, against the size of the fineparticle, for that filter. When purchasing a depth-type filter paper, one is usually able to acquire information on the average fineparticle captured by the filter and the penetration efficiency of that filter.

Membrane filters are very different in structure to fibrous depth filters. In one of the earliest methods for making a membrane filter, finely ground salt was mixed into a molten plastic that was then spread out into a very thin film. Hot water was then used to dissolve the salt fineparticles. The percentage of salt in the mixture had to be high enough for the dissolved fineparticles to leave continuous random paths through the thin plastic film. The resultant membrane film had a structure somewhat like a sponge. Modern membrane fineparticles are made by using a gelation process with plastics such as cellulose acetate and cellulose nitrate. The pore structure of a cellulosic membrane filter is shown in Figure 3.10.

The stopping power of a membrane filter has to be calibrated by a test aerosol because of the many variable aperture random paths throughout its structure. Thus the membrane filter shown in Figure 3.10*b* has a rating of 0.45 μm. This should be compared with the size of the holes in the adjacent picture, which, at the magnification of the two photographs, represents physical holes of 0.4 μm. The access pores on the surface of the membrane filter appear to be many times larger than the rating of the filter. This is because the stopping power of the filter depends on the minimum throat encountered in a traverse through the filter, the efficiency of fineparticle capture due to electrostatic forces, and also the effects of Brownian motion and Van der Waal's forces. Many different membrane filters with different pore ratings are available from several manufacturers [13–16].

Membrane filters are considerably thinner than random mat fibers; 150 μm is a typical thickness for a membrane filter. They usually have to be supported in special filter holders to avoid damage during the filtering

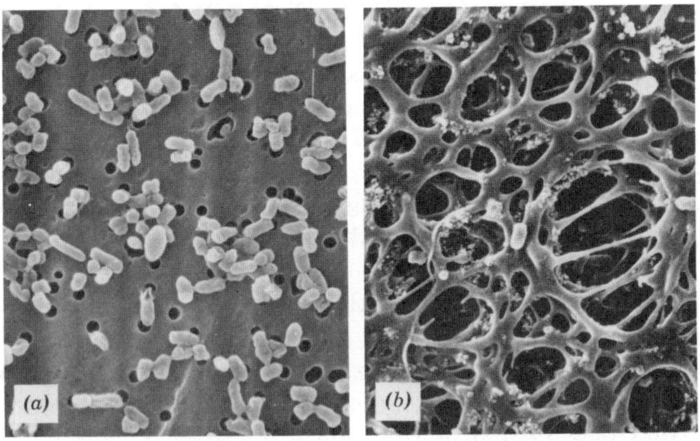

Fig. 3.10 Filtration technology for aerosol sampling has been revolutionized by the development of new types of filters: (*a*) the surface of a Nuclepore filter; (*b*) the structure of a cellulosic membrane filter. Both filters are photographed at the same magnification; the rated size of the Nuclepore filter is 0.4 μ and that of the cellulosic filter 0.45 μ. (Bacteria on the surface of the filters—*Serratin Marcescens*.)

process. These holders are available from the manufacturers of the filters. Membrane filters also tend to be of high dielectric constant and can easily pick up electrostatic charges when handled. Stray aerosol fineparticles attracted by the electrostatic forces distort the estimates of the fineparticle population of a subsequently sampled system, especially if the filter is being used to investigate a system with low fineparticle content such as a pharmaceutical solution or a clean room in a hospital or an aerospace laboratory. Unwanted fineparticles that wander into a measuring system are termed "tramp fineparticles." To avoid the pickup of tramp fineparticles on nonconducting filters, it is usual to use a low-power radioactive source to discharge the surface of the filter while it is being handled and mounted in its holder. The suppliers of membrane filters supply suitable low-cost static eliminators. Thus Biorad Laboratories sell a static eliminator containing 500 μCi of polonium 210 [15].

The filter shown in Figure 3.10*a* is called a *Nuclepore® filter*. This type of filter is also referred to as a *membrane filter*, but the Nuclepore filter is made by a process entirely different from that used in the manufacture of the sponge-type membrane filter. To make the Nuclepore filter, a thin film of plastic is irradiated with neutrons that pass through the film at right angles to its surface. The passage of a neutron through the film weakens the plastic along its track. Subsequent etching of the neutron track produces a smooth cylindrical pore through the plastic film. The flux of neutrons through the plastic is a completely random process, and hence the random pattern of holes in the Nuclepore filter (see Figure 3.10*a*). It can be shown from a study of the theory of random processes that the chance of two adjacent holes fusing to produce a larger hole becomes significant if the area occupied by the holes approaches 5% of the total area. For this reason, the available pore area of a Nuclepore filter is kept low to limit the chance of adjacent pores fusing to provide a larger hole passing through the filter. It is possible to produce a whole range of filters of various pore size simply by altering the etch time when processing the bombarded plastic. Standard Nuclepore membrane specifications are summarized in Table 3.2.

As can be seen from the photographs shown in Figures 3.10*a* and 3.10*b*, Nuclepore-type filters are very useful when the filtered fineparticles are to be examined by image analysis procedures. Extensive studies on the use of membrane filters and Nuclepore filters for the general study of aerosols have been reported by Spurný [17, 18]. Special filters for use with the scanning electron microscope, called *micropore filters*, are available from the Amicon corporation [19].

TABLE 3.2. Standard Nuclepore Membrane Specifications

Specified Pore Size, μm	Pore Size Range, μm	Nominal Pore Density, Pores/cm²	Nominal Thickness, μm	Typical Flow Rates at 10 psig ΔP—70°F: Water gpm[a]/ft²	N_2, cfm[b]/ft²
8.0	6.9–8.0	1×10^5	8.0	144.0	138.0
5.0	4.3–5.0	4×10^5	8.6	148.0	148.0
3.0	2.5–3.0	2×10^6	11.0	121.0	128.0
1.0	0.8–1.0	2×10^7	11.5	67.5	95.0
0.8	0.64–0.80	3×10^7	11.6	48.3	76.0
0.6	0.48–0.60	3×10^7	11.6	16.3	33.0
0.4	0.32–0.40	1×10^8	11.6	17.0	33.0
0.2	0.16–0.20	3×10^8	12.0	3.1	8.9
0.1	0.08–0.10	3×10^8	5.3	1.9	5.3
0.08	0.064–0.080	3×10^8	5.4	0.37	2.6
0.05	0.040–0.050	6×10^8	5.4	0.12	1.3
0.03	0.024–0.030	6×10^8	5.4	0.006	0.19

Source. Reference 17.
[a]Gallons per minute.
[b]Cubic feet (per) minute.

3.4 CAPTURE OF AEROSOL FINEPARTICLES BY ELECTROSTATIC PRECIPITATION

Many natural aerosols are electrically charged because of the way in which they are generated [20]. However, it is usually safer to charge the aerosol prior to the collection process if high collection efficiencies are desired. This is achieved by flooding the aerosol stream with ions from a subcorona discharge prior to the aerosol entering the electrostatic precipitator. A good discussion of the elementary theory of an electrostatic precipitator and the generation of subcorona discharge is given by Moore [21]. A more advanced discussion of the performance of electrostatic precipitators is that of White [22]. Several electrostatic precipitators for aerosol sampling are available commercially [14, 23].

3.5 THERMAL PRECIPITATION OF AEROSOL FINEPARTICLES

Fineparticles placed in a thermal gradient will tend to move in the direction of the lower temperatures, ultimately depositing on the cool surface. This principle is used in the "thermal precipitator" for sampling fineparticles from an aerosol cloud. In these instruments the cold and the hot surfaces are placed relatively closely together so that very high thermal gradients are achieved with temperature differences as low as 50°C. Thermal precipitators collect fineparticles between 0.01 and 5 μm with high efficiency. They are often used as a second stage in a total sampling system in which cyclones and/or impactors are used to remove the larger fineparticles before passing the residual aerosol stream through a thermal precipitator [20, 23].

NOTES AND REFERENCES

1. The derivation of these fluid dynamic formulas can be found in physics textbooks; see, for example, R. Stevenson and R. B. Moore, *Theory of Physics*, Saunders, Philadelphia, 1967.
2. H. E. Huntley, *Dimensional Analysis*, Macdonald, London, 1958.
3. A. Lieberman, "Automatic and Remote Monitoring of Airborne Particles," *Heating, Piping and Air Conditioning*, 117–120 (November 1968).
4. J. Happle and M. Brenner, *Low Reynolds Number Hydrodynamics*, Prentice-Hall, Englewood Cliffs, N.J., 1965.
5. R. Dennis (Ed.), *Handbook on Aerosols*, Technical Information Centre, Office of Public Affairs, U.S. Energy Research & Development Administra-

tion, available as document No. TID26608 from National Technical Information Service, U.S. Department of Commerce, Springfield, Virginia 22161, 1976. See Chapter 6, "Summary Tables and Graphs for Static and Dynamic Characteristics of Aerosol Systems."

6. N. A. Fuchs, *The Mechanics of Aerosols*, Pergamon, London, 1955.
7. C. N. Davies, "The Entry of Aerosols into Sampling Tubes and Heads," *Br. J. Appl. Phys.* (*J. Phys. D*) 1 (1968), 921–932.
8. For a review of nozzle design for sample intake tubes, see relevant sections of T. Allen, *Particle Size Measurement*, 2nd ed., Chapman and Hall, London, 1975.
9. B. W. Loo, J. M. Jaklevic, and F. S. Goulding, "Dichotomous Virtual Impactors for Large-Scale Monitoring of Airborne Particulate Matter," in *Fineparticles*, B. H. Liu (Ed.), Academic, New York, 1976, pp. 311–350.
10. T. G. Dzubay, R. K. Stevens, and C. M. Peterson, "X-ray Fluorescence Analysis of Environmental Samples," in *Application of the Dichotomous Sampler to the Characterization of Ambiant Aerosols*, Thomas Dzubay (Ed.), Ann Arbor Science Publishers, Ann Arbor, Michigan, 1978.
11. H. L. Green and W. R. Lane, *Particulate Clouds*, Van Nostrand, New York, 1964.
12. "Glass Fibre Filter Paper, A Versatile Laboratory Tool," *Am. Lab.* (April 1976).
13. Gelman Instrument Company, P. O. Box 1448, Ann Arbor, Michigan 48106.
14. B. G. I. Incorporated, 58 Guinan Street, Waltham, Massachusetts 02154.
15. Biorad Laboratories, 2200 Wright Avenue, Richmond, California 94804.
16. Millipore Corporation, Ashby Road, Bedford, Massachusetts 01730.
17. Nuclepore Corporation, 7035 Commerce Circle, Pleasanton, California 94566, provides comprehensive literature on the structure and the properties of Nuclepore filters.
18. K. Spurný, "The Investigation of the Filtration Process of Nonactive and Radioactive Aerosols by Means of Fibrous and Membrane Filters," Academia, Státní tiskárna n.p. závods, Praha 8, Czechoslovakia.
19. Amicon Corporation, 21 Hartwell Avenue, Lexington, Massachusetts 02173.
20. See discussion in R. D. Cadle, *Particle Size Theory and Industrial Applications*, Reinhold, New York, 1966.
21. R. D. Moore, *Electrostatics*, No. S57 in the Science Study Series, Anchor Books, Doubleday, New York, 1968.
22. H. J. White, *Industrial Electrostatic Precipitation*. Addison-Wesley, Reading, Massachusetts, 1962.
23. Thermo-Systems, Inc., 2500 North Cleveland Avenue, St. Paul, Minnesota 55113.

CHAPTER

4

SIEVE FRACTIONATION OF POWDERS

4.1 THE BASIC CONCEPTS

A sieve is defined as a vessel with a meshed or perforated bottom used to separate the fine part of anything from the coarse. The word "mesh" means an opening between the threads of a net or the threads and knots bounding the opening. Thus we see a sieve as a net used to separate the large fineparticles from small ones. A powder can be split into fractions of known size range by using a series of sieves of different mesh size. From the quantity of powder in each fraction, the size distribution function of the powder can be deduced. The mesh aperture of a sieve constitutes a go/no-go gauge used to split the powder into two groups.

In recent years the development of membrane filters of different pore size has made it possible to extend the concepts of powder sieving to the fractionation of aerosols using a series of membrane filters of different pore size (see the discussion on the use of membrane filters for aerosol sampling in Chapter 3). The fractionation of aerosols in this way is complicated by the important effects of surface and electrostatic forces. A detailed discussion of the technology and techniques is beyond the scope of this textbook, but those interested in this extension of the concept of sieving should consult the work of Spurný and co-workers [1]. The technique of sieving can be extended down to the fractionation of molecules by using substances known as *molecular sieves* [2].

In a characterization study the powder is placed on the coarsest sieve and the residue weighed. The process is repeated for each sieve of diminishing aperture size. From these residues, the cumulative percentage oversize weight distribution function of the powder is calculated.

The inaccuracies and the uncertainties of characterization by sieve fractionation arise from the experimental problems of determining the sieve residues and from the nonideal nature of the sieving surfaces. Characterization of a powder with sieves does not give information on the largest and smallest fineparticles present in the powder. As a result of this lack of information on the tail ends of the distribution function, conversion of sieve-generated size information into average parameters such as mean size and surface area can yield only approximate answers.

Theoretical aperture size, nominal size, and cut size of the sieve are equivalent terms for the size of holes in a sieve. The presence of a range of aperture sizes in any real sieving surface is a source of error in sieve-based characterization studies since the theoretical or nominal size of the sieve is taken to be the boundary limit for the sieve residue. The range of aperture sizes present in a sieving machine is referred to as the *sieve aperture distribution* or, in the case of wire-woven sieves, the *mesh aperture distribution*.

The effect of the mesh aperture distribution on the fractionation achieved by a sieve depends on the sieving technique. If sieving is carried out for an infinite time, the largest aperture present is the cut size of the sieve. With the finite times of real sieving experiments, the effective cut size lies between the largest aperture and the nominal size of the sieve. In many situations the presence of some apertures smaller than the nominal size delays the passage of some fineparticles able to pass the sieve, whereas the presence of the oversize aperture allows a few fineparticles through the sieve that should have been retained. These opposing effects tend to compensate each other, so that the magnitude, but not the constitution, of the residue is closer to the predicted residue using the nominal cut size than would be anticipated from a study of the frequency of oversize apertures in a sieve. There is very little information on the effect of mesh aperture distribution for various sieving actions on the effective residues retained on sieves. The effective cut size of a sieve is usually determined by using a standard powder.

The effective cut size of a sieve depends on the amount of material in a powder being characterized that lies within the size range of the mesh apertures. For other than spherical fineparticles, there is no simple way of correlating the size of a fineparticle determined by sieving to the size measured by a different physical procedure. The "kinetic residue" of a sieve is a powder characterization parameter related to the mesh aperture distribution of the sieve. The theory underlying the measurement of kinetic residues is discussed in Section 4.4.

Wear of a sieve surface can change its effective cut size. There is little quantitative information available on the type of damage occurring in a sieve surface during use. The cost of characterizing the very large number of apertures in a sieving surface usually means that criteria used in industrial plants for replacing sieves, or rejecting them for analytical purposes, depend on the arbitrary action of the plant manager. Newer methods of characterizing the apertures of a sieving surface, based on automated iconometrics, offer more information on the wear rates of sieving surfaces, and their effect on the sieve's fractionating capacity will become available in the not too distant future. Effects of sieve wear are currently a major cause of uncertainty in the characterization of powders by sieve fractionation [3, 4].

Coarse sieves are often made by punching holes in a metal sheet. For sieves of aperture smaller than a few millimeters, the sieving surface is usually made from woven wire cloth. Until the early 1940s, the practical limit for accurate wire-woven sieves was a 200-mesh, 76-μm-aperture sieve. For this reason, methods for characterizing fineparticles that are smaller than 76μm were referred to as *subsieve methods* of fineparticle characterization. Modern sieves can have apertures as small as 1 μm, and the term "subsieve size" is no longer useful and should be avoided.

The mesh number (or the "mesh") of a sieve is the number of parallel wires per inch in the weave of the cloth. The mesh number of a sieve is not a unique description of the size of the mesh aperture since this latter dimension depends also on the diameter of the wire and the type of weave. For this reason, any record of a powder characterization by sieving should specify the actual aperture size and the geometry or the specific type of standard sieve being used. Many types of weave are used in creating wire cloth for industrial fractionation of fineparticles, but in the making of cloth for sieves used in characterizing powders, a simple under–over weave on equally spaced warp threads is used. Fine wire-woven sieves used for characterizing powders are referred to as *analytical* or *standard* sieves. The standards set by many countries specify that the warp and weft wires for standard sieves are to be of equal diameter and of magnitude equal to the size of the aperture formed in the woven surface. The geometry of a sieve aperture made by this type of weaving is shown in Figure 4.1.

It was assumed a few years ago that as the art of weaving improved, the lower size limit of sieve characterization would move downward and that these future sieves would maintain the specification of wire width equal to aperture dimension. On this basis, powders were described by their theoretical mesh size. Thus a powder of an average size of 20 μm was referred to as 625-mesh powder and 5-μm average size powder, as 2500-mesh powder. The confusion has been compounded in the literature of fineparticle science by the fact that inevitably the term "theoretical" has been dropped from powder description. The scientist new to the field of fineparticle science reads about "2500-mesh" powder and can spend fruitless hours trying to locate the manufacturer of such a sieve. Wire-weaving technology was developed to produce a standard sieve of 400 mesh (37 μm); at this size a very wide range of aperture sizes had to be permitted in the sieving surface so that one could have a realistic specification for the wire cloth (see Table 4.1). Sieves that have apertures as small as 1 μm have been developed by electroplating and photoetching

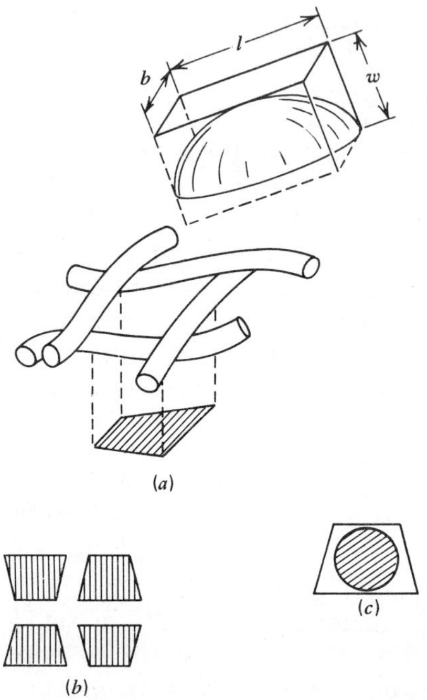

Fig. 4.1 Structural features of the apertures in a wire-woven sieve: (*a*) the "midpoint" is taken to be the aperture dimension; (*b*) repeating pattern of apertures created by a simple weave of equal diameter warp and weft threads.

techniques. Such sieves do not have solid strips between apertures equal to the aperture width. Therefore, the term "theoretical mesh" is not meaningful for such sieves, and the terminology should not be used [5].

One way of making sieves of aperture smaller than 37 μm is to deposit metal by electrolysis onto a drawn mask of the required sieving surface. The deposited metal is peeled off, and the resultant material is termed the *electroformed sieving surface*. The size of the aperture in the electroformed surface can be made smaller by prolonging the electrodeposition process. However, prolongation of the electroplating process means that the percentage of the sieving surface that is open to the passage of fineparticles decreases as the aperture size decreases. This can result in slow sieving rates. For example, some of the early 1-μm-aperture sieves made by this process had an open area as low as a few percent of the sieving surface. Most electroformed sieving surfaces have to be mounted on a coarse support grid. Electroformed sieves are expensive but precise. Manufacturers guarantee aperture dimension variation of ± 2 μm on the nominal size of the aperture. Wire-woven sieves remain the most widely used type of sieve, but electroformed sieve use is increasing. When electroformed sieves were first introduced, they were sometimes called *mi-*

TABLE 4.1. Permitted Variations in Aperture Dimensions for Wire-Woven Sieves Conforming to U.S. Sieve Series (Fine Series)

Width of Opening, μm	Screen No.	Permissible Variation, %: Average Opening (±)	Permissible Variation, %: Maximum Opening (+)[a]	Wire Diameter, mm
5660	3.5	3	10	1.28 –1.90
4760	4	3	10	1.14 –1.68
4000	5	3	10	1.00 –1.47
3360	6	3	10	0.87 –1.32
2830	7	3	10	0.80 –1.20
2380	8	3	10	0.74 –1.10
2000	10	3	10	0.68 –1.00
1680	12	3	10	0.62 –0.90
1410	14	3	10	0.56 –0.80
1190	16	3	10	0.50 –0.70
1000	18	5	15	0.43 –0.62
840	20	5	15	0.38 –0.55
710	25	5	15	0.33 –0.48
590	30	5	15	0.29 –0.42
500	35	5	15	0.26 –0.37
420	40	5	25	0.23 –0.33
350	45	5	25	0.20 –0.29
297	50	5	25	0.170–0.253
250	60	5	25	0.149–0.220
210	70	5	25	0.130–0.187
177	80	6	40	0.114–0.154
149	100	6	40	0.096–0.125
125	120	6	40	0.079–0.103
105	140	6	40	0.063–0.087
88	170	6	40	0.054–0.073
74	200	7	60	0.045–0.061
62	230	7	90	0.039–0.052
53	270	7	90	0.035–0.046
44	325	7	90	0.031–0.040
37	400	7	90	0.023–0.035

[a]For sieves from the 1000-μm (No. 18) to the 37-μm (No. 400) size, inclusive, not more than 5% of the openings shall exceed the nominal opening by more than one-half of the permissible variations in maximum opening.

cromesh sieves. Since electroformed sieves are now manufactured for apertures up to at least 100 μm in size, this is not a meaningful term and should be dropped from common usage. Some small-aperture sieves are made by photoetching techniques. They are sometimes referred to as *Dutch sieves* [6].

The aperture in a wire-woven sieve when projected into a two-dimensional plane is a trapezoid. The aperture size of a wire-woven sieve aperture as viewed through a microscope is the magnitude of the distance between the midpoints of the sides of the trapezoid (see Figure 4.1). For most purposes, this is equivalent to the effective diameter of the sieve as measured by passing a spherical fineparticle through the aperture. The equivalence is not exact because the maximum distance between the sides of the wire-woven aperture lies in a plane at a slight angle to the plane of projection of the trapezoid.

For an irregularly shaped fineparticle, the probability of passing through the sieve depends on the orientation at which it encounters an aperture. Thus the fineparticle shown in Figure 4.1 will pass the sieve if it is presented to the aperture end on but will not pass if it lies across the aperture. The orientation of a fineparticle at the sieving surface depends on the sieving actions. Traditional sieving actions include sifting, swirling, shaking, and jarring. Sifting describes any movement in which the fineparticles move sideways across the sieving surface. The term "swirling" is used to describe circular sifting; "shaking" describes vertical motion of the sieve; and "jarring" indicates that the rim of the sieve is rapped sharply on a hard surface to cause vibration of the sieve surface. In general, sifting actions result in the nonspherical fineparticles lying with a major axis along the surface, whereas vertical shaking and jarring randomize the orientation of the fineparticles. The sieving action affects the probability of a fineparticle encountering a few larger-than-nominal apertures. Thus sifting causes the fineparticles to track across the sieving surface, thus increasing the probability of encountering larger apertures.

"Blinding" of a sieving surface occurs when fineparticles become jammed into the apertures of the sieve. It is also called "pegging." A jarring element is often incorporated into a sieving action to help reduce blinding. The jarring action vibrates the surface of the sieve and shakes oversize fineparticles out of the apertures. Blinding can be severe with irregular fineparticles sifted on a wire-woven sieving surface because the adjacent sides of the wire-woven aperture act as a funnel to guide elongated fineparticles into the hole. A following large fineparticle moving in a horizontal action can then help jam the fineparticle into the hole. (Note in Figure 4.1 that the wires are not as large as the width of the hole; smaller wires were shown for the sake of clarity. The funnel shape of the aperture is quite noticeable in woven surfaces where the wire diameter equals the aperture width.)

The amount of powder placed on the sieve is the "powder bed." Heavy powder beds placed on the sieving surface can cause it to sag, a deformation known as "bellying." Gross bellying can create permanent damage to the sieve and increase the cut size of the sieve. The increased aperture size caused by mild bellying is not easily noticed because the sieve surface returns to its original state once the powder is removed. Powder falling onto the sieving surface during the final stage of a vertical sieving action can increase bellying for a few moments. This is called "transient bellying." As the mesh springs back, oversize, near-aperture-magnitude fineparticles entering the aperture during transient bellying can be trapped in such a way that they are difficult to remove from the aperture by gentle cleaning actions.

Many standard fractionation procedures specify that a sieve fractionation study is complete when a minimal rate of powder passage through the sieve has been attained. Undetected blinding of sieves can slow the rate of powder passage and result in premature termination of the sieving action and measured residues that are too large. Cleaning a blinded sieve is difficult. One useful technique is to invert the sieve and rap it sharply on a hard surface. Air jets are also useful to blow the trapped fineparticles out of the apertures. Some technologists have recommended the use of wire brushes to clean blinded sieves; this can result in damaged apertures. Ultrasonic cleaning baths have also been recommended for cleaning sieves. This technique should be used with caution. From the outside, the treatment received by a sieve in an ultrasonic cleaning bath appears to be a gentle agitation; however, cavitation occurring with the fluid of the cleaning bath can create severe local stress and local damage to the sieve apertures. Some sieve manufacturers give a chart indicating which of their more fragile, finer sieves may be damaged by ultrasonic cleaning [6].

Sometimes physical processes can interfere with the passage of fineparticles through the sieve and make measured residues higher or lower than the true expected values. The agitation of dry nonconducting powders on the surface of the sieve in low-humidity conditions can cause transient agglomeration of fineparticles due to triboelectric charges created on the fineparticle. Those electrostatic forces that interfere with passage of fineparticles through the sieve can usually be dissipated by carrying out the analysis in a controlled environment, with relative humidity adjusted to be higher than 30%. The powder can also be discharged by using a β-ray source. Some workers have claimed that a piece of metal chain or small metal balls placed in the powder on the sieve helps to discharge triboelectric charges. Provided the outer case of the sieve is electrically grounded, this can be useful; however, great caution should be exercised before taking such steps to aid powder movement through the sieve since these so-called sieving aids could damage the apertures of the sieve and grind the powder on the sieve. Any friable (low-strength) material may break down into smaller fineparticles if subjected to prolonged agitation on a sieve. Powders like tungsten carbide can cut through a sieving surface unless great care is exercised during the sieving process. Wet sieving, in which fineparticles are washed through the sieve, is often used to fractionate friable fineparticles and to overcome electrostatic problems.

The American Society for Testing of Materials (ASTM) specifications for standard sieves are summarized in Table 4.2. The range of apertures is such that the ratio of alternating members of the sequence are in root of 2 progression. This spacing of standard sieve apertures was decided on by early workers in the subject of fineparticle characterization so that if the oversize distribution function of a powder were to be plotted on a logarithmic scale, the size data points would be equally spaced on the logarithmic scale. (Industrial scientists in Germany in the late 1890s found experimentally that many crushed products such as crushed rock give a straight-line relationship if the oversize distribution function was plotted on log–log graph paper.) For many powders, this $\sqrt{2}$ progression of available apertures is not particularly convenient. Modern electroformed sieves are available in a much wider range of aperture sizes.

A "nest" of sieves consists of a stack of sieves with the base rim of one sieve fitting snugly into the top of another sieve. The nest is assembled with the largest-aperture-size sieve at the top. The finest sieve at the bottom is fitted with a container in which the fines are accumulated. In a nest of sieves the powder able to pass the finest sieve must diffuse down through the residues accumulating on each sieve in the nest. This requires hours of sieving time if loads of a few hundred grams are sieved on a column of 8-in.-diameter sieves. It is very difficult to determine when a nest sieving process is complete, and termination time is usually based on preliminary experiments using standard powders similar to the one to be fractionated in the nest of sieves.

The terms "screen" and "sieve" are sometimes used as synonyms. The term "screen" should be reserved to describe industrial processes for fractionating powders. Thus the screening of crushed coal would be the process of fractionating the coal using large screens. The subject of screening as distinct from sieving is concerned with fractionation as distinct from characterization.

A "riddle" is a coarse sieve. A "trommel" is a large circular sieve

TABLE 4.2. Comparison of Various Types of Standard Sieve

Tyler[a]	U.S.[b]		British[c] Standard		German DIN[d]		
Equivalent Mesh	Mesh No.	Openings, mm	Mesh No.	Openings, mm	DIN No.	Mesh/cm²	Openings, mm
3½	3½	5.66	...	...	1	1	6.000
4	4	4.76	...	...	...	...	...
5	5	4.00	...	...	...	...	...
6	6	3.36	5	3.353	2	4	3.000
7	7	2.83	6	2.812	...	...	...
...	...	...	...	...	...	...	...
8	8	2.38	7	2.411	2½	6.25	2.400
9	10	2.00	8	2.057	3	9	2.000
10	12	1.68	10	1.676	4	16	1.500
...	...	...	...	...	...	...	...
12	14	1.41	12	1.405	...	...	...
...	...	...	...	...	...	...	...
14	16	1.19	14	1.204	5	25	1.200
...	...	...	...	...	...	...	...
16	18	1.00	16	1.003	6	36	1.020
20	20	0.84	18	0.853	...	...	...
...	...	...	...	...	8	64	0.750
24	25	0.71	22	0.699	...	...	...
...	...	...	...	...	...	...	...
28	30	0.59	25	0.599	10	100	0.600
...	...	...	...	...	11	121	0.540
32	35	0.50	30	0.500	12	144	0.490
35	40	0.42	36	0.422	14	196	0.430
42	45	0.35	44	0.353	16	256	0.385
...	...	...	...	...	...	...	...
48	50	0.297	52	0.295	20	400	0.300
60	60	0.250	60	0.251	24	576	0.250
65	70	0.210	72	0.211	30	900	0.200
80	80	0.177	85	0.178	...	...	...
100	100	0.149	100	0.152	40	1600	0.150
...	...	...	...	...	...	...	...
115	120	0.125	120	0.124	50	2500	0.120
150	140	0.105	150	0.104	60	3600	0.102
170	170	0.088	170	0.089	70	4900	0.088
...	...	...	...	...	...	...	...
200	200	0.074	200	0.076	80	6400	0.075
250	230	0.062	240	0.066	100	10000	0.060
270	270	0.053	300	0.053	...	...	...
325	325	0.044	...	...	...	...	...
400	400	0.037	...	...	...	...	...

[a]Manufactured by W. S. Tyler Company, Cleveland, Ohio.
[b]National Bureau of Standards L.C.-584 and ASTM-E.11.
[c]BS-410 1943.
[d]DIN 1171 (German Standard Specification).

rotating on a horizontal axis used in industry to screen crushed rocks and other relatively coarse fineparticles.

4.2 SIEVE APERTURE CHARACTERIZATION AND STANDARD SPECIFICATIONS

There is very little information available on how the mesh aperture distribution of a sieve affects a measured sieve residue. There is also very little information on how the distribution of apertures varies over the area of an 8-in. sieve and how it changes with the wear of the sieve.

It has been shown that variations of mesh aperture distributions permitted by standard sieve specifications can give rise to significant differences in the kinetic residues deduced from the rate of sieving [7]. The way in which kinetic residues are measured is discussed later in this chapter.

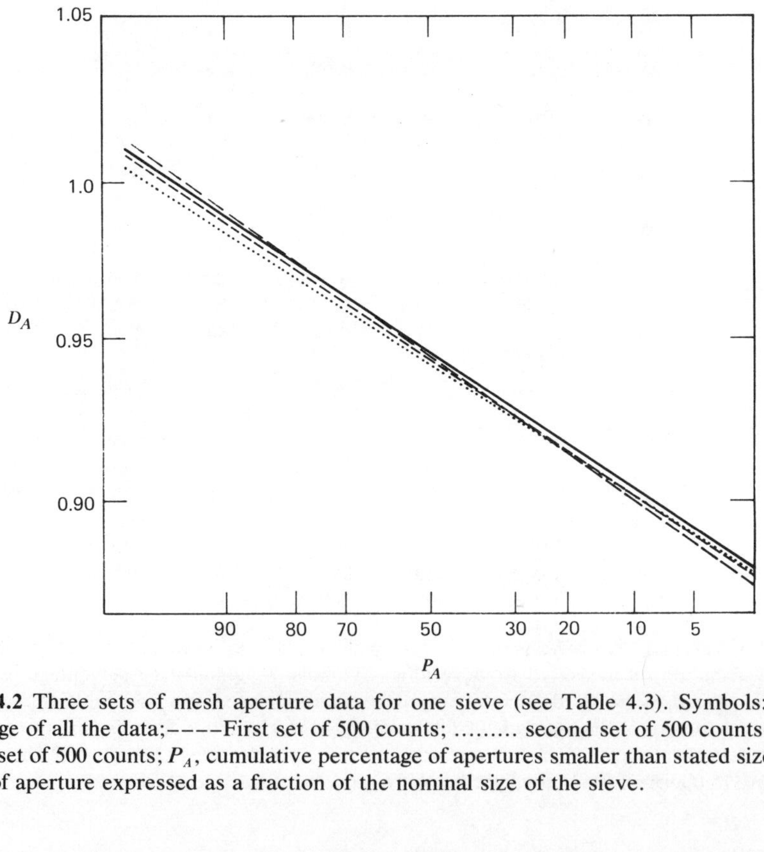

Fig. 4.2 Three sets of mesh aperture data for one sieve (see Table 4.3). Symbols: —— average of all the data;----First set of 500 counts; .... second set of 500 counts; ------ third set of 500 counts; P_A, cumulative percentage of apertures smaller than stated size; D_A, size of aperture expressed as a fraction of the nominal size of the sieve.

The data from this experiment are summarized in Table 4.3 and Figures 4.2 and 4.3. The aperture distributions of Table 4.3 for each sieve are based on data from 50 sets of 10 apertures in a straight line selected at random over the area of the sieve. The aperture size was taken to be the distance between the parallel wires at the midpoint of the aperture. Statistical tests established that the Gaussian distribution function could be used to describe the data. In Figures 4.2 and 4.3 the mesh aperture distributions are plotted on Gaussian probability paper. The data of Figure 4.2 show that the certainty of the various distribution curves of Figure 4.3 is high and that the differences between the distributions of mesh apertures in Figure 4.3 are real and significant. The sieves were $1\frac{1}{2}$-in.-diameter cylindrical wire-woven sieves conforming to the appropriate British Standards Institute specifications. They are referred to in Table 4.3 as *BSS sieves*. The sieves were purchased in two batches, as indicated

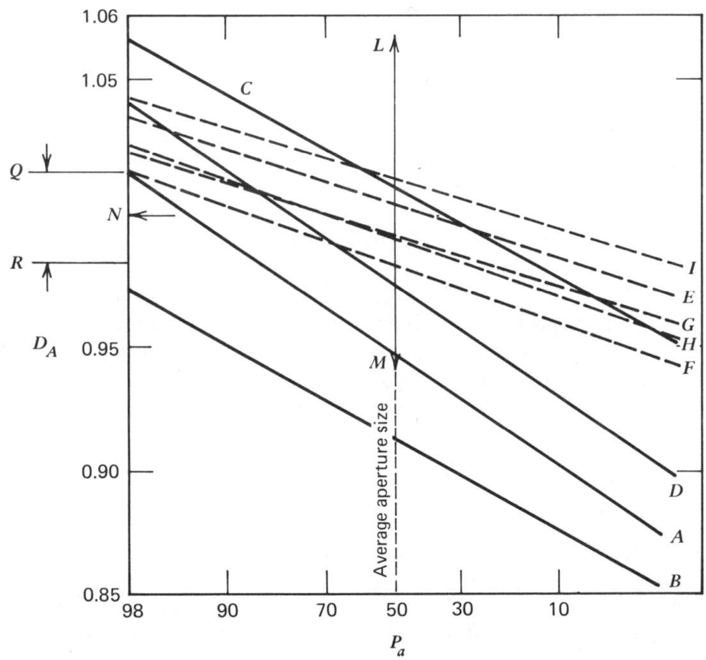

Fig. 4.3 Aperture distributions of wire-woven sieves used to investigate the effect of aperture distribution on measured sieve residue. Symbols: P_a cumulative percentage of apertures smaller than the stated size; D_A, aperture size expressed as a fraction of the nominal size of the sieve; N, nominal size of the sieve; $L \rightarrow M$, permitted average apertures of the sieves as set out in the British Standards Institute specifications [23]; $Q \rightarrow R$, permitted aperture range in electroformed sieves of the same nominal size.

TABLE 4.3. Data Showing Effect of Mesh Aperture Distributions on the Deduced Kinetic Residue for Spherical Bronze Powder

Aperture Distribution of the Nine Sieves
Cumulative Percentage Undersize of Apertures

Size of Aperture, μm	147	143	140	136	132	128	125	121	117
Batch 1									
A, 1st count	100	100	98.0	98.5	85.2	65.2	38.2	11.2	0.4
A, 2nd count		100	99.2	97.8	90.0	68.2	30.6	11.4	0.6
A, 3rd count			100	97.6	89.4	59.9	21.4	12.6	0.1
A, average		100	99.1	96.9	88.2	64.5	30.7	11.7	0.4
B						100	75.4	40.2	11.0
C	100	99.0	85.4	53.6	18.2	0.8	0		
D	99.6	99.2	97.2	87.6	71.0	40.8	12.2	2.2	0
Batch 2									
E		100	98.0	68.4	14.0	0.4	0		
F			100	97.4	60.0	11.8	0		
G			100	91.8	33.6	0			
H			100	91.6	45.8	3.8	0		
I		100	95.0	51.6	2.4	0			

Measured Sieve Residues on Nine Sieves of Same Nominal Aperture[a]

Sieve	Kinetic Residue Expressed as a Percentage of Original Powder Sample	Difference in Kinetic Residues Expressed with Respect to Lowest Measured Residue
A	75.6	3.0
B	76.4	3.8
C	72.6	0
D	75.1	2.4
E	74.1	1.5
F	74.4	1.8
G	74.4	1.8
H	74.3	1.7
I	73.8	1.2

[a]Percentage on the different sieves. Sample weight 23.88 g. Oversize and undersize powder recombined after each analysis to form the sample for the next sieve. Precision 0.1%.

in Figure 4.2. Only sieve B of the first batch of sieves failed to comply with specifications. This sieve had on average measured aperture size only 3% from the permitted range. To test the operational effect of the mesh aperture distributions of the sieves, the same powder was fractionated on each of the sieves in turn, and the kinetic residue was calculated from the rate at which powder passed through the sieve. The oversize and undersize fractions from one experiment were recombined and used on the next sieve.

The kinetic residues measured on each sieve (Table 4.3) are significantly different even though 90% of the sieves conformed to British Standards specifications. In Figure 4.4 the measured kinetic residues are plotted against the mean sieve aperture. It can be seen that the measured kinetic residues fall into two groups that are related to the batch to which the sieve belongs. One batch of sieves, which has a smaller standard deviation than the other batch, appears to result in kinetic residues closer to each other than the sieves that have larger measured apertures.

It is unfortunate that "long-sieving-time" residues were not recorded in the experiment summarized in Figures 4.3 and 4.4. It is probable that such data would have accentuated the differences between the maximum

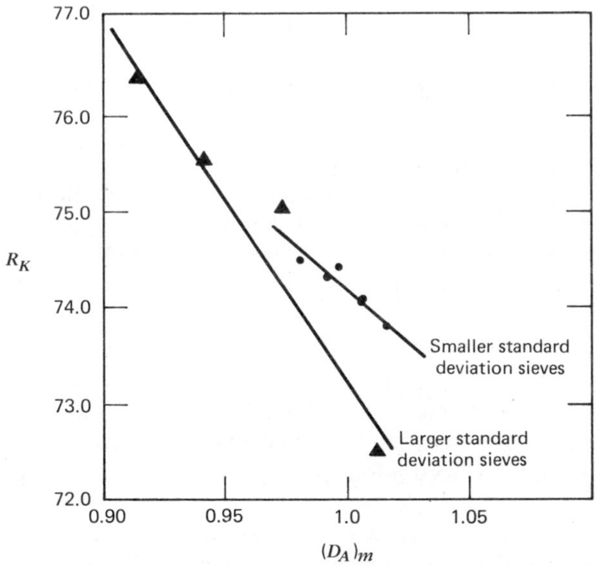

Fig. 4.4 Variation in the kinetic residue of a powder as determined on a series of sieves of the same nominal aperture. Symbols: R_k, kinetic-residue for a given sieve (percentage weight of original sample); $(D_A)_m$; mean sieve aperture (expressed as a fraction of the nominal aperture size of the sieves).

apertures present in each sieve. It has been well known for several years that even sieves meeting standard specification can give differences in residues for accurate work. Major companies such as the Tyler Sieve Company have offered to let purchasers select from among their standard sieves special sieves that when tested with the customer's standard powder yield residues within specified limits.

The availability of lower-cost data processing equipment and the redesign of sieving equipment will probably result in the wider use of kinetic residue data in sieving experiments. Figures 4.3 and 4.4 demonstrate that in many situations the analyst may be wise to use electroformed sieves.

The fact that significant differences in kinetic residues can be found with sieves meeting standard specifications indicates the need for rapid evaluation of the mesh aperture distribution and for techniques for finding damaged apertures so that they can be sealed and the useful life of the sieving surface prolonged. A technique for the rapid evaluation of aperture distributions has been reported by Kaye and Robb [4]. In their technique, the sieve surface is photographed by placing it in an ordinary photographic enlarger where a high-contrast film is held flat against the mesh and illuminated with 1-second exposure. The apertures appear as black objects on the negative and can be characterized with any of the automated iconometric systems such as the Omnicon (see Chapter 9). This very rapid method of evaluating the apertures of a wire-woven sieve makes it possible for the technologist to consider making detailed studies of sieve wear, to make rapid checks that new sieves are within specifications when purchased, and to record how sieves within specifications differ from one another [3].

Erosion of the wires of a sieve results in aperture enlargement. One can check for bellying damage to a sieve by measuring the dimensions of a representative sample of apertures near the center of the sieve and another set near the periphery. Serious bellying makes a sieve worthless for accurate work; however, the writer has seen many severely bellied sieves still in use in characterization studies, probably because distortion of the apertures by bellying is seldom immediately visible to the naked eye. Many electroformed sieves have stronger coarse support grids welded to their sieving surfaces to prevent their distortion under load. If this practice were to be extended to wire-woven sieves of finer mesh size, the life of many analytical sieves could be prolonged. The traditional 8-in. brass cylindrical sieve could be partitioned advantageously into many compartments with the sieve surface firmly attached to the sieve partitions. By partitioning the sieve in this way, not only is the mechanical strength of the sieve greatly increased, but the effects of any few very large apertures present in the sieving surface are isolated [8].

Naylor and co-workers have described a technique based on optical image processing for the location of damaged apertures [9–11]. In this procedure, the far-field diffraction pattern of a new sieve conforming as closely as possible to the appropriate specifications is photographed. The same type of diffraction pattern is generated with the sieving surface to be inspected. The negative of the diffraction pattern of the good sieve is placed in the focal plane of the lens generating the diffraction pattern of the sieve being inspected. If the sieve being inspected does not differ significantly from the new sieve used as the comparison reference, then all the light from the diffraction pattern of the sieve being inspected will be blocked by the negative of the diffraction pattern of the good sieve. When used in this way, the diffraction pattern of the good sieve is described in optical theory as a *spatial filter*. If there is a damaged aperture in the sieve being inspected, then some of the light diffracted by this aperture will move around the spatial filter. A second lens placed after the filtered diffraction pattern can now be used to construct an image of the sieving surface being inspected as shown in Figure 4.5. Because of the spatial filter, only damaged apertures will be imaged by the second lens. An electroformed sieve with various types of damage present in its surface is shown in Figure 4.5*b*. The reconstituted image produced by generating the diffraction pattern of this damaged sieve and filtering it with a spatial filter made using an undamaged sieve clearly shows the location of the various types of damaged and blinded apertures [9, 10].

Naylor and co-workers have incorporated the basic principles of their technique into an instrument that can be used to locate damaged apertures and display them on a television screen. This instrument is very useful when repairing an expensive sieve by locating the damaged apertures and blocking them with solder so that they no longer contribute to the fractionating power of the sieve [11].

It should be noted that the range of sieve apertures produced by various procedures is not necessarily described by the Gaussian distribution function (see Figure 4.6).

4.3 SIEVING EQUIPMENT AND PROCEDURES

In this review of commercially available instruments, only typical instruments that represent different classes of instrumentation are described. In the "Rotap" sieving machine a nest of sieves is swirled around in a horizontal plane while the column is rapped on the top by a hammer. The aim of the combined sieving actions is to simulate a type of motion that has proved to be successful when sieving shallow beds of powder by

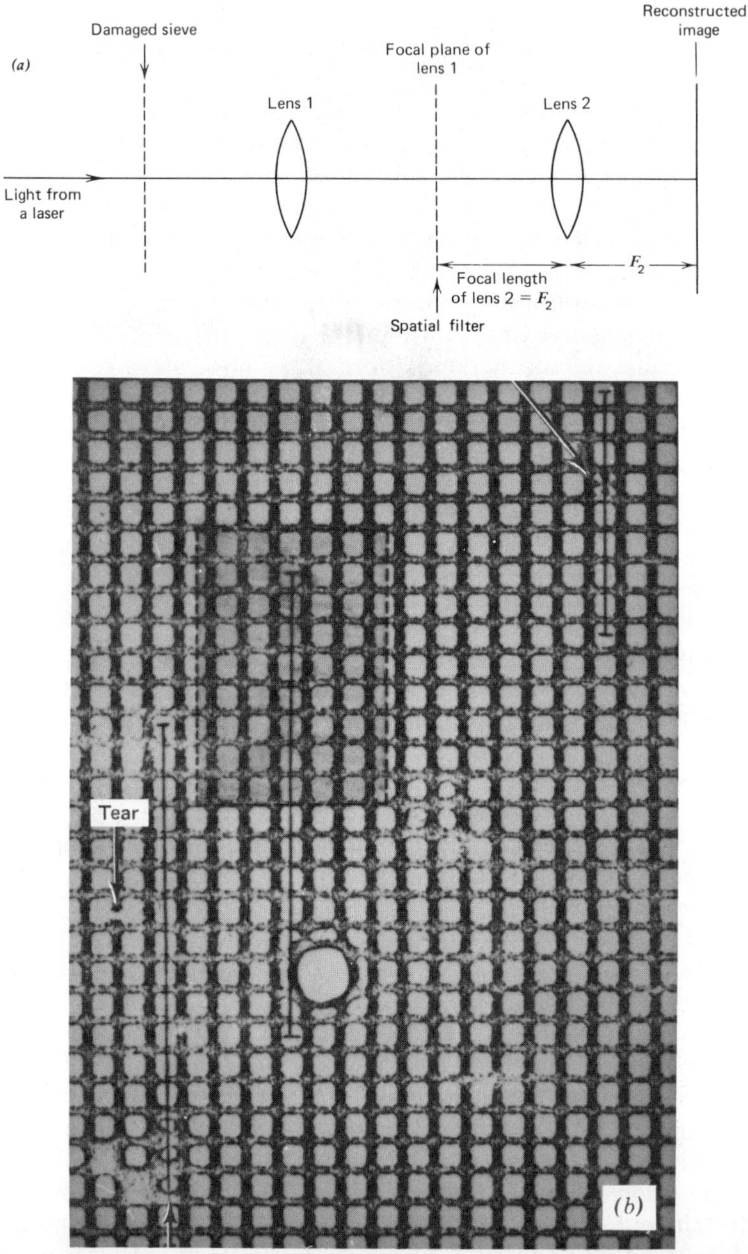

Fig. 4.5 Spatial filtering of an image facilitates the location of damaged or blocked apertures: (*a*) basic system for producing filtered image of damaged sieve; (*b*) photograph of damaged sieving surface.

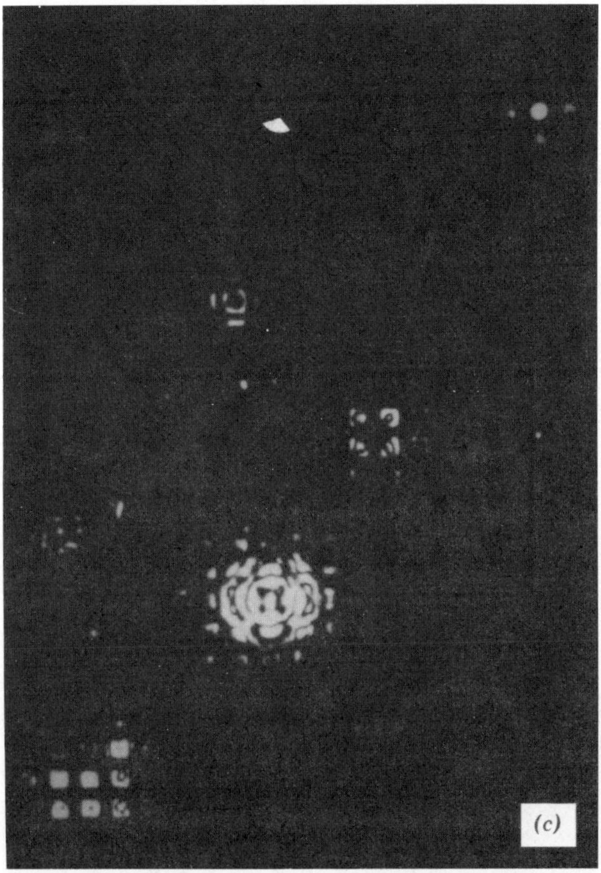

Fig. 4.5 (*c*) Filtered image of the sieve showing location of damaged apertures.

hand, that is, a horizontal swirling action combined with a vertical jarring action achieved by tapping the sieve on the bench top. In use, the Rotap equipment is noisy and is usually operated inside a sound-insulating cabinet [12]. In the Cenco sieving machine a movable platform for holding the sieves is mounted on four large helical springs. Vertical and lateral gyratory vibration are combined in the sieving motion. The device is self-balancing, as the moving parts comprise a free system that executes its motions about its own center of gravity [13].

In the Alpine Corporation's air-swept sieve, the sieve is sealed with an airtight plastic lid and fitted onto a cylindrical stand [14]. Air is sucked through the device by means of a vacuum pump. Air enters the base of the sieve through a set of nozzles drilled in a pipe that rotates just underneath

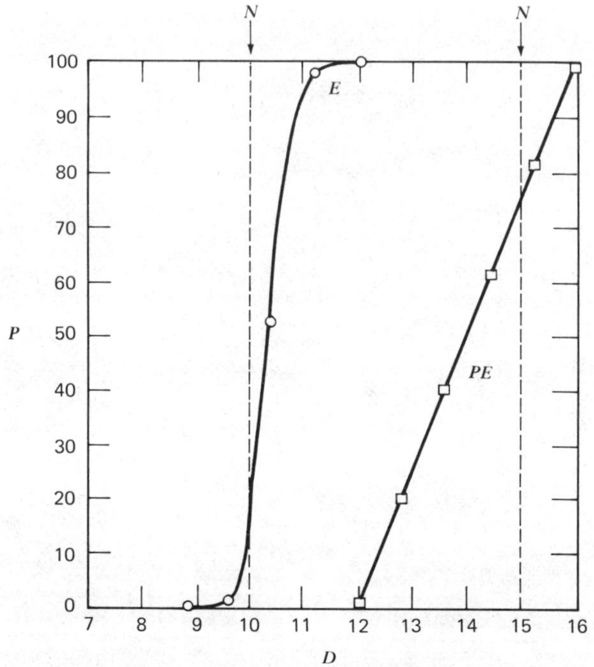

Fig. 4.6 Aperture distribution functions for two nonwoven sieve surfaces [16]. Symbols: *P*, percentages of apertures less than stated size; *D*, aperture diameter (Dutch sieves) in microns or length of side (electroformed sieves); *N*, nominal aperture size; *PE*, electroformed sieve surface; *E*, Dutch electroformed sieve surface.

the surface of the sieve. When the powder is placed on the sieving surface and the unit sealed, the rotating air jets clean the sieve apertures immediately above them and fluidize the powder. The return flow of air over other portions of the sieve aids the passage of fines through the sieve. The fines are collected by a filter inserted between the sieve stand and the vacuum pump. General turbulence at the sieving surface jets helps to break down agglomeration in the powder. This equipment can be used to fractionate powders down to 10 μm by using electroformed sieves. When very fine powders are sieved with dry air, electrostatic charges can cause problems. It may be necessary either to suspend a β-ray generating antistatic capsule inside the sieve or to treat the powder with an antielectrostatic type of flow agent. A flow agent is a substance added to a powder to improve its flow properties. Not all flow agents promote flow in the same physical manner. One way in which silica flow agents, composed of agglomerates of fine colloidal silica, appear to promote flow is that they give the surface of the fineparticles a conducting surface composed of OH groups tightly bonded to the silica molecules [16]. Another class of flow agents, represented by compounds such as magnesium stearate, appear to promote powder flow by agglomerating very small fineparticles into granules that roll over each other easily. These flow agents are waxy solids that make the fineparticles stick to each other. This type of flow agent should *not* be used as a sieving aid since it will alter the structure of the powder in a gross manner and result in distortion of measured residues. Stearates are sometimes added to metal powders during their manufacture to act as grinding aids. (The rapid coating of freshly created surfaces of fineparticles formed during the size reduction process with stearate prevents spontaneous welding together of newly created fineparticles.) Agitation during the sieving process of stearate-coated fineparticles will cause "balling," that is, agglomeration, of the powder. In such cases wet sieving may have to be used to characterize the powder [17]. Washing the powder with an organic solvent prior to the sieve study might remove the stearate in some cases.

In the "sonic sifter" the column of air inside a nest of sieves is vibrated at sonic frequencies by a rubber diaphragm at the top of the column. The base of the fines receiver is also a rubber diaphragm. To aid the movement of powder through the sieve, the equipment is jolted mechanically [15].

Heywood [18] reported that many smaller fineparticles cling to the larger ones retained on a sieve, thus increasing the measured residues. To overcome this problem, Heywood recommended wet sieving on the finest sieve before dry sieving the coarse portion of the powder. Such a procedure is time consuming, and the use of silica flow agents and/or β-ray radioactive treatment may help to overcome this type of problem. If the clinging of the smaller fineparticles to the larger ones is due to traces of moisture in the powder, the silica flow agents may help by dehydrating the system.

The aperture variation in the sieve cloth arises from (*a*) slight shifts in the weave of the cloth, (*b*) small variations in the wire diameter, and (*c*) subsequent shifting of a wire in the weave of the cloth during sieve manufacture. All these are essentially localized effects occurring at random over the whole sieving area. It seems reasonable to postulate that any subregion of the sieve surface can be described by a normal distribution of mean and standard deviations slightly different from those of the total sieving surface. If the larger fineparticles were to be prevented from searching out the few larger-than-nominal apertures by placement of a set of partitions on the sieve surface, with the sieve cloth firmly attached to the boundaries of the partitions, the sieve surface would be effectively transformed into a set of sieves of slightly differing mean aperture and standard distribution [8]. Most of these small sieve surfaces would have an average aperture size approximately equal to the nominal aperture of the sieve cloth, whereas a few would have a size lower than nominal and another few, greater than nominal. If the partitions were such that under subsequent sieving action the portions of the powder remained within their original compartments, the following statements about the fractionating power of the partitioned sieve could be made. In a partitionless sieve smaller-than-nominal size apertures only delay the sieving rate, whereas in partitioned sieves the portion of powder fractionated on a local area composed of lower-than-average apertures will have a slightly higher residue than will a portion of powder located on an area with higher-than-average aperture size. The net effect is that the higher-than-anticipated residue from the finer effective sieve will compensate for the lower-than-expected residue on the portion containing the larger apertures. Overall, the net effect will be to create an "averaged-out" residue on the partitioned sieve that will probably be closer to that predicted from the cut size of the sieve than the residue measured on the free-range sieve.

Theoretical considerations would indicate that the more partitions per unit surface area of the powder, the better the averaging effects of the various residues for each compartment of the sieving surface [8].

When considering the design of sieving equipment, it is pertinent to note that equipment designed many years ago may not constitute the optimum design in the current laboratory situation. From this perspective one may ask why do 8-in.-diameter wire-woven sieves dominate analytical practice in many fineparticle laboratories. Where the aim of sieving is separation into groups, the 8-in. sieve with its large surface area is useful; however, it is not necessarily useful in the analytical characterization of fineparticles. One can speculate that the 8-in. sieve came into its present dominant position in analytical studies of fineparticle systems because the large amounts of powder used with the 8-in. sieve facilitated the weighing process with the relatively simple equipment used in the field. In the modern industrial laboratory where rapid automatic weighing to thousandths of a gram is a routine matter, it is hard to understand the continual domination of laboratory procedures by the 8-in. sieve, particularly since the development of modern efficient sampling equipment makes it easy to efficiently obtain relatively small samples of powder. It is now quite feasible to use 0.5-in.^{2} sieves as basic units for the sieve characterization of powders using modern accurate equipment for sampling and weighing. If one wanted a larger sieving area to characterize larger samples of powder and to use simpler weighing equipment, then several of the 0.5-in.^{2} units could be clamped together to form 4, 6, 9, and 12 composite unit sieves. Such sieves would have all the advantages of a multipartitioned sieve just discussed, with the added advantages that damaged portions are easier to replace and that each unit could be examined rapidly for damage.

Widespread use of a basic sieve unit of 0.5-in.^{2} size would also offer an attractive alternative to traditional nest sieving. Whenever nest sieving is terminated, there is always uncertainty about the amount of powder having passed each layer of the nest. An alternative to nest sieving would appear to be small 0.5-in.^{2} sieves assembled in a matrix. Each small sieving area is well supported, and since only a small amount of sieve surface material is required, electroformed sieving surfaces could be used. The powder would be spread evenly between the various small sieves and then a container seal placed over the matrix of sieves. It would perhaps be convenient to mount such a matrix of sieves in a cylinder mounted on two circular supports so that the sieving machine could rotate on a modern support system like that used in the SORSI (self-organized sieve) sieve equipment (see Section 4.5). The sieve would be made to go through a given number of sieving actions, after which time each of the portions of powder on the sieve would be weighed. In this procedure the sieving action is carried out on each sieve size in parallel rather than in series (as in traditional nest sieving) and the errors from sieve to sieve are independent of each other.

The 8-in. circular wire-woven sieve continues to dominate laboratory practice in spite of modern alternative equipment, probably because industrial standard specifications for both sieves and their use are written largely around the use of the 8-in. sieve. Widespread shift to the use of alternative equipment would require new industrial standard specifications. Industrial laboratories with years of accumulated data are reluctant to switch to new equipment when existing equipment functions satisfactorily, especially since such a change may require the rewriting of customer and/or user specifications of a powder.

The Autosieve 5900, illustrated in Figure 4.7, completely abandons not only the traditional sieve shape and size, but also the traditional sequence followed in nest sieving of a powder [19].

Four different sieving surfaces are mounted around the periphery of a hollow plastic cylinder. Interchangeable cylinders with different sets of sieve cloths are available. One-fifth of the cylinder is an open area for the discharge of the oversize fineparticles. The Autosieve is an intelligent machine with its movements under the control of a small computer. Powder is fed into the cylinder through a central port on the axis of the cylinder. (In the commercial design of this instrument the cylinder is 8 in. in diameter and the surface carrying the sieve cloth is $3\frac{1}{2}$ in. wide.) At the beginning of the experiment the finest sieve cloth is at the bottom of

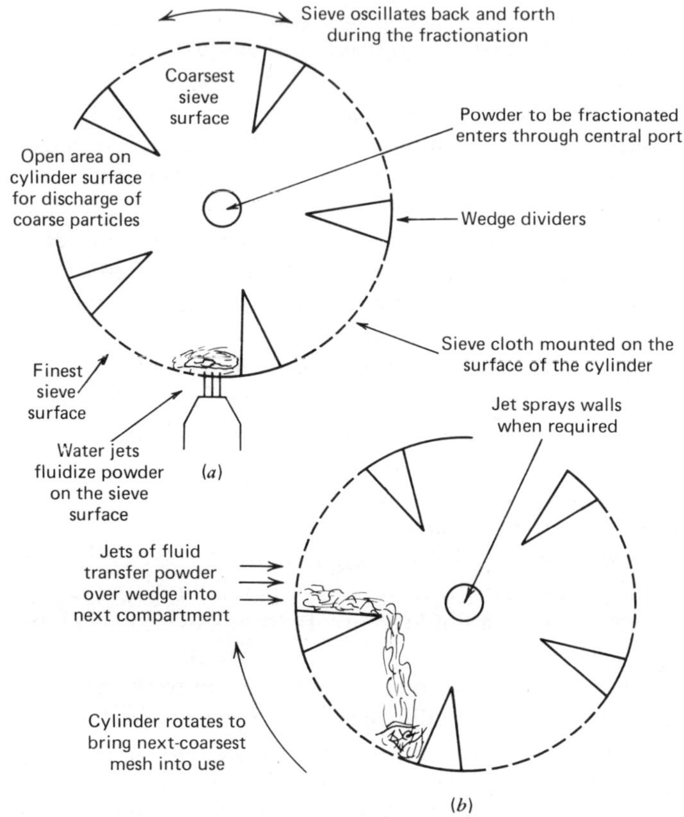

Fig. 4.7 Basic system employed in the Autosieve developed by Orr and co-workers [19]: (*a*) configuration during sieving; (*b*) configuration during powder transfer.

the cylinder (Figure 4.7*a*). When the powder is in position on the sieve, the computer initiates sieve oscillation over approximately one-fifth of an arc. At the same time, fluid jets directed through the sieving surface fluidize the powder, overcoming any problems due to aggregation of the powder, electrostatic charge, and "dustiness." Fluid draining back helps to clear the fines from the sieve. The draining slurry of powder is filtered through a special filter hanging from an automatic weighing device. The fines retained by the filter are automatically weighed each time the cylinder oscillates. The entire weighing operation is carried out under fluid. The control computer stores information on successive weighings and calculates the average of eight sequential weighings. It arbitrarily decides that a sieving fractionation is complete when the running average of two sets of eight weighings do not differ by more than 1%. This arbitrary decision-taking process can be programmed to achieve higher or lower levels of scrutiny of the sieving process. When fractionation is complete on one sieving surface, the computer rotates the sieve cylinder into the next position. At this stage ancillary jets of fluid positioned at the side of the equipment spurt through the sieve cloth and wash the oversize fineparticles retained on the first sieve down the dividing wedge onto the next-finest mesh. These transfer jets also clean the mesh of the first sieving compartment. Next the computer activates the fluidizing jets in the bottom position and the sieve cylinder oscillates, and the fines are weighed. Once more the computer decides when the sieve fractionation is complete and initiates a move to the next sieving position of the cylinder. The computer also controls the calibration and cleaning procedures to be applied to the filter used to collect and weigh the fines.

4.4 KINETIC RESIDUES

The fact that all the apertures of a sieving surface are not identical is the major cause of uncertainty present in the characterization of fineparticles with sieves. It has been pointed out that the probability of a fineparticle of a given size encountering a hole through which it can pass is related to the aperture distribution function as well as the method of agitation. There have been many extensive studies of the rate at which powder passes through a sieve, but these are seldom of general value because of the varied sieving actions of the several machines used in the studies [20, 21]. In practice, the technologist often has to decide arbitrarily that sieve fractionation is complete at some experimentally determined low rate of passage of fineparticles through the sieve [22].

In past studies of the movement of powder through a sieve, not enough distinction has always been made between studies aimed at fractionating powders in the industrial scene and the needs of the technologist wishing to characterize fineparticles with a sieve. When characterization of the powder is the main aim of the study, it is useful to arrange the kinetics of fineparticles on a sieve so that probability theory can be used to relate the distribution function of the powder and the aperture distribution function of the sieve to the rate of movement of powder through the sieve, even if the resultant rate of passage of fineparticles is lower than in other more efficient shaking techniques. In studies carried out by the author, a type of sieving action that enabled probability theory to be applied to the study of the rate of powder movement through the sieve consisted of the following sequence of events. First, a sieve fitted with a tight lid and containing powder is shaken vertically. The sieve is jarred on a hard surface at the end of the vertical movement. As a result, the powder's lateral movement across the sieve is discouraged by restraining the movement of the sieve immediately after impact on the surface. Then the sieve is lifted carefully, inverted, and the powder shaken vigorously to randomize the position of the fineparticles remaining on the sieve. Once more, the sieve is inverted and shaken to bring the powder vertically onto the sieve surface, and lateral motion is again discouraged. For this method of shaking the sieve, the probability that a fineparticle will encounter an aperture of a given size is directly proportional to the frequency of that sized aperture in the sieve surface. The second stage of the shaking routine ensures that the probability of passage through the sieve of any fineparticle capable of doing so is directly related to the number of such fineparticles remaining in the sieve residue. When the sieve is shaken in this manner, all fineparticles smaller than the smallest aperture present are eliminated very rapidly, but fineparticles in the size range of the aperture size range will pass the sieve at a rate governed by the following equation:

$$R_t = R_{KE} + Ae^{-\lambda t} \tag{4.1}$$

(In the original publication describing this investigation, the constant labeled in this equation R_{KE} was called R_∞. This was an unfortunate notation because it implied that the residue involved in such a calculation was equivalent to the residue that would be retained by the sieve after infinite sieving time. In fact, the residue after infinite sieving time, assuming that the powder is sufficiently robust that it does not disintegrate in the sieving action, will be smaller than the residue R_{KE} of this equation because after infinite sieving, all the fineparticles less than the largest sieve aperture present will have infinite time to find that hole and pass through. Prolonged sieving times accentuate the effect of the larger apertures, whereas R_{KE}, which is described as the kinetic residue of the sieve, is a measure of the fractionating power of the sieve when the effect of the presence of a few large apertures has been diminished because of the arranged kinetics of the fineparticles.)

In the original publication, a graphical method was used to solve equation 4.1 to find R_{KE}. The method was based on the transformation of equation 4.1 to the form

$$\log_e (R_t - R_{KE}) = A - \lambda t \tag{4.2}$$

If the value of R_{KE} is known, a plot of $\log(R_t - R_{KE})$ against R_{KE} will be a straight line of slope $-\lambda$. To solve equation 4.2 graphically, a series of values of R_t are obtained, where it is assumed that t is great enough so that equation 4.1 can be used to describe the sieving rate. The next step is to make an estimate of R_{KE}, denoted as R'_{KE}. A graph of $\log_{10}(R_t - R'_{KE})$ is plotted against t on semilog graph paper. Consider the effect of an error in R'_{KE} on the graphical plot. If the estimate is larger than R_{KE}, the values of $\log_{10}(R_t - R'_{KE})$ will rapidly approach $-\infty$; the curve will be asymptotic to the $\log_{10}(R - R_{KE})$ axis. If, however, $R_{KE} - R'_{KE}$ is a small positive quantity, then $\log_{10}(R_t - R'_{KE})$ will tend to a constant value, and the curve will be asymptotic to the time axis. By taking various estimated values of R_{KE} and plotting $\log_{10}(R_t - R'_{KE})$ against time, the correct value of R_{KE} that gives a linear relationship can be found.

The shape of the initial curve indicates the necessary modification to R'_{KE} in order to arrive at a more accurate value. This method will give a linear relationship only when t is greater than T, which is the time that must elapse before the assumptions made in the derivation of the equation are valid. Prior to this time, probability of passage for the various fineparticles is a complex function of time, and $\log_{10}(R_t - R_{KE})$ against time cannot be linear.

A typical graphical solution for a kinetic residue is shown in Figure 4.8. Experiments have shown that the kinetic residue deduced in this way is independent of sample weight, sieve dimension, and type of powder used [22]. The data in Figure 4.5 demonstrate that the magnitude of the kinetic residue is related to the mesh aperture distribution of the sieve surface.

To demonstrate that the kinetic residue is independent of sample size and to compare results obtained by this technique with those obtained following a national standard procedure, one of the experiments described in the original scientific publication is discussed in some detail. A glass powder was sieved on an 18-mesh sieve (nominal aperture size 0.853 mm).

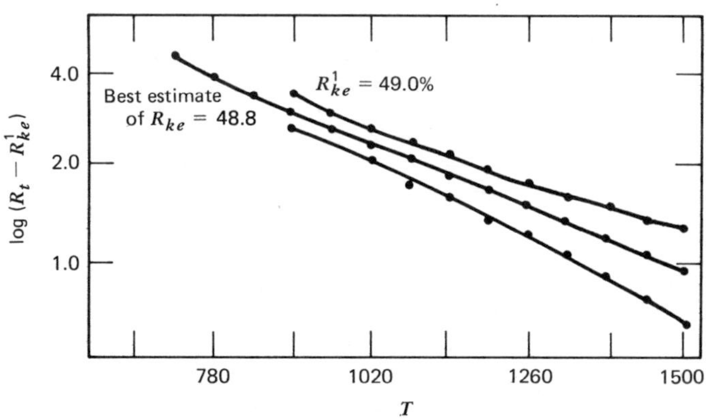

Fig. 4.8 Graphical solution of rate data for the estimate of a kinetic residue. Symbols: T, total number of taps; R_t, residue after t taps of the sieve; R^1_{ke} = estimated kinetic residue; 35 g of spherical bronze powder sieved on a $1\frac{1}{2}$ inch, 120 mesh, BSS sieve [22].

Eight samples of approximately the same weight were characterized following procedures specified by the British Standards Institute [23] and by the rate method described previously. Experimental data are summarized in Table 4.4.

The oversize and undersize powders for each sample were recombined and the samples amalgamated in pairs. The kinetic residue of the amalgamated samples was measured and compared to the residue predicted from the measured kinetic residues of the constituent samples. The British Standards Institute recommends procedures for sieving that specify a small sample size; thus the residue of the combined samples cannot be measured by this procedure, but residues for the larger sample sizes predicted from the values of R_{BS} for the constituent samples can be calculated. The data for this series of experiments are summarized in Table 4.5, where it can be seen that the kinetic residue measured is independent of the sample weight used in the characterization study.

A basic assumption made in the derivation of the equation used in deducing kinetic residues is that when the number of fineparticles still capable of passing the sieve is small and the total number of fineparticles remaining on the sieve is large, the probability of passage of a particular-sized fineparticle is independent of time and size. Once sufficient time has elapsed for this assumption to become operative, successive fractions of the powder passing through the sieve should have the same size distribution. Thus the physical validity of the assumptions made in the derivation of the equation underlying the calculation of kinetic residues can be tested by studying the size distribution functions of the various fractions of powder passing the sieve at different times. Consider the powder sieving

TABLE 4.4. Sieve Fractionation of Glass Powders (All Residues Quoted as Percentages of Original Sample Weight)

Sample Number	Sample Weight	R_{BS}[a]	R_{KE}[b]
1	80.7	9.0	7.3
2	87.7	8.7	7.3
3	81.2	12.9	12.0
4	88.5	6.3	5.3
5	84.7	8.8	7.8
6	84.3	9.2	7.8
7	80.2	9.2	8.3
8	83.4	9.4	8.1

[a]Residue obtained using BSS procedure.
[b]Kinetic residue deduced by the rate method.

TABLE 4.5. Residues of Amalgamated Samples of Glass Powder

Sample Combination	Weight	$R_{P(BS)}$[a]	$R_{P(KE)}$[b]	R_{KE}[c]
1 + 2	168	8.8	7.3	7.1
3 + 4	170	9.5	8.5	8.5
5 + 6	169	9.0	7.8	7.8
7 + 8	164	9.3	8.2	8.2
1 + 2 + 3 + 4	337	9.2	7.9	7.8
5 + 6 + 7 + 8	332	9.1	8.0	8.0
All Samples	669	9.2	7.9	7.9

[a]Predicted residue using residues measured according to British Standards procedure.
[b]Predicted residue from measured kinetic residues of component sampling.
[c]Kinetic residue of combined sampling measured directly.

experiment with the total amount of glass powder (669 g) summarized in Table 4.5. The information on the rate of sieving for this powder is detailed in Table 4.6. The graphical processing of the data yielded an estimate of R_{KE} of 7.9% and indicated that equation 4.1 was a valid description of the rate of sieving after 16 minutes of sieving time.

The size distribution of the powder passing the sieve in the time intervals 0–2, 2–4, 4–8, 8–12, 12–16, 16–20, 20–28, 28–36, and 36–44 minutes was measured by looking at the fineparticles through a microscope equipped with a linear eyepiece reticule. (It was necessary to combine powder for the longer time intervals in the later stages of the sieving process to collect sufficient powder to carry out a size determination.) The cumulative percentage undersize distributions calculated for the various fractions of powder are summarized in Figure 4.9.

The data in Figure 4.9 illustrate that as predicted, the smaller fineparticles are rapidly eliminated in the early stages of the sieving process and that when the linear region of the kinetic residue decay curve has been reached (i.e., after 16 minutes), the successive fractions have virtually the same size distribution. The median size of the different fractions is plotted in Figure 4.9*b* to emphasize this fact.

The glass beads used in this experiment were very close to exact spheres; a maximum divergence of 5% between diameters measured at right angles was observed [22]. Therefore, the data in Figure 4.9 can be used to compare the data obtainable from sieve fractionation and direct dimension measurement. To facilitate this comparison, the size distribution curves in Figure 4.9 have been plotted as two sets in Figure 4.10 with the size of the fineparticles expressed as a fraction of the nominal aperture

TABLE 4.6. Sieving Data for 669.49 g of Glass Beads Sieved on an 18-Mesh (0.85-mm Nominal Sieve Opening) Sieve[a]

Time, minutes	4	8	12	16	20	24	28	32	36	40	44	48	52
R_t[b]	23.0	16.3	14.0	12.8	11.9	11.3	10.8	10.4	10.1	9.7	9.5	9.3	9.1

[a]Kinetic residue calculated as 7.9%. Linear relationship established after 16 minutes.
[b]Percentage weight of powder retained on the sieve after time *t*.

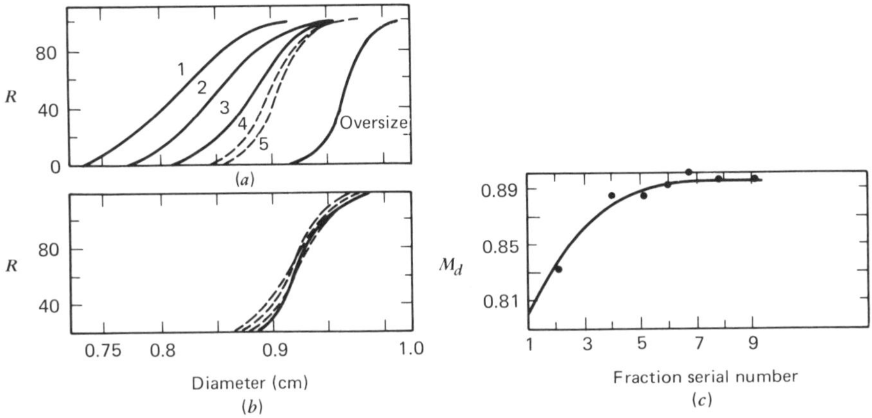

Fig. 4.9 Size distribution functions of successive fractions of glass beads passing an 18-mesh sieve: (*a*) cumulative undersize distributions; (*b*) size distributions of fractions generated during the linear region regime; (*c*) median size of successive sieve fractions. Symbols: F, fraction serial number; M_d, median diameter in centimeters; R, percentage weight undersize.

of the sieve. Only the first three fractions passing the sieve, that is, powder passing the sieve in the first 8 minutes, contain any significant amounts of powder smaller than the nominal size of the sieve. This demonstrates that prolonged sieving time is necessitated not by the need to achieve fractionation of powder smaller than the nominal aperture of the sieve, but because of the much slower passage of fineparticles of diameter range equal to the mesh aperture distribution function of the wire-woven sieve. This indicates that a shift to electroformed sieves at higher-aperture sizes would not only give greater aperture precision, but could significantly reduce the time of sieve characterization because there is a much smaller range of aperture sizes within a given sieve.

It is unfortunate that the studies summarized in Figure 4.9 did not report the sieve aperture distribution; however, the data in Figure 4.10*b* seem to indicate that the maximum aperture present was about 1.08 of the nominal aperture, well within specification for an 18-mesh British Standard series sieve. The zone indicating the size distribution functions of the linear portion of the kinetic residue demonstrates that after 16 minutes of sieving, the powder passing the sieve in the next 28 minutes is all within the aperture distribution range. Furthermore, the data in Figure 4.10*b* would indicate that up to 80% of the powder remaining on the sieve after 44 minutes of sieving can still pass the sieve if sieving is prolonged indefinitely. This would bring the residue for the sieve down to 2.0% of

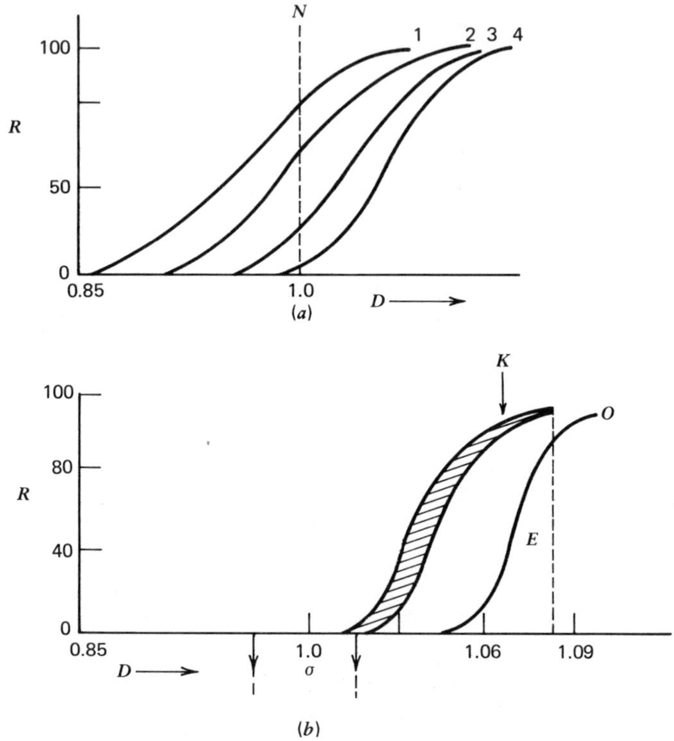

Fig. 4.10 Size distribution data for successive sieve fractions of glass beads. Numerals 1, 2, 3, and so on denote serial number of sieve fraction; (*a*) size distributions of early sieve fractions; (*b*) size distribution of fractions passing for linear portion of kinetic residue rate curve all fell within band denoted by *K*. Symbols: 0, size distribution of oversize fraction; *R*, percentage weight of fineparticles less than stated size; *D*, diameter of fineparticle expressed as a fraction of the nominal size of the sieve aperture; *N*, nominal size of the sieve aperture; *E*, fraction of oversize powder that would eventually pass the sieve as determined by maximum aperture present in the sieve; σ, permitted variation in electroformed sieves.

the original sample weight. This figure differs by more than 300% from the kinetic residue and 400% from the residue measured following BSS procedures and would take hours of sieving to achieve. The data in Figure 4.10 also indicate that if powder is fractionated on electroformed sieves of the same nominal aperture as a wire-woven sieve, very big discrepancies between measured residues on the two types of sieve can be anticipated.

More data from the original publication describing the rate method for calculating the kinetic residue of sieves are summarized in Figure 4.11. These experiments were carried out with an aluminum powder consisting of highly irregularly shaped grains. The powder was sieved with the ap-

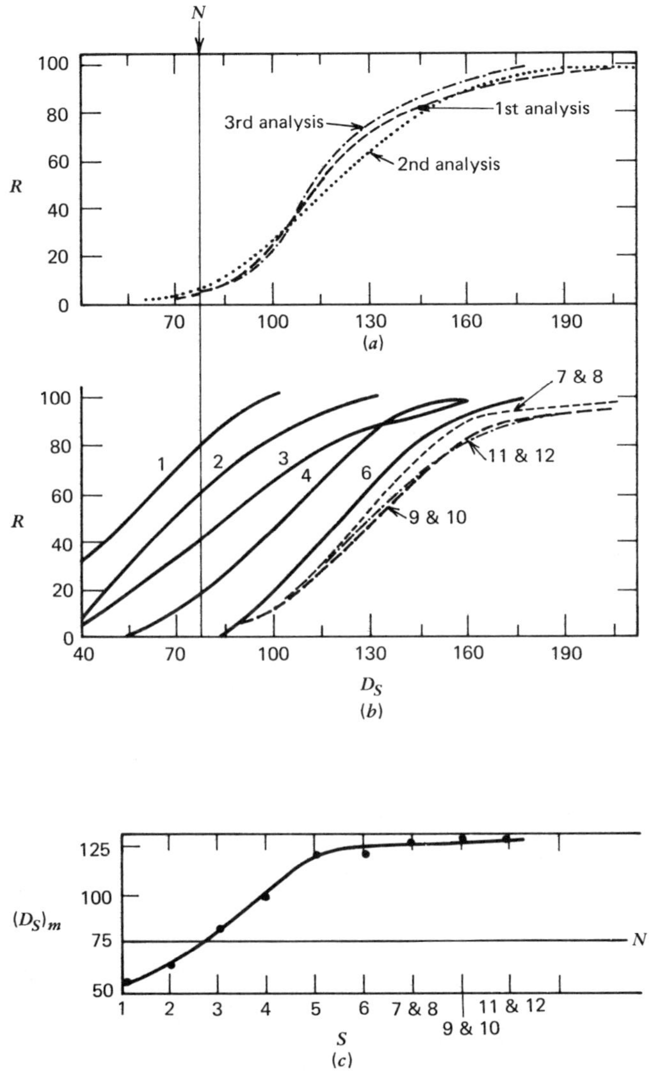

Fig. 4.11 Experiments carried out with fine-metal powder having highly irregular shape characteristics: (*a*) repeat characterizations of fraction 5; (*b*) size characterizations of successive fractions; (*c*) mean size of successive fractions.

Note: Graphical processing of rate data indicated rate equation valid after fraction 5. Symbols: N, nominal aperture of the sieve; S, serial number of sieve fraction; D_S, Stokes diameter; R, percentage by weight less than stated size; $(D_s)_m$, mean Stokes diameter.

propriate shaking technique on an 8-in. circular 200-mesh BSS sieve. The successive fractions of powder passing the sieve were collected and weighed. Graphical processing of the rate data established that the linear region of the sieving rate curve was established when fraction 6 was passing through the sieve.

The size characterization of the fractions was carried out with a Boundbrook photosedimentometer. The Stokes diameter distribution of the various fractions is summarized in Figure 4.11. The average size of the fractions numbered 6 and higher is virtually constant, again confirming the validity of the physical assumptions underlying the rate method for deducing kinetic residues.

For irregularly shaped fineparticles, there is no direct correlation between the effective sieve size and the Stokes diameter. However, even making allowance for this lack of direct correlation, it is interesting to note that after fraction 5 there are no fineparticles passing the sieve with a Stokes diameter less than the nominal aperture of the sieve. The Stokes diameter of the largest fineparticle passing the sieve is more than twice the magnitude of the nominal aperture size of the sieve. The average Stokes diameter of the fractions passing the sieve in the linear region of the rate curve is 65% greater than the nominal aperture size of the sieve. These data strongly suggest that the linear region of the rate curve, when a wire-woven sieve is used, is a function of the aperture distribution of the sieving surface.

Although the deduced kinetic residue is independent of the operator and the sample size taken for characterization studies, the graphical solution for the kinetic residue is time consuming and tedious. Furthermore, with traditional equipment it is not easy to obtain a running series of weights as the sieving proceeds. Therefore, it is not surprising that the rate method of sieving and the kinetic residue deduced from the rate at which powder was passing through the sieve remained an academic curiosity for many years. However, in the late 1970s Kaye and Robb explored the possibility of using small computers to automate the calculation of the kinetic residue. They suggested that sieve equipment such as that shown in Figure 4.12 could be used with advantage in automated sieve procedures for characterizing powders [24]. This sieve rotates about the axis shown. Powder falls onto the sieving surface by bouncing off many randomizing rods positioned in the body of the sieving equipment so that the powder arriving at the sieve surface has very little lateral movement. Powder passing the sieve is caught and weighed. The sieve is then inverted. On the return transit, the positions of the fineparticles are randomized as they move to the base of the container. The sieve chamber is again inverted, and the powder falls through the randomizing rods to

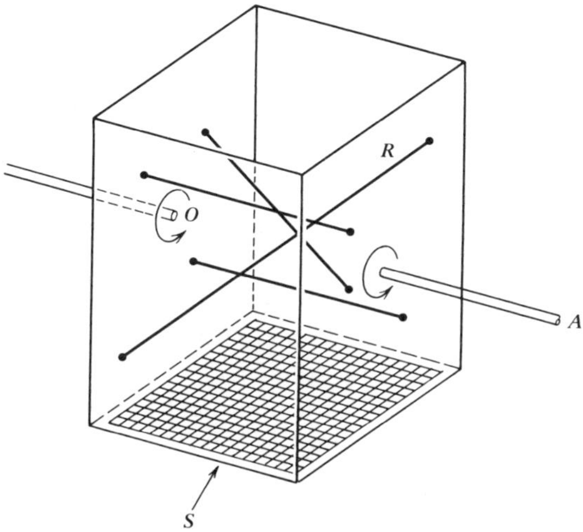

Fig. 4.12 The Tipit sieve for measuring kinetic sieve residues (R, denotes randomizing rods). For the sake of clarity, only a few are shown in the diagram. Symbols: A, axle; S, sieving surface.

encounter the sieving surface again. This type of sieving action embodies the theoretical postulates underlying the derivation of equation 4.1 so that the rate of powder movement through the sieve should be describable by equation 4.1. Kaye and Robb reexamined this basic equation and suggested that it be rewritten as follows. Let

$$R_{(J)} = R_{KE} + Ae^{-\lambda J} \tag{4.3}$$

where R_{KE} is the kinetic residue and $R_{(J)}$ is the amount retained on the sieve after the J rotation or pass at the sieve. From equation 4.3, we obtain

$$R_{(J+1)} - R_{(J)} = \Delta(J)$$

Thus

$$\Delta(J) = R_{KE} + Ae^{-\lambda(J+1)} - (R_{KE} + Ae^{-\lambda J})$$

$$\Delta(J) = Ae^{-\lambda J}[e^{-\lambda} - 1]$$

But

$$Ae^{-\lambda J} = R_{(J)} - R_{KE}$$

Hence

$$\Delta(J) = [R_J - R_{KE}][e^{-\lambda} - 1] \tag{4.4}$$

This equation can be written in the form

$$\Delta(J) = cR_J + d \tag{4.5}$$

where

$$c = [e^{-\lambda} - 1] \quad \text{and} \quad d = -R_{KE} \cdot c.$$

It follows from this equation that a plot of $\Delta(J)$, the amount of powder passing in the Jth sieving action, against R_J is a linear relationship when equation 4.1 is a valid description of the process. Equation 4.5 can be used as the basis for an algorithm for the calculation of R_{KE}. In essence, a series of values of $R_{(J)}$ is assumed to be governed by a linear relationship, and if a "least squares" method is used to find best values of c and d, then an estimate for R_{KE} is calculated from the relationship

$$R_{KE} = -\frac{d}{c} \tag{4.6}$$

In the early stages of sieving, before the kinetic residue equation is a valid description of the rate of sieving, the very rapid elimination of fineparticles smaller than the nominal aperture of the sieve gives sieving rates that, if extrapolated over long sieving times, would give very small sieving residues. Therefore, the estimate of R_{KE} from the beginning of the sieving process based on, for instance, running sets of eight values of R_J, gradually increases until the valid region is reached and then stays steady for several running sets of $R_{(J)}$. Thus a sequence of invariable estimates of R_{KE} for successive running sets of $R_{(J)}$ indicates achievement of the linear region of the kinetic residue decay curve, and the steady value can be taken to be the best estimate of the kinetic residue. Kaye and Robb have automated this algorithm and demonstrated the logic flow for a small computer to initiate the sieving action, monitor the rate of powder movement through the sieve, establish when the rate equation is valid as a descriptor of the rate of powder movement through the sieve, deduce the kinetic residue for the powder system, terminate the sieving action, and initiate processing of the next powder sample.

4.5 THE SORSI SIEVING SYSTEM

The SORSI system developed by Kaye and co-workers [25] can be used to fractionate powders by sieving on a batch or continuous basis. When operating in the batch mode, it can be used to measure kinetic residues. When operating in the continuous mode, it can be used to characterize powders and can be readily programmed to function as an on-line instrument for feedback loop design in quality control procedures. The basic components and operating system of SORSI (the name is an acronym derived from the phrase "self-organized sieve") are shown in Figure 4.13. The sieving chamber used to fractionate the powder is a hollow icosahedron. The sieve surface is a triangular-shaped area that clips into place on the wall of the sieving chamber. The active sieving area can be several

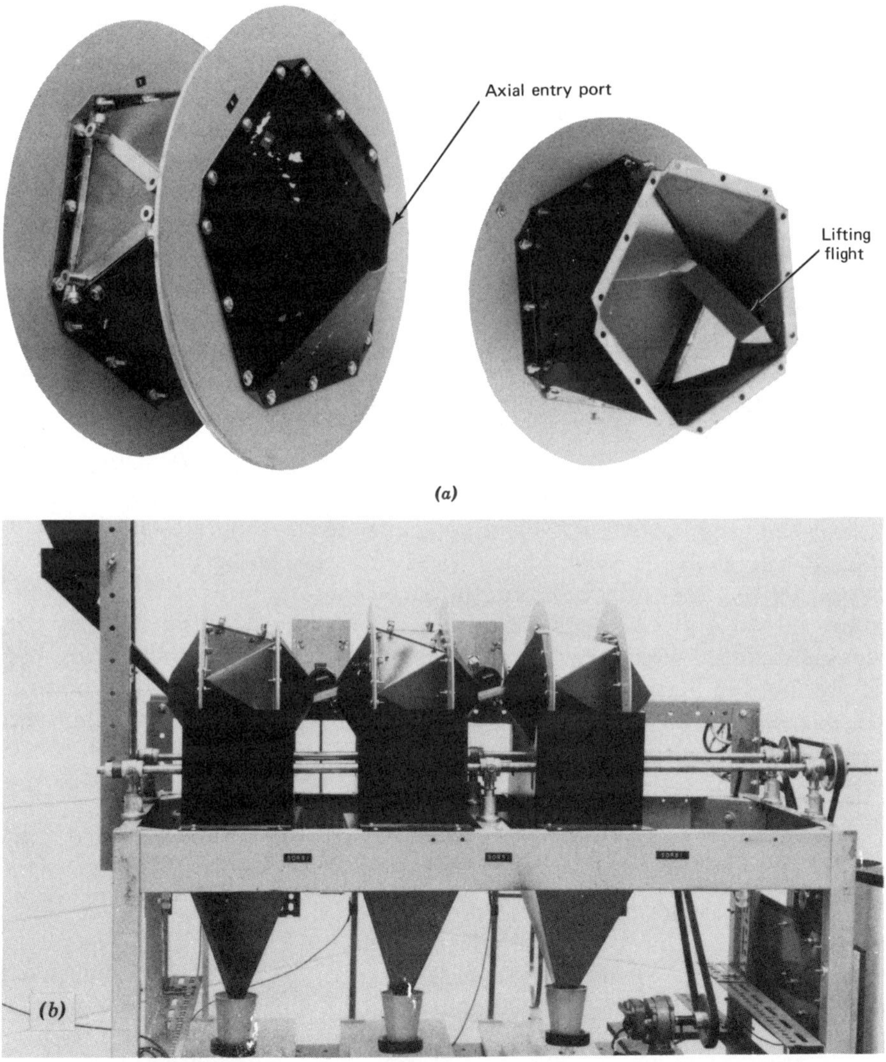

Fig. 4.13 The SORSI sieving system: (*a*) details of sieving chamber construction; (*b*) assembly of sieving chambers.

triangles of the wall of the icosahedron up to a maximum of 10 triangles for equipment used in fractionating powders rather than in characterizing powders. A minimum equipment for SORSI experiments consists of a single sieving chamber with a whole range of interchangeable sieve surface triangles representing the range of sieve apertures required. To carry out equivalent of nest sieving, several chambers are used in cascade format with the coarsest sieve mesh on the first sieving chamber and the next coarsest on the immediate neighbor sieving chamber. The sieving chambers are mounted on circular wheels that fit to the pentagon on the sides of the icosahedral chambers, and the individual sieving chambers are placed on rollers. The powder to be sieved is introduced into the left-hand side of the sieving chamber. The sieving chamber rotates as the rollers turn and the powder moves across the sieving surface. If the sieve is being used in the batch mode, the transfer chute between sieving chambers slopes down to the left. In Figure 4.13*c* the transfer chutes between the three sieving chambers are in this batch position, keeping the powder in each chamber. The equipment shown in Figure 4.13 is provided with automatic devices for altering the slope of these transfer chutes. When in the transfer position a flight inside the sieving chamber (marked by the letter A in Figure 4.13*b*) can pick up the powder as the chamber rotates and pour it onto the chute so that it moves into the next sieving chamber.

Theoretically, it is reasonable to postulate that maximum sieving rates should be achieved when a single layer of fineparticles moves slowly across the sieving surface. When powder is fed at this rate, all fineparticles able to pass the sieve should pass the sieve in very short traverse across the sieving surface. Even if the feed rate is a little higher, fractionation should still be complete within a very short journey across the sieving surface. When powder is fractionated under these conditions, the lifting flight is placed immediately after the sieving surface to scoop up the oversize fineparticles and dump them in the exit chute. Note that if the particles move at too high a speed over the sieving surface, kinetic energy carries them over the sieve apertures and error will be introduced into the fractionation study. Some preliminary experiments are usually necessary before optimum feed rates and rotation rates can be designed for any specific powder. The probabilities of encountering a larger-than-nominal aperture are less for a direct linear traverse of a fineparticle across a sieve than the probabilities for multiple random tracking across the sieving surface; therefore, residues measured using SORSI with monolayer powder feed rates will be higher than those using 8-in. cylindrical sieves with wire-woven sieves and traditional sieving equipment. For some powders, it may be necessary to use β-ray electrostatic discharge capsules inserted through the axial entry port. For others, it may be necessary to add silica flow agents to the powder to ensure smooth transition across the sieving surface.

If the SORSI equipment is used in the batch mode to measure kinetic residues, the flight is positioned on the opposite side of the equipment in such a way that it scoops up the powder and pours it down onto the sieving surface; the transfer chute acts as a randomizing rod on the downward-moving powder. When fine powders are used in the batch mode, it may be necessary to place rubber stoppers in the entry and exit ports.

As the sieving chamber rotates, it can be inspected by shining a laser through an inspection port on the side opposite the sieve surface that generates a diffraction pattern that may be used to monitor sieve damage and blinding. When the sieve is in the top position of the rotation cycle, the sieve apertures can also be cleaned by air jets.

The powder passing through the sieving surface is collected by a chute and funneled into a cup that rests on one arm of a device for automatically weighing the powder that passes through the sieve. The system shown in Figure 4.13*c* can weigh quantities of powder automatically with an accuracy of 0.1 g over a range of less than 1 g up to several hundred grams. The rotation time of the system shown in Figure 4.13*c* is 30 seconds. The weighing cycle is complete before the next rotation delivers powder down the collection chute. These limits and time schedules can be readily varied. The equipment is easily converted to wet-sieving equipment by mounting a spray pipe through one end of the icosahedral sieving chamber and directing the spray to the base of the chamber. For wet sieving, either the collection cup can be provided with a porous base so that only retained solids are weighed, or a larger container can be used for subsequent processing of the slurry collected.

The sieve surface in the SORSI equipment is much better mechanically supported than that of traditional sieving equipment; therefore, any damage due to bellying is virtually eliminated. Since only a small triangle of material is required, it is now possible to consider using electroformed sieving surfaces for all aperture sizes. In addition, SORSI equipment is quiet and needs only a low-power motor for operation. The flexibility of design of the SORSI equipment now makes it possible to link it with a small computer and operate it as an intelligent machine.

4.6 FELVATION

The basic system used in a technique for fractionating powders with electroformed sieve surfaces that combines elements of sieving technol-

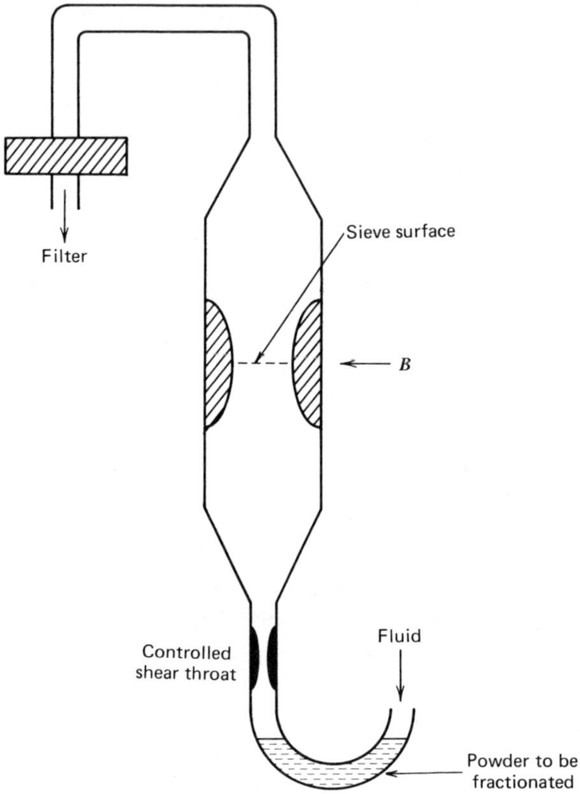

Fig. 4.14 Felvation column (*B*, body constriction to accelerate flow in the region of the sieve surface).

ogy, elutriation, and fluidized bed technology is shown in Figure 4.14. The procedure is known in industry as *felvation* [26, 27]. As can be seen from Figure 4.14, the sieve surface used in the separation of the fineparticles is mounted across the middle of the column. The powder to be fractionated is placed in the supply sump and fluidized into the felvation column. The state of dispersion in the felvation column can be controlled by varying the dimensions of the shear throat, a constriction in the flow system situated just below the main column.

In the initial phase of a felvation fractionation, the operating fluid is fed into the column at a low rate, which is increased slowly. The first stage of felvation is achieved when the powder becomes a fluidized bed at the base of the conical portion of the column. During this stage a circulation pattern becomes established, with larger fineparticles dropping back down the sides of the conical portion back into the jet of high-velocity suspension emerging from the shear throat. The second stage of the felvation process involves a further gradual increase in the flow rate through the column. At this stage the fineparticles that can pass easily through the sieve are swept up through the apertures without having to compete with oversize fineparticles as in traditional sieving techniques.

The flow constriction near the sieving surface increases the flow at the sieving surface, thus ensuring that fineparticles fluidized up to the surface are quickly assisted through the apertures. It also keeps the fluid region above the surface free of fineparticles so that they cannot fall back into the apertures of the sieve and interfere with the sieving process. Flow conditions under the sieve are generally turbulent, and it is possible for the odd large fineparticle to move up to the sieving surface prematurely. If this happens, the aperture becomes blocked, thus causing the flow rate in its immediate vicinity to fall. The large fineparticle can no longer be supported by the fluid, so it falls back into the turbulent fluid. Fineparticles that drift toward the sides of the column slide downward back into the high-shear region and are redispersed into the fractionation column. The final stage of felvation is reached when fineparticles just larger than the sieve apertures are carried up to the sieve surface. At this stage the fluid leaving the sieve becomes clear and the pressure drop across the sieve surface increases suddenly. The end of the felvation process can be detected immediately by monitoring either of these physical phenomena.

In the felvation process the sieve surface is covered with fineparticles for only a short time. As soon as the fluid flow is stopped, the fineparticles fall back into the sump of the felvation column. Because of the gentle and segregated way in which the fineparticles are presented to the sieving surface, there should be very little blinding of the sieve apertures. In essence, felvation utilizes a fluidized bed to deaggregate the powders and achieve primary fractionation. Then the fineparticles are elutriated up the column in sequence with the smallest first. Here they encounter the sieve surface, which acts as a go/no-go gauge to fractionate the fineparticles. It should be noted that the use of elutriation to feed the particles efficiently to a go/no-go gauge is different from the use of a relatively coarse grid placed across an elutriator in an attempt to smooth out the parabolic fluid flow profile present in an ordinary elutriator (see Chapter 7).

If a fluid such as kerosene is used in a felvation experiment, then, since it is a nonconductor of electricity, problems can be encountered with electrostatic effects unless an additive is placed in the kerosene to make it electrically conducting.

NOTES AND REFERENCES

1. K. Spurný, "The Investigation of the Filtration Process of Nonactive and Radioactive Aerosols by Means of Fibrous and Membrane Filters," Academia, Státní tiskárna n.p. závods, Praha 8, Czechoslovakia.
2. C. K. Hersh, *Molecular Sieves*, New York, Reinhold, 1961.
3. B. H. Kaye, N. I. Robb, and A. G. Naylor, *Methods for Studying and Minimizing Wear in Sieves Used in Characterizing Powders*, proceedings of Powder Technology Conference, Chicago, May 1977; published by Powder Advisory Center, 222 West Adams Street, Chicago, Illinois.
4. N. I. Robb and B. H. Kaye, "Sieve Aperture Characterization," *Powder Technol.* (Report in preparation.)
5. Electroformed sieves are available from Buckbee-Mears Company, 245 East Sixth Street, St. Paul, Minnesota 55101.
6. Photoetched sieves are available from Veco-Stork International, 4925 Silabert Avenue, Charlotte, North Carolina 28205.
7. B. H. Kaye, "Physical Problems in Particle Size Analysis," Ph.D. thesis, London University, 1961.
8. B. H. Kaye, "Potential Advantages of Multipartitioned Sieve Surfaces for Characterizing Powders," *Powder Technol.*, 19 (3) (1978), 121–123.
9. H. Konowalchuk, A. G. Naylor, and B. H. Kaye, "A Rapid Method for Assessing the Quality of Sieves," *Powder Technol.*, 13 (1) (1976), 97–101.
10. H. Konowalchuk, M.Sc. thesis, Laurentian University, Sudbury, Ontario, Canada.
11. A. G. Naylor, B. H. Kaye, and P. E. Legault, "An Instrument for Assessing Sieve Aperture Damage," *J. Powder Bulk Solids*, 1 (1977), 9–12.
12. W. S. Tyler Company, Cleveland 14, Ohio.
13. Cenco Instrument Corporation, 1700 Irving Park Road, Chicago, Illinois 60613.
14. Alpine American Corporation, 3 Michigan Drive, Natick, Massachusetts 01760.
15. ATM Corporation, Sonic Sifter Division, P. O. Box 2405, Milwaukee, Wisconsin 53214.
16. B. H. Kaye, "Silica Flowagents and the Movement of Powder," in Proceedings of the PowTech Conference, Chicago, May, 1977; published by Powder Advisory Center, 222 West Adams Street, Chicago, Illinois.
17. F. J. Colon, "Sieve Analysis to 5 Microns with Micro-mesh Sieves," *Chem. Ind.* No. [6] (1965), 263.
18. H. Heywood, "A Study of Size Analysis by Sieving," *Transact. Inst. Min. Met.*, 55 (1945–1946), 373.
19. C. Orr, R. W. Camp, and D. K. Davis, *Particle Size by Automatic Sieving*, PowTech Conference, May 1977; published by Powder Advisory Center, 222

West Adams Street, Chicago, Illinois. Literature on the Autosieve 5900 available from Micromeritics Instrument Corporation, 800 Goshen Springs Road, Norcross, Georgia 30071.

20. A. W. Fahrenwald and S. W. Stockdale, "Effect of Sieve Motion on Screening Efficiency," U.S. Bureau of Mines Report Investigation No. 2933, 1929.
21. K. T. Whitby, "The Mechanics of Fine Sieving," A.S.T.M. Special Technical Publication No. 234, 1959, p. 3.
22. B. H. Kaye, "Investigations into the Possibilities of Developing a Rate Method of Sieve Analysis," *Powder Met.* (1962), 199–217.
23. British Standards Institute Publication Nos. 410 (1943) and 1796 (1952).
24. B. H. Kaye and N. I. Robb, "A New Algorithm for Calculating Kinetic Residues for the Characterization of Powders Using Sieves," *Powder Technol.*, 24 (1979), 125–128.
25. B. H. Kaye, "SORSI and Digest. Two Intelligent Machines for Characterizing Fineparticles," in *Proceedings of the Powder Technology Conference*, Chicago, Illinois, March 10–12, 1977.
26. B. H. Kaye and M. R. Jackson, *A New Technique for Fractionating Powders, Powder Technol.*, 1 (1) (1967), 43–50.
27. M. W. G. Burt, *Proceedings of the Society for Analytical Chemistry*, Vol. VI (1970), pp. 156–168.

CHAPTER

5

GRAVITY SEDIMENTATION STUDIES OF FINEPARTICLES

5.1 INTRODUCTION

In gravity sedimentation methods for characterizing fineparticles, the falling speeds of fineparticles settling in a viscous fluid under the influence of gravity are measured. The measured speeds are transformed by using Stokes' equation into a size parameter known as the *Stokes diameter*. The various sources of uncertainty in the measured Stokes diameter distribution of a fineparticle system arise from the various compromises that have to be made in the design of systems for studying the group dynamics of the fineparticles.

In the published literature of sedimentation procedures for characterizing fineparticles, there is considerable confusion over the interpretation of sedimentation data for irregularly shaped fineparticles. It is sometimes claimed that "increasing error" occurs if sedimentation procedures are used to study highly irregular fineparticles. This is a misleading statement. If irregular fineparticles are allowed to sediment in a fluid, there is an uncertainty range for the Stokes diameter of the fineparticle because different orientations of the fineparticle are possible. For example, if an irregular fineparticle falls with its largest dimension perpendicular to the direction of motion, it will fall more slowly than if its maximum dimension is aligned with the direction of fall. Davies [1] has shown that under laminar flow conditions the initial orientation of an irregular fineparticle falling under gravity is maintained throughout the complete trajectory of the fineparticle. Therefore, an irregular fineparticle will have a range of settling speeds that will be interpreted as a range of Stokes diameters. Furthermore, the average Stokes diameter of an irregular fineparticle will differ from the diameter of the dense smooth sphere of the same material because the irregular fineparticle drags stagnant fluid with it during its fall. The difference between the measured Stokes diameter of a fineparticle and the theoretical magnitude of the Stokes diameter calculated by considering the dynamics of a sphere of equal volume is a measure of the shape of the fineparticle. Before the physical reasons for this discrepancy were fully appreciated, this discrepancy was sometimes classified as an "error"; it is now seen to be information concerning the structure of the fineparticle.

Problems of dispersing fineparticles in a fluid are discussed in Section 5.2. Problems of maintaining the thermal stability of the suspension of fineparticles are discussed in Section 5.3. The containing vessel walls can interfere with the free fall of a fineparticle. The presence of other fineparticles in a suspension can also interfere with the free fall of a fineparticle. Disturbances caused by the presence of other fineparticles are referred to as *concentration effects*. Published studies concerning the concentration levels at which mutual interference between sedimenting fineparticles becomes negligible are confusing, as shown by the data in Table 5.1 [2, 3]. The effects of the containing vessel and the suspension concentration on the accuracy of characterization studies are studied in detail in Section 5.4. The remaining sections of this chapter are concerned with the details of the various procedures and instruments for characterizing fineparticles by gravity sedimentation.

5.2 DISPERSING FINEPARTICLES IN A FLUID

To achieve good contact between a fineparticle and a liquid in which it is to be dispersed, it is common practice to start off by adding a few drops of the fluid to the dry powder, followed by the application of high-shear

TABLE 5.1 Recommended Upper Limit Concentration to Be Used in Sedimentation Methods for Characterizing Fineparticles

Year of Publication	Observer	Volume Concentration of Solids
1941	Martin [4]	3.0
1942	Carey and Stairmand [5]	0.01
1947	Stairmand [6]	0.2
1951	Amstein and Scott [7]	2.0
1951	Hawksley [8]	0.05
1954	Jarrett and Heywood [9]	1.0
1959	Orr and Dallavalle [10]	2.0
1963[a]	Irani and Callis [11]	0.05

Source. Reference 2.
[a]Added to Boardman's list.

forces to bring the fluid and the fineparticle into good mechanical contact. At this stage the technologist should avoid applying greater forces to the fineparticles in the viscous fluid than those to which the fineparticle will be subjected when in the operative system being studied. Failure to match the severity of dispersion used prior to a characterization study with that experienced by the fineparticles in the system of interest results in a mismatch of information. Thus, if the fineparticle under study is a pigment that is to be dispersed in a viscous oil medium using very severe shear forces, the technologist should use equally severe forces to disperse the fineparticles before a sedimentation characterization study. Again, if a pharmaceutical powder is to be ingested by a patient, where the dispersal technique is gentle stirring of the powder after it is poured into a glass of water, then equally mild dispersion shear forces are appropriate when characterizing such a powder.

If the fluid carrying the sedimenting fineparticles is a gas, it is common practice to disperse the powder in the gas using high-velocity gas jets. The moisture content of the gas and/or the powder can make large differences in the state of dispersion achieved by this technique. If the moisture content of the fineparticles and the gas is very low, this type of dispersion technique can result in the fineparticles carrying a high electrostatic charge that can interfere with the settlement dynamics of the fineparticles and lead to deposition of fineparticles on the walls of the sedimentation vessel (see discussion of the Micromerograph® technique described in Section 5.5).

When attempting to disperse a powder in a liquid, one should always check the history of the powder to see if it may have suffered surface contamination that will make dispersion difficult in the given fluid. For example, many fine-metal powders are manufactured by a grinding process in which a stearate is added to the powder to prevent spontaneous welding of the fineparticles to each other as they are being formed. Such a powder is likely to disperse easily in an organic fluid but will not be wetted by water. Simple rinsing of such a powder in an organic solvent is unlikely to completely remove the stearate layer, and it will usually be necessary to use a surface active agent to aid dispersion of the powder. Many metal powders have an oxide film on their surface so that their surface physical properties are those of the oxide, not of the bulk material. It has been suggested that this accounts for the fact that the fineparticles of many conducting metals can be treated as nonconductors when studying them in resitazone stream counters (see Chapter 8).

A distinction should be made between a dispersing agent and a surface-active agent (in modern technical literature the latter is often referred to as a *surfactant*). A dispersing agent is any general additive placed in a suspension to maintain maximum interparticle distance, whereas a surface-active agent is a particular kind of dispersing agent, usually designed to achieve intimate contact between the surface of the fineparticle and the body of the fluid. Particularly in industrial science, the term "dispersing agent" is sometimes applied to materials that are colloids, that is, gums and/or gels that achieve and maintain maximum fineparticle separation by increasing the local viscosity within the suspension. In some early publications on fineparticle science it was indicated that a measure of a good dispersion was a bulky sediment at the end of the settling experiments. This is not a safe rule to follow since bulky sediment may result from a gelatinous structure in suspension achieved by a gelling or colloidal agent.

A particular combination of a sedimenting fluid and a surface active agent may result in the fineparticles carrying ionic charge that can create a potential gradient within the settling suspension. This can give rise to electroviscous effects [12]. Effects due to electroviscosity are generally neglected in characterization studies of fineparticles but can be of significant magnitude. In the general discussion of characterization procedures given in this book it is assumed that the technologist has selected the fluid and dispersing agents so that electroviscous forces can be neglected.

If a nonconducting fluid is used, electrostatic effects can interfere with sedimentation dynamics. It is possible to minimize this type of problem by adding chemicals in very low quantities to fluids such as kerosene and aliphatic hydrocarbons to make them electrically conducting.

The problem of determining the compatibility of a surfactant–fineparticle–fluid system is complex [13, 14]. In the experience of the author, the choice of an appropriate additive to aid a fineparticle–fluid system dispersion is usually so specialized that one has to rely on the people requesting the fineparticle characterization for advice. It does not seem useful to give a list of surfactants compatable with a series of powders and fluids since such a list will not be of general interest. Technologists facing a problem of choosing a surfactant or other dispersing agent should consult the specialist literature in this area [13, 14].

In all sedimentation characterization studies it is difficult to decide when the fineparticles can be treated as having begun a steady fall at their terminal velocity. Agitation of the suspension immediately before a sedimentation study results in a residual random motion of the fineparticles. This motion damps out relatively quickly but is a source of uncertainty in the accuracy and precision of the characterization procedure. Allen [14] has shown that for most purposes the time period required for a fineparticle to achieve its terminal velocity under laminar flow conditions in a viscous fluid is negligible compared to the general precision and accuracy of sedimentation procedures.

It is often useful to place a suspension to be used in a characterization study in a vacuum environment for a short period of time to degas the system because small air bubbles clinging to the fineparticles can be a major source of error in the measurement of the falling speeds of the fineparticles. Any subsequent stirring should not introduce gas bubbles into the suspension. In some published procedures for sedimentation characterization studies, a final violent turbulent stirring of the suspension is achieved by bubbling compressed gas through the system. This procedure might be acceptable in a quality control situation but should never be used in precise research level characterization studies because minute bubbles might become attached to the small fineparticles.

When a paddle stirrer is used to agitate a suspension, the blades of the paddle should be designed to ensure general upthrust of the fluid being stirred. A poorly designed stirrer can impart a strong rotary motion to the suspension, thus centrifuging the fineparticles so that a density gradient will exist across the suspension vessel. When stirring ceases, denser suspension will move down the walls and cause an upwelling along the axis of the container. This convection current will seldom be visible from the outside of the container. The rapid rotation of the shaft of a paddle–stirrer can create a vortex at the surface of the suspension, thus resulting in the entrainment of air bubbles. The use of a magnetic stirrer at the base of the sedimentation column can prevent problems due to air entrapment at the surface of a sealed suspension. A simple and effective stirrer for generating general turbulence and mixing in a suspension consists of a perforated disk attached to a rod that is moved up and down in the suspension. As the disk is moved, the suspension is forced through the holes in a turbulent manner, thus generating a uniform dispersion of the fineparticles.

Sometimes in published descriptions of sedimentation characterization studies there is a vague reference to ultrasonic agitation. There are many different kinds of ultrasonic dispersion device that operate at different power levels. If the ultrasonic dispersion device is based on the use of standing acoustic waves in a fluid, that is, an ultrasonic "cleaning" bath, then the geometry of the vessel containing the suspension and its position in the ultrasonic bath can affect the acoustical coupling and the power transmitted to the suspension. Many ultrasonic treatments of a suspension involve a process known as *cavitation*. The local forces within a cavitating zone of a fluid being treated with ultrasonic waves can be very severe and can result in the rapid degradation of a fineparticle system. The use of an ultrasonic probe to treat a suspension can result in a rapid temperature rise of that suspension because of the power absorbed by the suspension. Any ultrasonic treatment of a fineparticle suspension should be carried out circumspectly, and the details of the treatment should be recorded in detail when reporting the results.

5.3 THERMAL STABILITY OF A SUSPENSION

To maintain the thermal stability of a sedimenting suspension, it is common practice to surround the sedimentation column with an outer jacket through which constant-temperature fluid is circulated. This is not sufficient thermal protection of the suspension if the top surface of the sedimentation column is open to the atmosphere. In such a system there is a vapor loss from the surface of the fluid, thus causing surface cooling of the fluid. This heat loss results in convection currents moving down from the surface of the fluid into the body of the suspension [15]. When the sedimenting column fluid is water, one way of preventing evaporative cooling is to float onto the top of the column a thin layer of immiscible low-vapor-pressure mineral oil immediately after the last stirring of the suspension. Davies and Kaye [16] described a procedure to prevent evaporation by covering the top of the surface with a thin layer of microballoons of glass. Because of the ready availability of styrofoam and other expanded plastics, probably many different kinds of small plastic beads of low density can now be used to achieve the same thermal protection of the open surface.

Thermal control of the outside of a sedimentation column is inadequate when radiant energy is incident on the sedimentation system. For example, the author once visited a laboratory where glass bottles containing the sedimenting suspension were stood in a constant-temperature water bath. The temperature bath was maintained steady at 25°C, yet it was obvious that convection currents were occurring within the sedimenting suspensions. The constant-temperature bath was stood on a bench in strong sunlight. Sunlight penetrating the glass cover of the sedimentation column was being absorbed by the fluid. As a consequence, the fluid was considerably warmer than its surroundings. The instability problem was prevented by shielding the temperature bath from direct sunlight. Bright laboratory lights adjacent to a temperature-controlled environment when using transparent containers can result in the same type of thermal instability. Inexpensive Dewar flasks can be used to thermally isolate the sedimentation system.

5.4 FLUID DYNAMIC SOURCES OF ERROR IN SEDIMENTATION METHODS

The retarding effect of the containing walls on the fall of a sphere moving axially in a cylindrical vessel has been studied extensively [23, 24]. If D is the diameter of the containing vessel and d the diameter of the sphere, then for D large compared to d, the magnitude of the retardation effect can be described by Landenburg's equation, which is usually written in the form

$$V_\infty = V_m\left(1 + 2.4\frac{d}{D}\right) \tag{5.1}$$

where V_∞ is the falling speed of the sphere in an infinite sea of fluid, that is, the theoretical Stokes velocity calculated from inserting the known characteristics of the sphere in Stokes' equation; and V_m is the measured velocity in a vessel of diameter D.

Experimental data on the terminal velocity of a sphere in a cylinder of viscous fluid, taken from the work of Fidleris and Whitmore [24], are summarized in Figure 5.1. Even when the cylinder is 20 times bigger than the sphere, there is still a 10% reduction in the falling speed of the sphere.

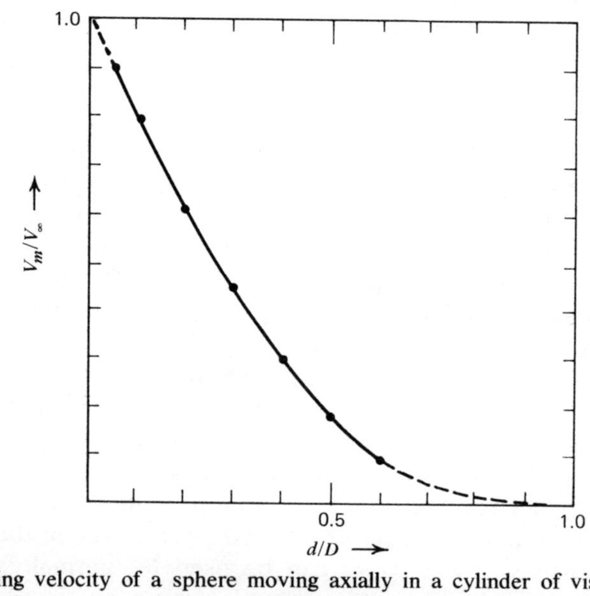

Fig. 5.1 Falling velocity of a sphere moving axially in a cylinder of viscous fluid [24]. Symbols: D, diameter of container; d, diameter of sphere; . . . extrapolation region; V_m, measured velocity; V_∞, ideal Stokes velocity.

As the sphere approaches the dimensions of the cylinder, the flow of the fluid around the sphere, rather than the fall of the sphere, becomes the dominating physical phenomenon. In a similar way, it has been established that if the settling behavior of highly concentrated suspensions containing solids of the order of 0.20 solids fraction (20% by volume) is studied, it is found that the major physical phenomenon occurring is the draining of the fluid from among the sedimenting fineparticles. The settling behavior of this type of high-concentration suspension is known as *hindered settling*. Such systems have been studied extensively by chemical engineers interested in the settling of slurries and "tailing" ponds from mining operations [25]. The equations for describing hindered settling are related to permeability equations for fluid flow through packed powder beds and are described in the companion volume, *Indirect Methods of Fineparticle Characterization*.

When studying the sedimentation of the fineparticle of the order of 50 μm in diameter, a sedimentation vessel of 1 cm in diameter constitutes a 200-diameter cylinder. Therefore, retardation effects predicted by Landenburg's equation are small. Most sedimentation procedures for characterizing fineparticles use sedimentation vessels larger than 1 cm, so that the effect of the bounding walls on the trajectories of the fineparticles can usually be neglected. The development of techniques for monitoring the dynamics of fineparticles using laser Doppler effects will make it possible in the not too distant future to consider carrying out experiments to measure the velocity of fineparticles moving near the wall of a containing vessel in the presence of other fineparticles at a series of distances from the wall. This should make it possible to clear up uncertainty concerning the effect of the walls on the sedimenting fineparticles.

In general, however, it can be stated that in sedimentation characterization studies interfineparticle effects are usually an order of magnitude greater than the effects of the containing vessel on the trajectories of the fineparticles. To gain an appreciation of the distances between fineparticles in a suspension at a given concentration, consider the ideal case of a suspension of spheres of equal size in which individual spheres are located at the intersection points of a three-dimensional square lattice. For such an idealized system, the relationship between the volume concentration of the suspension α and the distance between the centers of the spheres d expressed in terms of the diameters of the spheres D is

$$\frac{d}{D} = \left[\frac{\pi}{6}\,\alpha\right]^{\frac{1}{3}} \tag{5.2}$$

From this formula it can be calculated that even for the idealized suspension, the between-center distance of the fineparticles is four diameters at

0.01 volume fraction (i.e., at 1% by volume concentration). Although the suspension considered in developing equation 5.2 is for a very idealized suspension, it yields an impression of the order of magnitude of interfineparticle distances in a real suspension.

Wall effects on the behavior of a sedimenting suspension can be severe if a sloping wall occurs within the sedimentation system [26]. In the rest of this section we are concerned with fluid dynamic effects in containers that have parallel vertical walls.

As part of a study of the interaction of sedimenting fineparticles, Kaye [27] studied the motion of four contiguous spheres falling in a very large tank of viscous fluid. It was noted that the assembly of spheres fell as a group while rotating around each other. Although they are not physically connected, they maintain a tight configuration as they move downward through the fluid. This interaction of the spheres illustrates a phenomenon that has been described as dynamic clustering in which groups of fineparticles moving through an infinite sea of fluid do not move apart but act as a cohesive drop of suspension that has internal circulation. Dynamic clustering is an important aspect of the behavior of relatively dilute suspensions of fineparticles.

The interaction of two spheres falling under laminar flow conditions at various separations and orientations with respect to each other in a viscous fluid has been studied extensively [2, 20]. Data for spheres forming a horizontal and vertical doublet for various separations are summarized in Figure 5.2. It should be noted that even at seven diameters apart, the falling speeds of the doublets are greater than for the single sphere by approximately 4 and 10% for the horizontal and vertical doublets, respectively. This increase in falling speed arises from the fact that the liquid can move around the spheres more easily than between them. Therefore, the drag force on the two-sphere system is less than what would be experienced by the two spheres separately. To illustrate the confusion in the fineparticle literature with respect to the reality of fluid particle dynamics, consider the following quotation [21]:

> It is obvious that if two identical particles separated by only a few diameters move through a viscous liquid, the fluid passes between in such a way that the resulting viscous force is greater than that acting on a single particle. The velocity is thus less than would be predicted by Stokes' law. Experimentally, it has been shown that the effect is sufficiently small to be within the region of other uncertainties at volume concentrations below 1–3%. . . . The error introduced by particle interactions under these conditions is in the direction of a smaller indicated size and it is less than 1%.

This comment is 14 years later than the thesis of Hall, who definitely

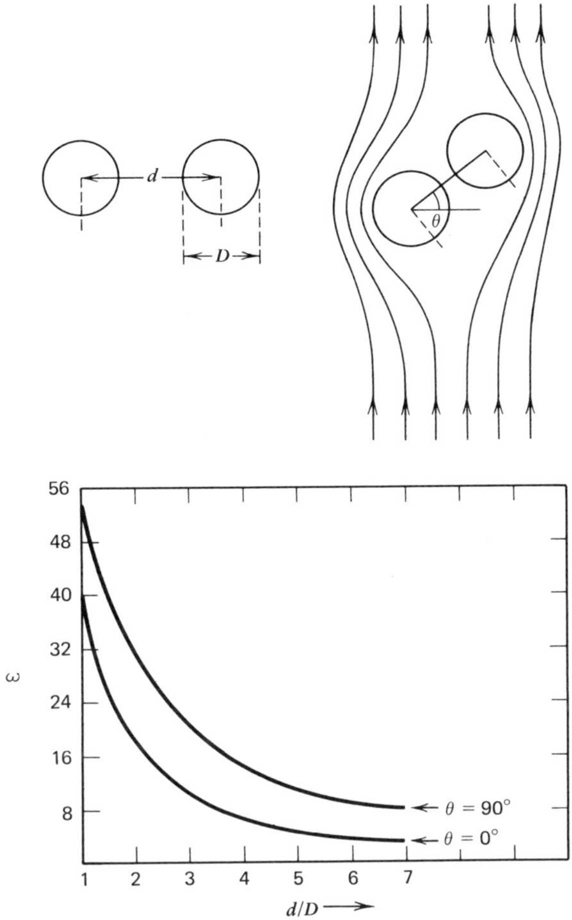

Fig. 5.2 The falling speed of a pair of spheres [20] (streamlines shown indicate the general nature of fluid flow and are not definitive of fluid element trajectories): ω, percent increase in falling speed of doublet with respect to that of a single sphere.

established that the flow conditions in the vicinity of a doublet of spheres is *around* and not *between* them! It is also diametrically opposed to the empirical fact that the net effect is a higher velocity for the doublet, *not* a reduced velocity. The experimental evidence referred to in the quotation is that of Martin, carried out in 1941 [4]. It is shown later in this discussion that the experimental data of Martin are capable of a completely different explanation and cannot be used to justify such high levels of solids concentrations in sedimentation studies.

When measuring the Stokes diameter distribution of a sedimenting suspension, two basic suspension systems can be used. In the simplest system the fineparticles to be characterized are placed at the top of a column of fluid and the movement of the fineparticles down the column studied (see Figure 5.3). Historically, this system has been referred to in the literature as a *two-layer sedimentation system*. This name is misleading since it implies that there are two layers in the column whereas there is only one layer of suspension on top of a column of fluid. The modern term for this type of system is *line start procedure*.

The movement of the fineparticles down the column can be monitored as they pass through a given zone at a certain depth, a technique

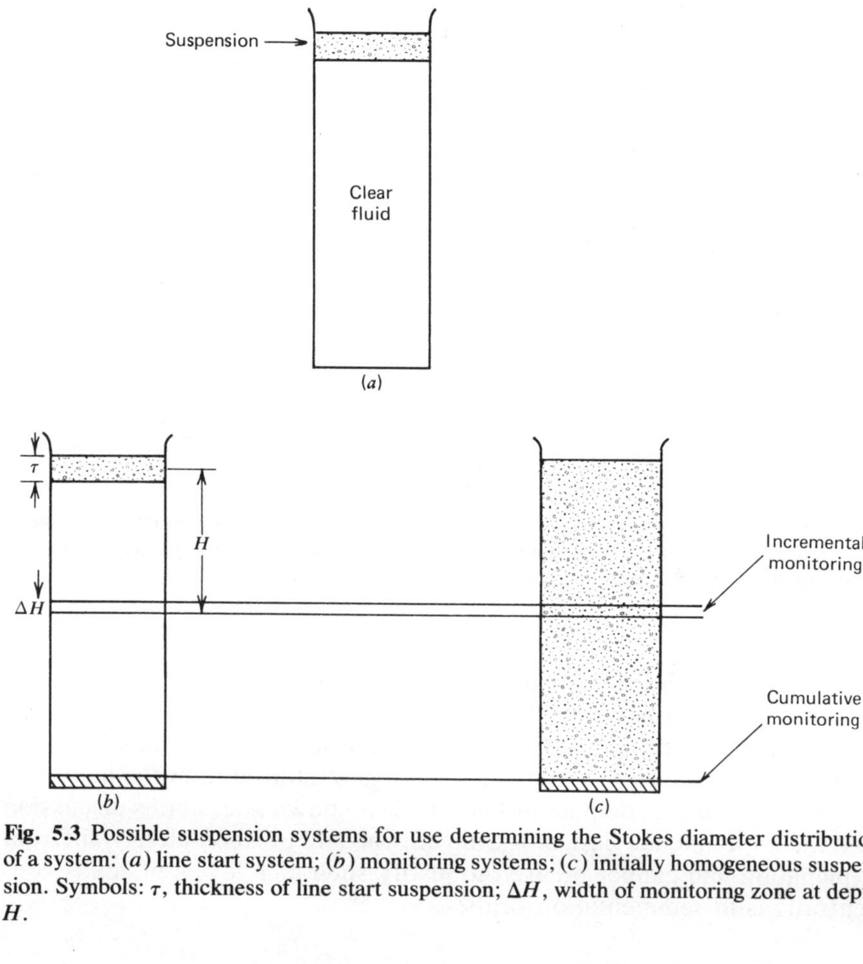

Fig. 5.3 Possible suspension systems for use determining the Stokes diameter distribution of a system: (*a*) line start system; (*b*) monitoring systems; (*c*) initially homogeneous suspension. Symbols: τ, thickness of line start suspension; ΔH, width of monitoring zone at depth H.

referred to as *incremental sampling of the suspension*. Characterization procedures that monitor the arrival of the fineparticles at the base of a column of fluid are known as *cumulative sedimentation methods*.

The line start procedure has the advantage that each size of fineparticle presents itself to an incremental measuring zone or a cumulative rate-measuring device in decreasing order of magnitude, thus enabling the output of the instrument to be interpreted as a size distribution function with minimum data manipulation. The disadvantage of the procedure is ensuring the stability of the layer of suspension used in the line start sedimentation dynamics.

In the early 1960s Kaye and Boardman investigated the stability of the line start sedimentation system. They found that for a suspension of monosized spheres, the concentration level at which interfine particle effects that caused instability in the line start system were negligible was of the order of 0.05% by volume (i.e., a volume fraction of 0.0005). This represents an interfineparticle distance calculated from equation 5.2 of seven diameters. This indicated that many concentration levels used in sedimentation characterization studies could result in dynamic clustering of fineparticles leading to error in the measured Stokes diameter distribution. Because of the significance of the Kaye-Boardman data for sedimentation methodology, their experiments are briefly reviewed [2, 3].

Kaye and Boardman measured the terminal velocity of individual red glass spheres falling in the presence of other glass spheres of various size ranges. The red spheres, termed *marker spheres,* had the same Stokes diameters as measured empirically with an accuracy of 1%. (They were selected from among a larger population of red glass beads by allowing them to fall individually under laminar flow conditions down a column of clear viscous mineral oil.) Measured under the microscope, these spheres had a diameter of 900 μm. The dynamics of these marker spheres in suspension of clear glass spheres in mineral oil were investigated experimentally. By using 50 marker spheres, random chance could be relied on to provide at least one sphere moving down along the axis of the sedimentation column. The terminal velocity of an axial marker sphere was measured several times at a series of concentrations in three different glass sphere powders. The glass powders were "ballotini," manufactured in large quantities for making reflecting road signs. The size distribution function of the three powders determined by microscopic examination is shown in Figure 5.4.

In the first series of experiments the red marker spheres were studied in various suspension concentrations of powder X, which had a narrow range of sizes close to that of the marker spheres. The experimental data from this series of experiments are summarized in Figure 5.5. At the time

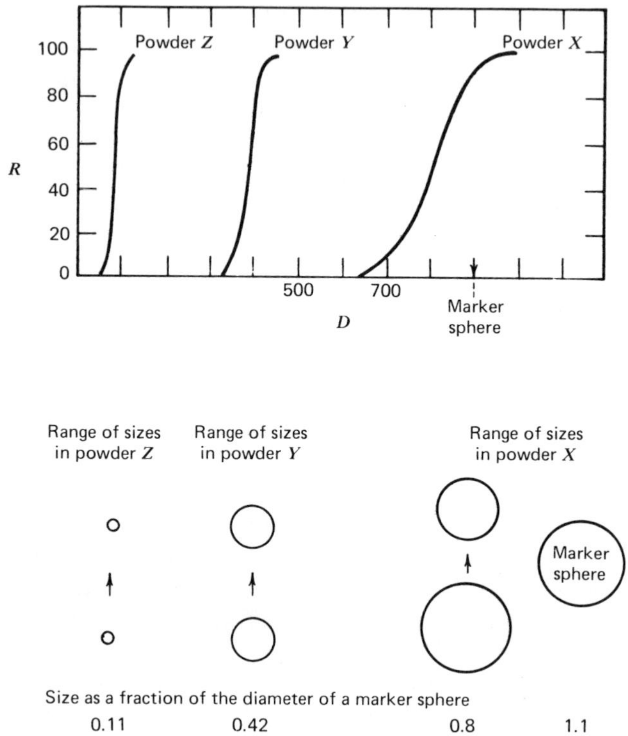

Fig. 5.4 Characteristics of powders and marker spheres used by Kaye and Boardman. Symbols: R, cumulative percentage by weight less than stated size; D, sphere diameter (microscope examination) in microns.

that these data were reported, they seemed to contradict the findings of all previous studies of sedimentation dynamics. In particular, the existence of a concentration range in which terminal velocities in excess of Stokesian velocities were measured became a center of controversy. This phenomenon has been the subject of later in-depth experimental work by Johne [28], Koglin [29], and Scarlett and Cowlam [30, 31]. Their studies confirmed the existence of a range of suspension concentration for which hyper-Stokesian velocities exist. Johne and Koglin actually reported higher effective terminal velocities than did Kaye and Boardman. Their observed data apparently differ from those of Kaye and Boardman for three reasons: (1) they used more closely size ranged powders, so that their system was a much better indication of effects to be anticipated as operative in a truly monosized suspension; (2) Kaye and Boardman used a relatively narrow sedimentation vessel, and Koglin has shown that re-

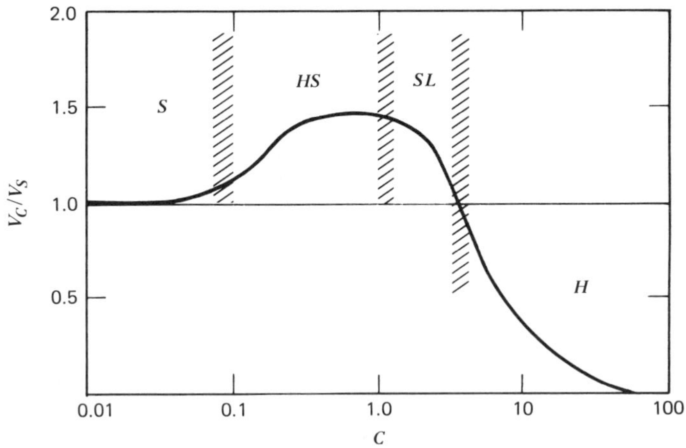

Fig. 5.5 Concentration-dependent behavior of a settling suspension of approximately monosized spheres [3]. Symbols: V_C, average velocity of marker spheres moving down the center of the column; V_S, Stokes velocity of marker sphere under free-fall conditions; C, volume of solids (percentage of powder X).

sidual wall effects probably reduced the magnitude of their measured excess Stokesian velocities [29]; and (3) electrostatic effects in the Kaye–Boardman system could have further lowered the measured terminal velocities [30]. Kaye and Boardman made measurements only on spheres moving along the axis of the sedimentation column, whereas Johne and Koglin recorded data for spheres across the horizontal plane of the sedimentation vessel. However, in broad terms, the data of Kaye and Boardman have been substantiated.

Kaye and Boardman postulated that the dynamic behavior of settling suspensions should be divided into the four regions shown in Figure 5.5. The locations of these boundaries shown in the diagram are only approximate and will depend on the size ratio of fineparticles present in any particular suspension. The very low concentration range corresponds to Stokesian or free-fall behavior. In the original data of Kaye and Boardman, the approximate range of the sizes present in the suspension with respect to the marker spheres was 0.8–1.1. With this system, the Stokesian region seems to terminate at approximately 0.05% by volume. On the other hand, Johne and Koglin report that with their closer-sized range of powder they observed significant departure from Stokesian conditions at concentrations lower than this. When Kaye and Boardman carried out interaction studies with powders Y and Z, the distribution functions of which are given in Figure 5.4, the interactions were of a much smaller magnitude. The Stokesian region of effectively free-fall behavior appeared to extend up to concentrations of approximately 0.1% when the marker spheres: suspension spheres ratio was 2.5:1. When the marker spheres: suspension spheres ratio was 9:1, the free-fall zone extended up to 1% by volume. These experiments are summarized in Figure 5.6. The second region of suspension behavior, described in Figure 5.5 as the dynamic clustering region, is one in which the dominant behavior inside the sedimenting suspension is the formation by random juxtaposition of dynamic clusters that fall at greater speeds than the individual fineparticles. Dynamic clusters formed in a suspension are less stable than those formed by a group of monosized spheres in a sea of fluid. Also, in a suspension the general movement of fluid around the dynamic clusters results in the breakup and the formation of various clusters on a random basis, with any one fineparticle participating in several clusters in its overall trajectory down through the suspension. Boardman and Kaye

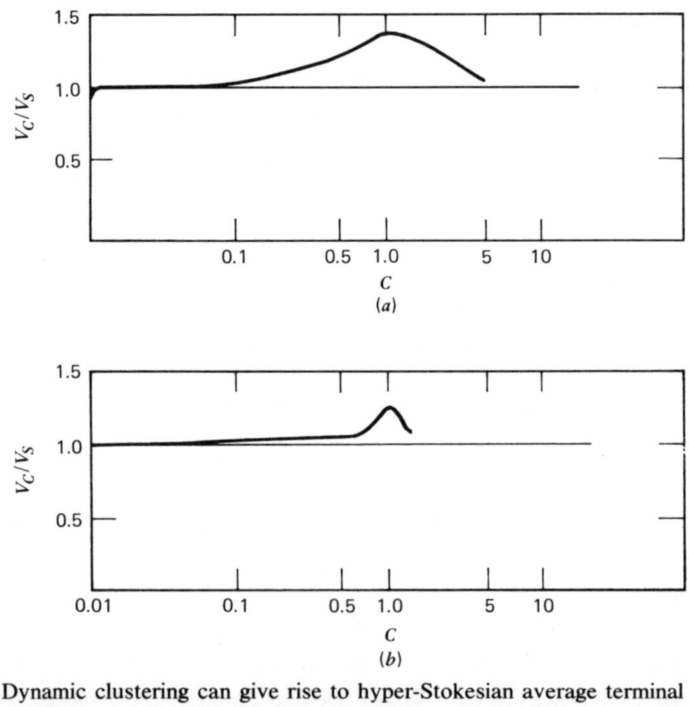

Fig. 5.6 Dynamic clustering can give rise to hyper-Stokesian average terminal velocities: (*a*) marker spheres in a suspension of powder *Y* (see Fig. 5.4) (*b*) marker spheres in a suspension of powder *Z* (see Fig. 5.4). Symbols: *C*, percentage (by volume) concentration of powder. V_C, velocity of marker spheres in concentration *C*; V_S Stokes velocity of marker spheres.

report that their marker spheres moved upward at certain times in their overall trajectory, and the data in Figures 5.5 and 5.6 represent average velocities over a continuous extensive trajectory down through the suspension.

As the concentration of the suspension increases, there is a higher probability of the formation of larger clusters that fall with greater speed to be interpreted as larger pseudofineparticles. However, further increase in concentration also results in interaction between the clusters with the movement of the fluid out of the way of the sedimenting fineparticles becoming the most important aspect of the suspension behavior. As pointed out earlier, when fluid drainage is the dominant physical behavior of a suspension, the process is known as *hindered settling*. The third region of concentration-dependent behavior shown in Figure 5.5 represents competition between dynamic clustering and hindered settling. As the concentration is increased still further, hindered settling becomes the dominant type of behavior in a suspension, as indicated in Figure 5.5. For the approximately monosized system of Kaye and Boardman, hindered settling overpowered dynamic clustering to reduce average terminal velocities below their theoretical velocities at approximately 5% by volume.

The general conclusion drawn by Kaye and Boardman from their data was that in the absence of prior knowledge concerning the size range of fineparticles present in a suspension, the safest procedure to follow when setting out to characterize the Stokes diameter distribution of a fineparticle system was to utilize volume solids concentrations less than 0.05% by volume. This conclusion has been justified by many experimental investigations reported since the date of their original investigation.

Experiments that illustrate that concentration effects are virtually eliminated in line start procedures with the use of solids concentrations of less than 0.05% by volume were reported by Kaye et al. [32] and Kaye and Jackson [33], who measured the Stokes diameters of monosized latex particles in the Illinois Institute of Technology (IIT) Research Institute version of the centrifugal disk photosedimentometer. This equipment is described in detail in Chapter 6. It is sufficient for our purposes to recognize that in this technique the fineparticles are sedimented outward from the center of a centrifuge and are monitored as they pass a beam of light at a certain distance from the center of rotation. In the line start procedure employed, a small amount of suspension is injected into the centrifuge to create a line start system. With this type of equipment, when using a monosized system, there should be no movement of the pen recorder beam used to follow the movement of fineparticles through the monitoring beam of light until the fineparticles have traveled the full distance from the line start system. If, however, there is a breakdown of the layer due to dynamic clustering, the arrival of pseudofineparticles having larger terminal velocities will be recorded by the characterization system.

Sets of pen recorder traces obtained by studying 2.965-μm latex spheres of polyvinyl toluene sedimenting in water at various concentrations of injected suspension are shown in Figure 5.7. The data demonstrate that for solids concentrations of less than 0.03% by volume in the line start suspension, fluid dynamic interfineparticle effects were negligible. However, at 0.3% by volume, pseudofineparticles of larger effective size were recorded, thus indicating dynamic clustering and the breakdown of the line start suspension. [The pen recorder sensitivity is different in the case of the three traces of Figure 5.7 (32).] At the high resolution

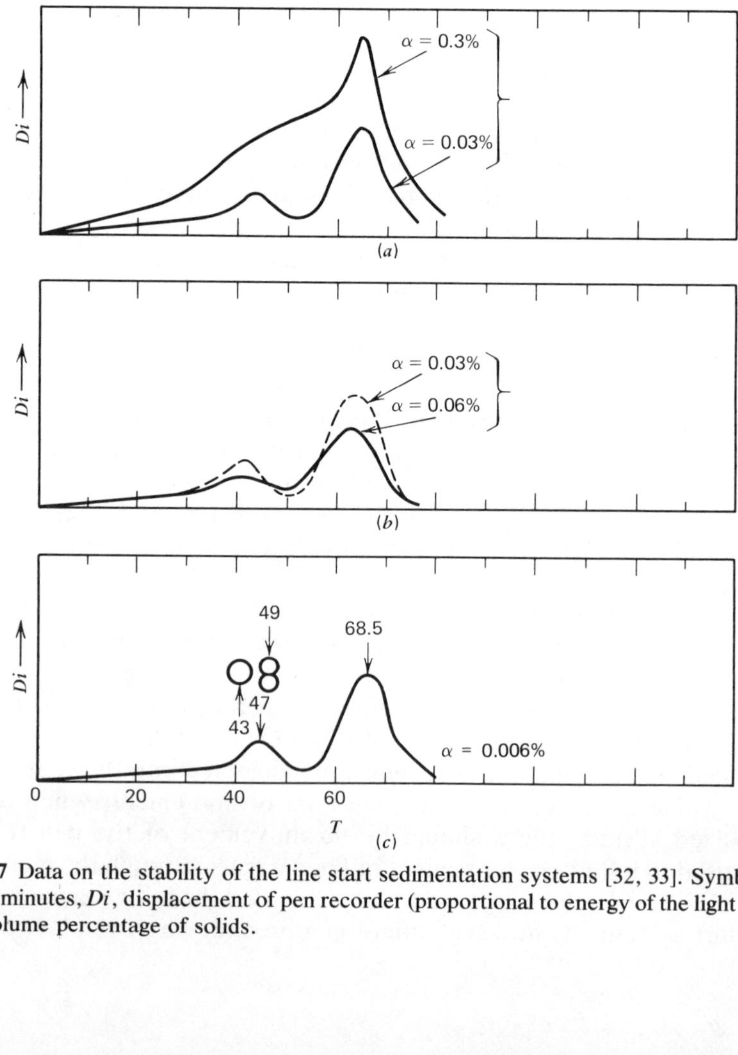

Fig. 5.7 Data on the stability of the line start sedimentation systems [32, 33]. Symbols: T, time in minutes, Di, displacement of pen recorder (proportional to energy of the light beam); α = volume percentage of solids.

obtainable by using 0.006% by volume solids, Kaye and Jackson [32] demonstrated the presence of permanent doublets in the nominally monosized latex as demonstrated by the data in Figure 5.7. If the time of arrival of the prime peak is taken to represent the arrival of the single spheres (68.5 minutes in Figure 5.7), then the predicted time of arrival of two spheres completely fused to form a single sphere is 43 minutes and a doublet moving with its axis along the direction of travel 49 minutes. The secondary peak in Figure 5.7*c* is at 47 minutes, indicating that this peak probably represents rigid doublets in various orientations. In later experiments Kaye, Jackson and Kahrun [32] were able to differentiate the arrival of doublets in the monitoring zone traveling along the direction of centrifugal force from the arrival of those traveling at right angles to this direction.

There is one exception to the general rule that stability in a line start system can be achieved only when the solids concentration in the line start suspension is less than 0.05% by volume solids. This is the special case when the thickness of the line start suspension is so thin that it is virtually one fineparticle thick. The equipment used by Kaye and Jackson was of sufficient sensitivity that this special case could be explored. They were able to utilize line start suspensions of only a few microns in thickness. Using such very thin line start suspensions, they were able to obtain stable layers with concentrations of up to 1% by volume solids. It should be stressed, however, that these extremely thin layers are a very special case and do not affect the general recommendation that in line start systems solids concentrations of less than 0.05% should be used wherever possible.

Kaye and Boardman report that interaction effects are reduced in suspension of smaller glass powders because of the inability of the marker spheres to participate in stable clusters with the smaller fineparticles as they moved through the fluid. Thus the clustering that gave rise to the hyper-Stokesian velocity in glass powders *Y* and *Z* were far more transient than the clusters noticed with glass powder *X*. Therefore, in an initially homogeneous suspension containing a wide range of different sizes of fineparticles, the effective concentration of fineparticles that can participate in dynamic clusters to give the hyper-Stokesian velocities may be much smaller than the overall concentration of the suspension. This probably explains why various workers have reported that they could not detect a range of hyper-Stokesian velocities at concentrations below 0.2% by volume solids with homogeneous suspensions. Thus Stairmand has reported the data summarized in Figure 5.8. The Phototrack equipment used to generate some of the data in Figure 5.8 was developed by Stairmand and Carey. In this instrument the trajectories of the fineparticles in a

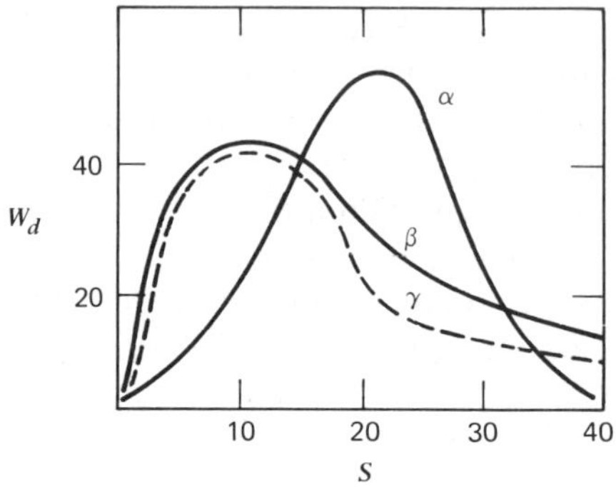

Fig. 5.8 Concentration effects on the measured Stokes diameter distribution of a powder [6]. Symbols: α, Andreasen 0.4% solids (4%); β, Phototrack (0.0005% solids); γ, Andreasen 0.2% solids; S, Stokes diameter in microns; W_d, weight in size range.

very dilute suspension were photographed by using long-exposure photography. The length of the track of each individual fineparticle was used to calculate the velocity and hence the Stokes diameter of the individual fineparticle. With this equipment, Carey and Stairmand were able to work with suspension volume concentrations as low as 5 ppm. They compared the Stokes diameter distributions of a powder measured at very low concentrations, using the Phototrack equipment with the Stokes diameter distributions determined by the Andreasen pipette [22] sedimentation procedure at various suspension concentrations. (This latter method for characterizing the Stokes diameter for distribution, discussed in Section 5.6, uses an initially homogeneous sedimenting suspension.) The data in Figure 5.8 show that the shift in the measured Stokes diameter distribution as the concentration changes from 0.2 to 0.4% is very large but that at 0.2% by concentration the distribution as measured by Andreasen and Phototrack are quite close. This explains the difference between the two concentration levels in Table 5.1 as recommended by Stairmand and Carey and then later by Stairmand. The Carey–Stairmand limit suggested in 1942 is probably an ideal limit based on work carried out with Phototrack equipment, whereas the higher limits suggested by Stairmand in 1947 is a more practical limit determined by extensive experiments using the Anreasen equipment. It should be noted that the shift in Stokes diameter distribution function of the data of Figure 5.8 as the concentration level goes from 0.2 to 0.4% solids is very drastic and much higher than any reported by any other workers. One is tempted to believe that there is a typing error in the original graph and that the curve should be labeled 4% solids curve (see discussion following the research paper as given in the proceedings of the symposium in reference 3). Such a figure would place that set of data in the maximum interaction zone postulated by Kaye and Boardman (note Stairmand's concentrations are by weight, not by volume).

In the opinion of the writer, for many systems encountered in practice when initially homogeneous suspensions are used, concentration levels of up to 0.2% by volume of solids probably represent a concentration level at which dynamic clustering and the consequent errors in Stokes diameter determination are negligible.

It should be stressed that all the recommendations in Table 5.1 are for initially homogeneous suspensions and do not take into account the fact that suspension concentrations used in line start systems must be much lower than the 0.2% level that the writer recommends as the safety limit for systems by using initially homogeneous suspensions. The difference between the stability of the two different suspension systems is due to the fact that when a line start system is used, any dynamic clusters formed move out into a clear fluid where there is no return flow from adjacent clusters to interfere with the stability of the dynamic cluster.

Many workers have attempted to improve the stability of line start systems by using a different fluid in the line start suspension to that of the main column of fluid so that the overall density of the line start suspension is less than that of the main fluid. If the movement of a dynamic cluster moving out into clear fluid is considered, it would appear that the fact that the cluster contained a fluid different from that into which it is moving would probably increase the stability of the cluster. The flow around the cluster would be unable to sweep the interstitial fluid out from between the fineparticles participating in the cluster. Line start systems using a different fluid in the suspension will probably result in the measurement of large pseudofineparticles from the persistence of stable dynamic clusters and will probably also show a deficiency in the finer fineparticles because they are stripped out of the initial layer by the stable dynamic clusters moving into the main sedimentation liquid [25].

In Table 5.1 the highest concentration recommended for use with sedimentation characterization procedures is that of Martin, given in 1942. His recommended value of 3% is greatly in excess of the limits suggested by Kaye and Boardman and Stairmand. Since Martin's work has been quoted as recently as 1970 as a justification for accepting higher solids concentrations in sedimentation systems used to characterize fineparticles, it is worth reviewing his data briefly and assessing the validity of his recommendations against the perspective of the more recent work on the interaction of fineparticles in a sedimenting suspension. Martin reached his conclusions concerning the safe concentration level in sedimentation systems by carrying out a centrifugal sedimentation study of a titanium dioxide pigment at six different volume concentrations. His data are presented in terms of the measured surface area of the pigment. Thus, if his measured fineparticle velocities are greater than their real

Stokesian value, these data would be interpreted as indicative of a reduction in the measured surface area of the powder. At the highest concentrations used, Martin reported that hindered settling was obviously decreasing the velocities of the fineparticles, and high values for the measured surface area of his powder were recorded. The measured surface area dropped continuously with reduced solids concentrations up to a concentration of 3% by volume. Thereafter there was a slight increase in the measured surface area of the powder as the solids concentration of the suspension decreased. At the time Martin carried out his work, hindered settling was considered to be the only effective mechanism interfering with the movement of fineparticles. Therefore, the apparent decrease in the sedimenting velocities of the fineparticles on going from 3 to 0.5% by volume was difficult to explain. From a modern perspective it appears that this decrease in surface area represents a decrease in measured velocities as the measurements move out of the peak velocities occurring in the dynamic clustering region of sedimentation dynamics and into the Stokesian region. If Martin had carried out investigations below 0.5% by solids, he may have recorded an even more marked reduction in the surface area of the powder as Stokesian sedimentation dynamics became established. In retrospect, it appears that Martin based his conclusions on inadequate data in the absence of independent information on the true Stokes diameter distribution of the fineparticle system. In fact, Martin's experimental data can be interpreted as being in accord with the data of Kaye and Boardman (see discussion in references 2 and 3).

Perhaps one of the reasons why the phenomenon of dynamic clustering has had difficulty in becoming widely accepted is that in a standard book on fluid fineparticle dynamics by Happel and Brenner, the data of Kaye and Boardman are contrasted with some apparently contradictory empirical data reported by McNown and Lynn as shown in Figure 5.9 [34, 35]. The apparent contradiction of the two sets of data in Figure 5.9 arises from the different experimental techniques of the different workers. Kaye and Boardman observed the behavior of fineparticles in the *body of a suspension* of approximately monosized fineparticles. McNown and Lynn observed *the movement of the boundary* between fluid and suspension in a settling suspension containing monosized fineparticles. If a set of monosized fineparticles in suspension moved at their Stokesian velocities, a clear boundary would move down through the column of suspension at the Stokes velocity of the individual fineparticles. However, when dynamic clustering takes place, the internal dynamics of the suspension results in above Stokes velocity sedimentation of most of the fineparticles in suspension until dynamic clustering has stripped the internal concentration of the suspension below that at which dynamic clusters can form.

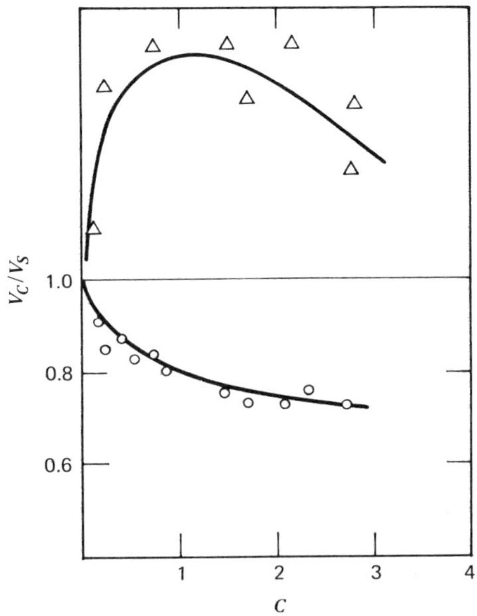

Fig. 5.9 Experimental data on fluid fineparticle interaction. Symbols: Δ, Kaye and Boardman [3]; ○, McNown and Lynn [36]; C, volume percentage of solids; V_C, velocity measured in concentration C; V_S, Stokes velocity of an individual sphere.

The dilute residue of fineparticles will then fall at the Stokes velocity of the individual fineparticles. However, the general turbulence created by the dynamic clustering will prevent settling of this residue of fineparticles until the effective concentration in the upper regions of the suspension falls to concentration levels in which free-fall conditions are operative. It will be recalled that Kaye and Boardman actually observed upward-moving fineparticles adjacent to dynamic clusters in their original experimental work. Thus dynamic clustering will delay the movement of the boundary down through the column, and this boundary will be delayed more seriously the greater the dynamic interaction of the fineparticles; that is, they will be slowed down more at higher solids concentrations. Thus it can be seen that there is no contradiction between the data of Kaye and Boardman and that of McNown and Lynn, only a difference in the observational strategy adopted for investigating the group dynamics of a settling suspension.

In summary, it can be stated that intensive studies of fineparticle dynamics during 1960–1977 have established that some of the earlier recommendations for concentration levels that can be used in the characterization of the Stokes diameter distribution of a fineparticle system are not valid. Furthermore, this work indicates that in line start systems one can use the same fluid in the line start suspension as the sedimentation if one stays below 0.05% be volume solids. If at all possible in a general sedimentation system, this upper limit of 0.05% by volume solids should be adhered to. However, for practical systems employing initially homogeneous suspensions, using gravimetric assessment of fineparticles, it may be permissible in certain cases to go as high as 0.2% by volume solids. The effect of using any higher solids volume concentration should always be investigated and experimental procedures standardized.

5.5 LINE START SEDIMENTATION METHODS

Werner's line start sedimentation system is shown in Figure 5.10*a* [36]. An external water jacket (not shown) around the sedimentation column is used for thermal protection of the equipment. The sedimentation column has a tapered base, the final length of which is a narrow-bore tube. The suspension of fineparticles to be characterized forms a narrow layer on top of the column. The amount of powder arriving at the bottom of the column at different periods of time is monitored by measuring the height of the sediment in the narrow-bore tube. It is assumed that the height of the sediment is proportional to the weight of powder that has sedimented at any given time. There are several problems associated with this equipment. First, some of the fineparticles sediment onto the shoulders of the tube. When sufficient quantities have accumulated, they slide down into the base of the tube, and the exact time of their arrival in the column is not related to their size. Sludge dumping from the shoulders creates convection currents, thus delaying the arrival of some of the fineparticles caught in the upward-moving portion of the convection currents.

The justification for relating the height of the column of sediment to the amount of the fineparticles arriving at the base is that along the length of the sediment column fineparticles in contact with each other are approximately the same size, and it has been shown that, in general, monosized fineparticles randomly pack to the same porosity irrespective of their absolute size for a wide range of sizes. The accumulation of sediment in the base of the column can consolidate the lower levels of the sediment, thus causing error in the estimate of powder quantities.

If the fluid used to prepare the suspension is the same as that in the column, the layer of suspension will have a higher density than the column fluid. In this situation the assembled line start system in the column will be essentially unstable and tendrils of suspension will tend to "bleed" down the walls of the tube. This gross breakdown of the suspension layer is described as "streaming." Allen comments on the Werner instrument, "Some of the suspension settles en masse in the form of pockets of fineparticles which fall rapidly through the clear liquid leaving a tail of

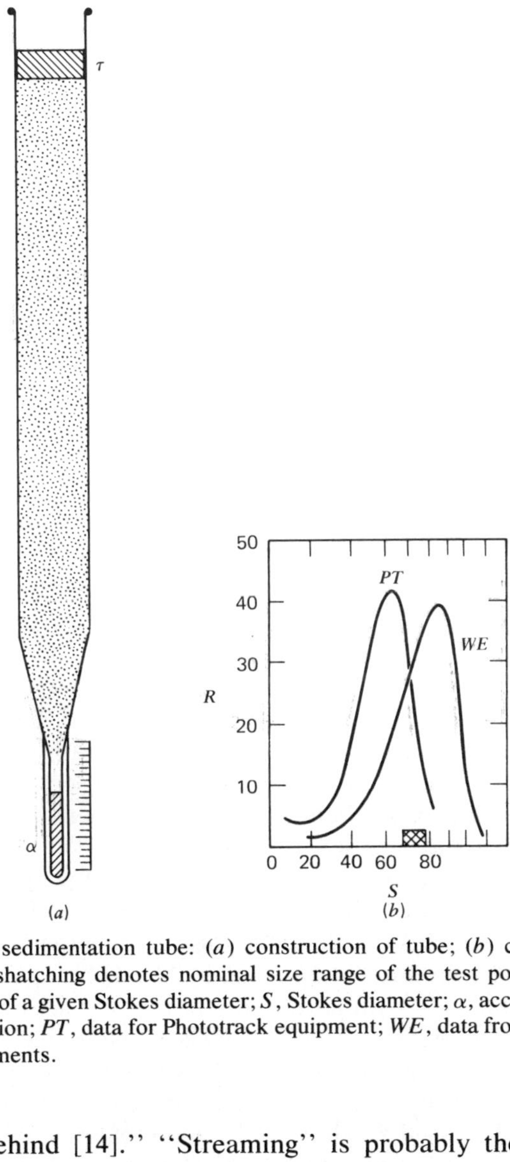

Fig. 5.10 Werner sedimentation tube: (*a*) construction of tube; (*b*) comparative performance data. Crosshatching denotes nominal size range of the test powder. Symbols: *R*, percentage weight of a given Stokes diameter; *S*, Stokes diameter; α, accumulated sediment; τ, layer of suspension; *PT*, data for Phototrack equipment; *WE*, data from Werner sedimentation tube experiments.

fineparticles behind [14].'' ''Streaming'' is probably the movement of massive dynamic clusters down the sedimentation column. It causes a coarsening in the estimated Stokes diameter distribution of the fineparticle system being characterized. Many attempts have been made to eliminate streaming from line start sedimentation studies by preparing the suspension layer with a lower-density liquid that is completely miscible with the column liquid so that the overall density of the suspension is lower than that of the column liquid (see discussion previously in this chapter). This technique eliminates gross streaming, but the use of a different liquid in the suspension layer can create a general coarsening in the estimated distribution function as compared to the use of line start procedures using suspensions of very low solids concentration in the same liquid as that employed in the sedimentation column [16]. This is shown by the data of Stairmand summarized in Figure 5.10*b*. He took a sand fraction that passed through a sieve of aperture 76 μm and was retained on one of 66 μm. He found that the Phototrack equipment generally gave a smaller Stokes diameter for this powder than did the nominal size of the sieves used in its preparation, whereas the Werner size distribution apparently described fineparticles much larger than the sieve apertures through which they had passed [69].

Although the Werner tube is not widely used in modern fineparticle science, the Mine Safety Association (MSA) particle size analyzer developed by Whitby and co-workers uses a miniature version of the Werner sedimentation tube that can be used in a centrifuge tube [38, 45, 46]. The MSA equipment is used to characterize fineparticles by a combination of gravity and centrifugal sedimentation, and problems similar to those encountered when using the Werner sedimentation tube can occur with the MSA equipment. The details of the equipment and procedure are given in Chapter 6.

To test the various claims that the use of two different but miscible fluids eliminated instability problems associated with line start sedimentation procedures, Kaye and Davies carried out a series of investigations with a Cahn sedimentation balance. This equipment is shown in Figure 5.11. The Cahn balance is a sensitive servo controlled electrobalance that has very small movements of the balance pan as it weighs objects placed on the pan [37]. It is capable of the continuous recording of small items falling onto the pan. It is thus suitable for weighing the fineparticles arriving at the base of a sedimentation column. The open-ended sedimentation column in Figure 5.11 is supported in an outer cylinder of liquid. At the beginning of an experiment, the light-weight disposable balance pan can be held firmly against the base of the sedimentation column. When the line start procedure is used, the pan is left suspended freely in the liquid; however, when it is used to study the behavior of an initially homogeneous suspension, the pan is clamped to the base of the column at the beginning of the experiment. When a line start procedure is used, the suspension column and the outer cylinder are filled with liquid and the suspension floated onto the column of liquid. To minimize uncertainties in measured sedimentation times, the floating of the suspension layer onto the column of fluid must be carried out quickly and smoothly. To facilitate the floating

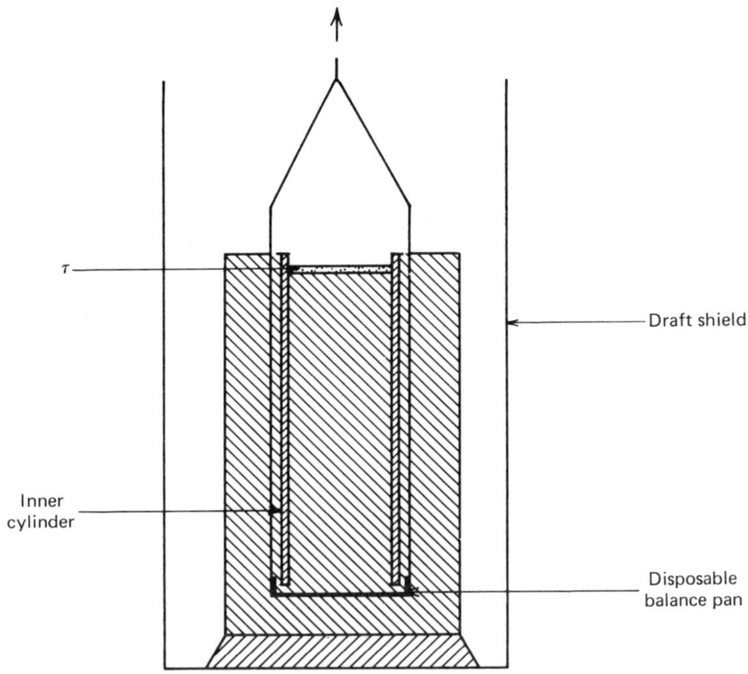

Fig. 5.11 Sedimentation balance equipment: τ, layer of suspension.

of the suspension layer onto the column of fluid, Kaye and Davies designed a device to be placed on top of the cylinder. It consisted of a lucite annulus, with an internal diameter of 3.5 cm and a depth of 0.7 cm. A wire mesh screen was placed on the bottom of the cell. The mesh helped to spread the suspension poured into it and absorbed most of the kinetic energy associated with the act of pouring the suspension onto the column fluid. Kaye and Davies used the Cahn sedimentation balance to characterize fineparticles with both the line start procedure and the initially homogeneous suspension procedure. They used two test powders, designated *A* and *B*, that had no fineparticles greater than the apertures of a 300 standard mesh sieve, the nominal aperture size of which is 53 μm. The sieve cloth used in constructing the feeding device for use with the sedimentation column was 300 mesh. Test powders *A* and *B* were produced by ball milling glass shards. The grains of the powder were irregular in shape and had a density of 2.7 g/cm^3. The Stokes diameter distribution of the test powders as determined by the Andreasen pipette method, using a suspension of 0.37% by volume concentration are given in Figure 5.17. (At this solids concentration, distortion to the Stokes diameter distribu-

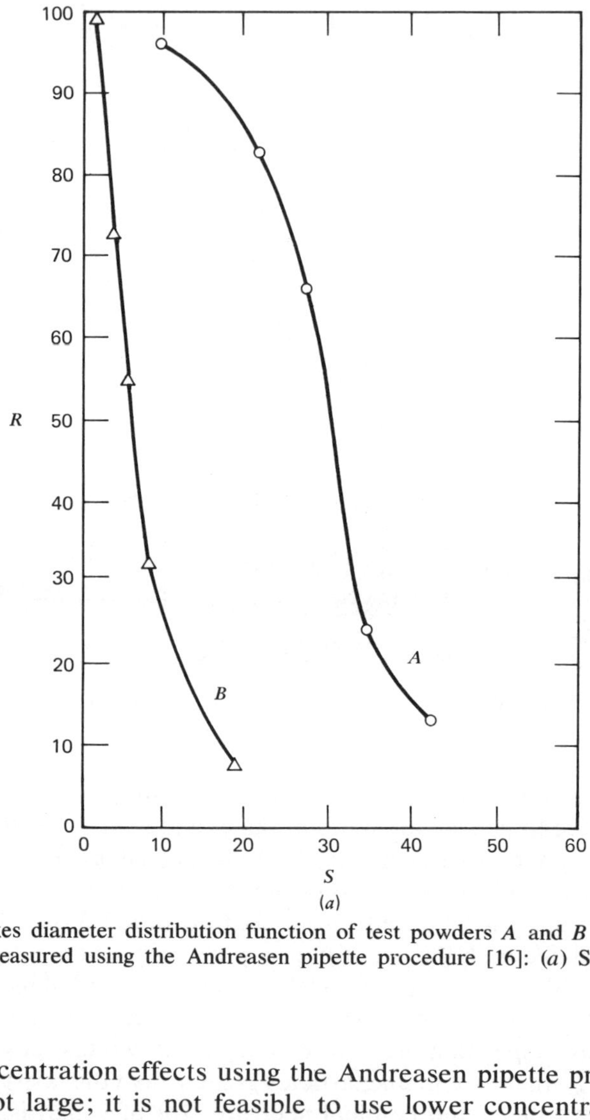

Fig. 5.12 Stokes diameter distribution function of test powders *A* and *B* used by Kaye and Davies measured using the Andreasen pipette procedure [16]: (*a*) Stokes diameter distribution.

tion by concentration effects using the Andreasen pipette procedure are probably not large; it is not feasible to use lower concentrations.) The distributions determined by the Andreasen pipette method are not more fundamentally "correct" than those determined by any other sedimentation procedure, but the technique has been widely used because it is relatively easy and inexpensive to implement in the average laboratory. There is a great deal of reference data generated by the method, and the Andreasen-determined distribution function of the test powders is given here as a set of reference data. The data in Figure 5.12*a* shows the relative

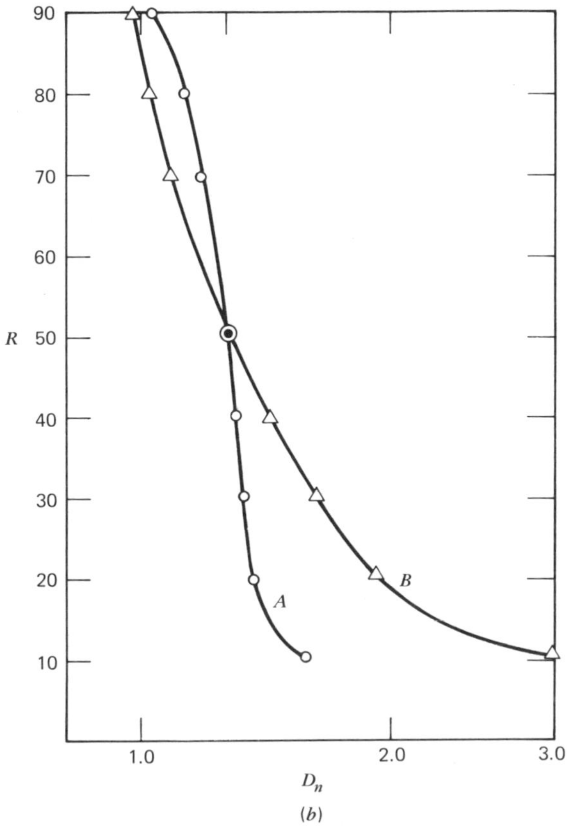

Fig. 5.12 (*b*) Normalized size distribution. Symbols: S, Stokes diameter; D_n, normalized diameter; R, cumulative percentage greater than stated size (by weight).

order of magnitude of the Stokes diameter distribution function of the two powders. It was pointed out in Section 5.4 that fineparticle interaction effects depend on the relative size range of fineparticles in a suspension. A dimensionless way of expressing the range of fineparticle diameters present in a powder is to express all sizes present as fractions of the mean diameter of the size distribution. The size distributions of the test powders expressed in this way are given in Figure 5.12*b*. Powder *A* has a narrower size range than does powder *B*; therefore, concentration effects should be more marked with test powder *A*.

To investigate the possibility that the use of a lower-density fluid in preparation of the initial suspension overcomes streaming problems associated with the line start procedure, two series of experiments were carried out. The first series of experiments were carried out by using distilled water plus 0.1% by weight of Atlas 1086 dispersing agent for the sedimentation column fluid and in the preparation of the suspension layer. Three different suspensions of 0.05, 0.50, and 1.00 percentage solids by volume were used. The Stokes diameter size distributions for test powder *A* measured with these suspensions using the line start procedure are summarized in Figure 5.13. It can be seen that the measured size distribution obtained using 0.05% solids, at which level it was predicted in Section 5.4 that fluid dynamic interaction effects would be negligible, agrees tolerably well with the Andreasen data. Thus the median diameter for the two sets of data are 28 and 29 μm, respectively. The two sets of data for suspension concentrations of 0.5 and 1.00 percentage by volume solids demonstrate a marked coarsening in the measured Stokes diameter distribution. Thus the shift in the median diameter for the three sets of data is $28 \rightarrow 37 \rightarrow 45$ μm. There is also a large increase with solids concentration in the measured percentage of powder greater than 50 μm from approximately 3% through 22–30%. This marked increase in the measured number of apparent large fineparticles is obviously an artifact of the measurement procedure. It represents the effect of increased dynamic clustering, resulting in the registering of pseudofineparticles at the pan.

Kaye and Davies [25] also carried out a series of experiments using distilled water plus dispersing agent to prepare the suspension and a brine solution of density 1.18 g/cm^3 in the sedimentation column. The data

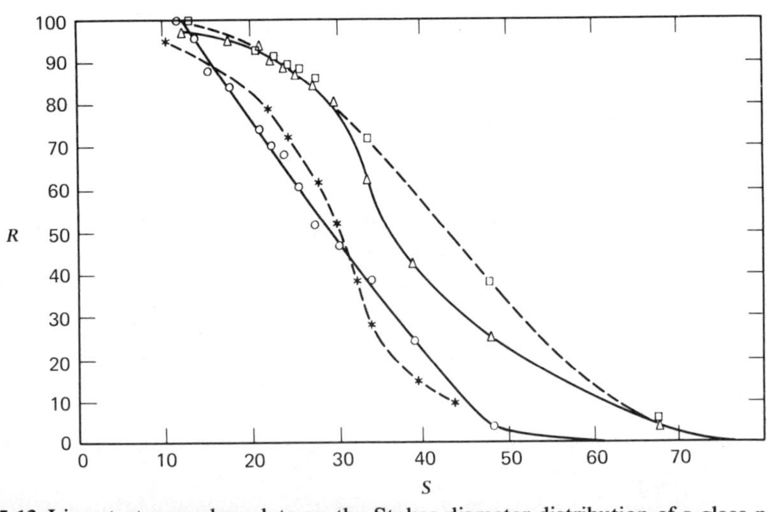

Fig. 5.13 Line start procedure data on the Stokes diameter distribution of a glass powder [16]. Water suspension floated onto water column. Symbols: S, Stokes diameter in microns; R, cumulative percentage oversize. Line start data: ○, 10 mg or 0.05% by volume; Δ, 100 mg or 0.50% by volume; □, 200 mg or 1.00% by volume; *–*–* Andreasen.

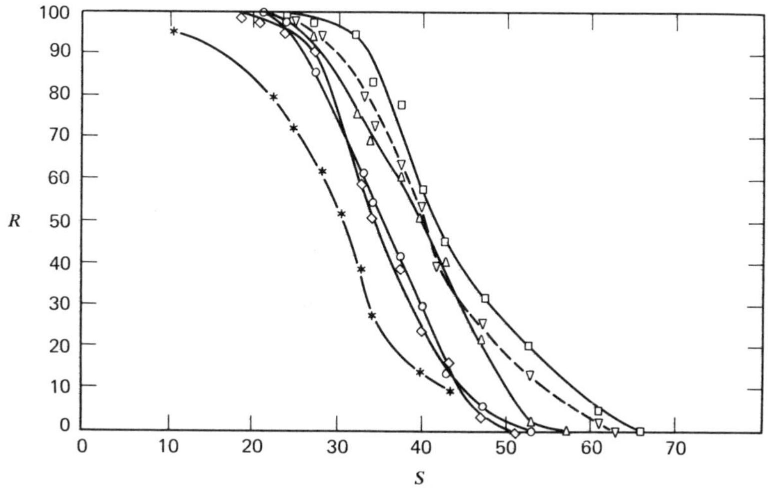

Fig. 5.14 Line start data obtained using glass powder in water placed on a column of brine [16]. Symbols: S, Stokes diameter in microns; R, cumulative percentage oversize by weight. Concentration of glass in water: ◇, 10 mg or 0.05% by volume; ○, 20 mg or 0.10% by volume; Δ, 50 mg or 0.25% by volume; □, 100 mg or 0.50% by volume; ▽, 200 mg or 1.00% by volume; * = *–*, Andreasen data.

obtained in these experiments are summarized in Figure 5.14. It will be noticed that the Stokes diameter distribution measured at the lowest suspension concentration of 0.05% is coarser than for the simple line start procedure using water in the sedimentation column shown in Figure 5.13. The largest effective size measured using a different liquid to prepare the layer of suspension is lower than that for the all-water system. This is consistent with the hypothesis put forward in Section 5.4 that the use of a different fluid in the line start layer suppresses gross streaming instability of the dynamic clusters that break away from the layer. The measured distributions of Figure 5.14 are generally deficient in fines compared to those in Figure 5.13. This is indicative of the fact that the dynamic clusters have a fluid medium different from that of the fluid through which they move and probably retain trapped smaller fineparticles that circulate with the larger members of the cluster. There has been some criticism of the water–brine work of Kaye and Davies because they did not take into account electroviscous effects, which could explain the absence of the large pseudoagglomerates; however, in the broadest sense their data are certainly compatible with the hypothesis in Section 5.4 concerning interparticle effects and illustrates the danger of using two fluids in the characterization procedure.

It is interesting to note that all the volume concentrations used in the line start procedures of both Figures 5.13 and 5.14 are below the safety level recommendation of several of the authorities quoted in Table 5.1. The work of Kaye and Davies definitely strengthens the case for limiting volume concentrations utilized in line start methods for characterizing fineparticles to less than 0.05% by volume. In the early days of the development of sedimentation balance characterization procedures, the weighing of the very small quantities associated with initial suspension levels of 0.05% solids was difficult and expensive. However, the data of Kaye and Davies show that such measurements are well within the routine performance of commercially available electronic balances. Technologists should give serious consideration to the possibility of a wider use of line start procedures in sedimentation balance characterization studies because of the ease with which the data can be interpreted, as compared with the interpretive theory for transforming data obtained using the sedimentation balance in the cumulative mode (see Section 5.9).

The data in Figure 5.14 indicate that when using 0.05% by volume in the line start suspension, stability has been achieved by virtually eliminating all interfineparticle fluid dynamic interaction. This suggests an interesting variation on the characterization procedure in which a shield can be used to eliminate wall effects. Thus, if the sedimentation column were to be fitted with a device for feeding a layer of suspension onto a column in such a way that the edge of the suspension layer were at least 1 cm in from the wall of the sedimentation column, then fineparticles entering the sedimentation column would be free from wall effects. Such a system is illustrated in Figure 5.15. In such a system there should be virtually no loss of fineparticles to the wall of the cylinder, provided that the line start system was free of dynamic clusters. The use of a line start procedure in sedimentation balances particularly with the advent of minicomputers would make it relatively easy to adapt this procedure for quality control purposes [75].

There are no reported studies in the fineparticle literature of gravitational systems in which line start procedures have been used in conjunction with a light beam to monitor incrementally the movement of fineparticles. This is somewhat surprising in view of the fact that such systems have been used very successfully in centrifugal characterization studies (see Chapter 6). The availability of low-cost laser systems and the possibility of line start procedures with incremental monitoring makes it possible to consider using instrument designs such as that shown in Figure 5.15. In this instrument the power drop in the straightforward beam can be measured, as in traditional photosedimentation equipment, or the diffracted light could be monitored. This latter measurement procedure has

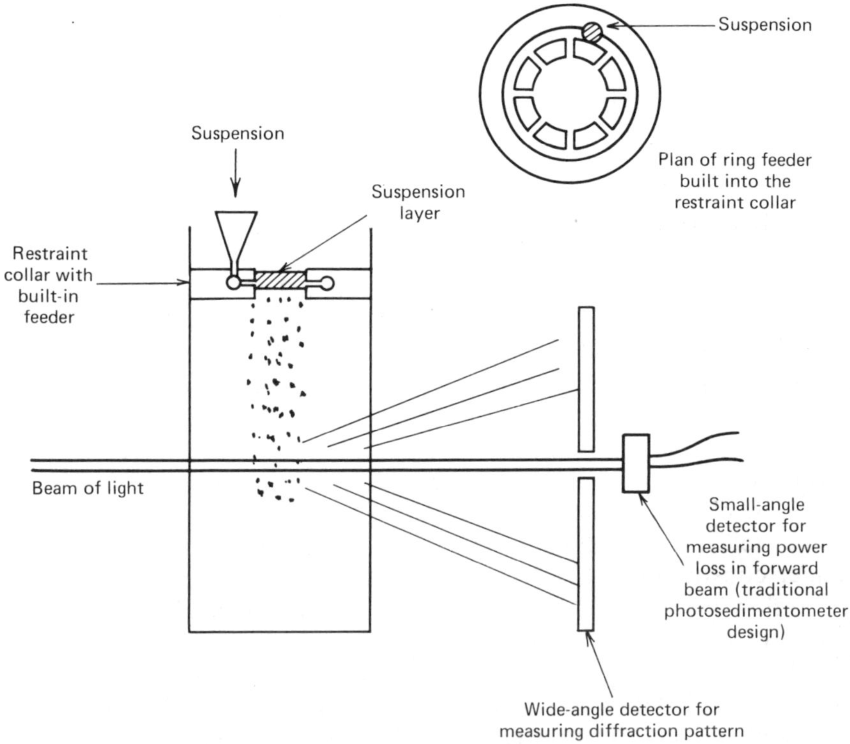

Fig. 5.15 Conceptual design of a differential photosedimentometer for use in conjunction with the line start sedimentation procedure for measuring Stokes diameter distributions.

the advantage that each size of fineparticle would be in the monitoring zone at a given time, and the diffraction pattern of a given Stokes diameter fineparticle could be measured directly. This information would considerably aid the interpretation of data from such equipment as the eriometers available from CILAS and the Leeds Northrup Company [47, 48]. If an instrument such as that of Figure 5.15 were to be used with white light, it would facilitate calibration procedures for use with ordinary white light photosedimentometers.

It should be noted that the photosedimentometer shown in Figure 5.15 would be useful in quality control situations because once the fineparticles have moved past the monitoring zone, they can be allowed to accumulate at the bottom of the sedimentation column without interfering with subsequent characterization studies carried out with the same column. One could aspirate fluid from the top of the column back through the feeder once the smallest fineparticle had cleared the interrogation zone and the next suspension to be characterized could be added to the top of the column. Alternatively, the restraint collar could be elevated by a small amount before inserting the next amount of suspension with the appropriate adjustments being made in the interpretation of the sedimentation data because of the slight change in the height through which the fineparticles are sedimenting.

The Micromerograph® uses the line start procedure in conjunction with a servo controlled electronic balance to weigh fineparticles arriving at the base of a nitrogen filled sedimentation column [39, 40]. The equipment has been widely used in quality control situations. (The name derives from the fact that at one time the term "Micromeritics," literally, "small measurements," from the Greek words *micros,* "small," and *metron,* "measure," was used to describe some aspects of fineparticle science [41].) The basic system of the Micromerograph is shown in Figure 5.16. The powder to be characterized is dispersed at the top of the sedimentation column by blowing a jet of high-velocity gas through powder placed in the sample holder. Theoretically, the cloud of fineparticles formed in this manner constitutes a thin layer of suspension at the top of the column. The airflow through the sample holder is designed in such a way that the powder is deagglomerated by the aerodynamic shear forces in the narrow feed channel. As the fineparticles are collected on the balance pan, a slight rotation of the balance arm occurs, thus activating a control circuit that, in turn, applies a current to a restoring force coil on the other side of the beam balance. The current required to keep the balance in equilibrium is a continuous measure of the weight of the settled fineparticles.

Kaye et al. [42] studied the configuration of the injected cloud using high-speed photography. They report that the cloud moved down one-seventh of the column before a stable cloud was obtained. This failure of the initial cloud to remain in a narrow region at the top of the column would cause the real height of fall to be less than the column height. This factor would cause the measured size distribution to appear coarser than it should be.

A source of difficulties that arise in interpreting the experimental data obtained from a Micromerograph is that the total weight eventually recorded by the balance beam does not always equal the weight of powder dispensed into the column. It is usually assumed that the missing powder is attached to the walls of the sedimentation column. Although the settling column is made of aluminum, aluminum surfaces are usually covered by a thin layer of oxide so that in essence the surface of the settling column has insulating properties with respect to the settling dust fineparticles. Therefore, electrostatic charging could cause them to adhere to the sides of the sedimentation column. It has been suggested that the use of anti-

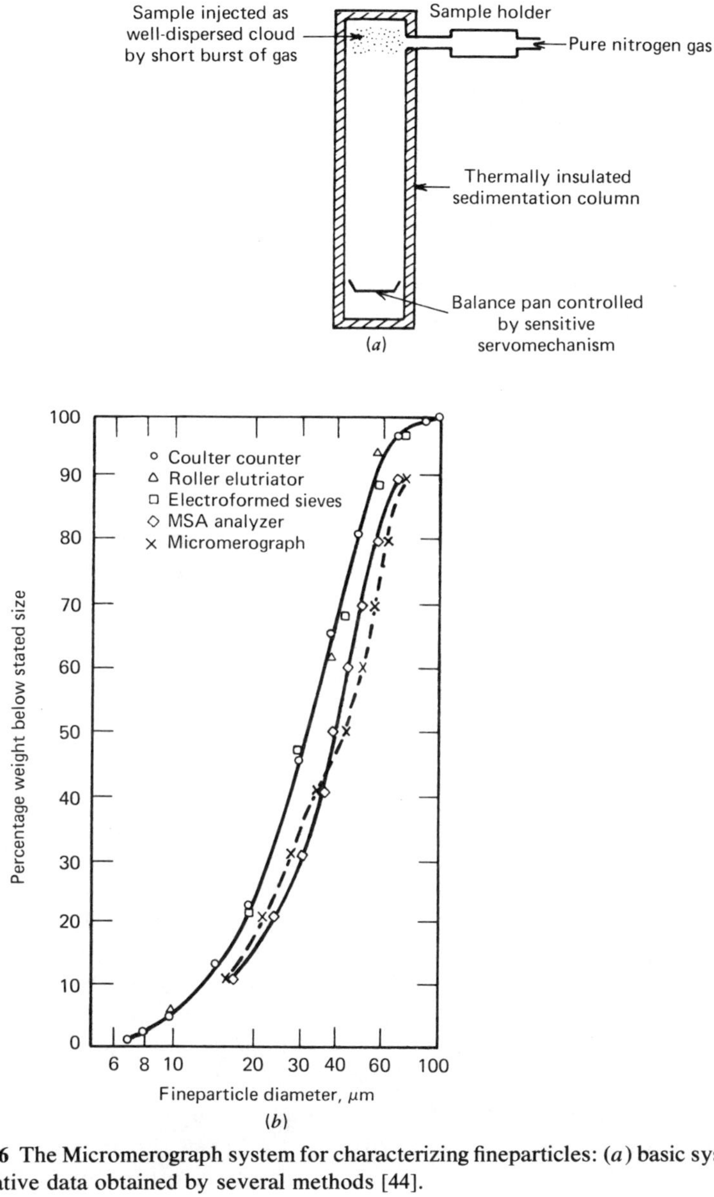

Fig. 5.16 The Micromerograph system for characterizing fineparticles: (*a*) basic system; (*b*) comparative data obtained by several methods [44].

static agents improves the accuracy attainable with the Micromerograph [43].

If the difference between the powder that settles onto the balance pan and the initial sample weight is small, it can be assumed that the powder lost is typical of the whole powder. However, if the weight discrepancy is greater than 5% by weight, any assumptions made concerning the size distribution of the missing powder should be checked at least once by an examination of feed powder and that lost to the walls of the equipment by microscopic examination. If it is assumed that the powder lost is representative of the entire powder, then the total weight of powder retained ultimately on the balance pan is used to calculate the percentage weight on the pan at any specific time representing the portion of powder greater than a given Stokes diameter.

Sometimes the increase in weight on the pan decreases exponentially, and experimenters are tempted to take a short cut and extrapolate to the ultimate weight retained at infinite time by graphical techniques. If such an extrapolation is coupled with assumptions concerning the missing powder, the absolute accuracy of the ultimately derived Stokes diameter distribution of the powder is questionable.

An operational disadvantage of the Micromerograph is that the sedimentation column is usually cleaned out at the end of a characterization study by a blast of gas. If the powder has any toxic attributes, it is very difficult to contain the emergent cloud. In specific instances this operational difficulty can be overcome, but the modified experimental procedure adds expense to the basic characterization study.

Ullrich [44] has compared the performance of the Micromerograph with other methods of characterizing fineparticles. He used a spherical aluminum powder so that the data from the various characterization procedures using different physical principles could be correlated. His data, summarized in Figure 5.16*b*, show that data from the Coulter counter, the Roller elutriator, and the electroformed sieves correlate very well, considering the very different physical parameters measured in these three methods of characterization. The data obtained with the MSA analyzer and the Micromerograph differ markedly from the other three sets of data. The MSA equipment is essentially the same as the Werner sedimentation tube and utilizes a line start procedure to study fineparticles sedimenting in a viscous fluid. If there had been no systematic error involved in the use of the line start procedure, the MSA data and the Micromerograph data should have been directly comparable to the elutriator-generated data, whereas they exhibit a shift of the order of 30% on the measured median diameter compared to that from the Roller elutriator. Moreover, the Micromerograph data compare directly with the MSA equipment, even though the two characterization studies differ in that one was carried out in a liquid and the other in a gas. The data of Ullrich again indicate that the use of a line start system in a sedimentation procedure for measuring the Stokes diameter distribution of a fineparticle system can cause an apparent coarsening of the distribution if truly stable conditions are not achieved at the commencement of the characterization study. The data also illustrate that data cannot be compared on the Stokes diameter distribution of a powder from various characterization procedures unless the sedimentation system and the volume concentration of the suspension used in the procedure are specified.

Connor and co-workers [53] developed a version of the Werner sedimentation tube in which accumulated sediment was measured by treating the fineparticles so that they became radioactive. Thus thoria powder was irradiated with neutrons to produce β-active thorium 233 with a β energy of 1.23 MeV. The thoria powder was then sedimented in a lucite sedimentation tube. The β activity of the sediment could be measured with a β scintillation counter and related to the mass of the accumulated sediment [53]. Connor and co-workers [53] also investigated the use of an external β-ray source for use with powders that could not be readily transformed to a radioactive species.

A line start sedimentation balance system for characterizing coarse fineparticles has been described by Brezina [67]. A line start procedure in which monitoring of fineparticles arriving at the base of the column is achieved by measuring hydrostatic changes at the base of the column has been described by Schlee [68].

5.6 HOMOGENEOUS SUSPENSION INCREMENTAL MONITORING SEDIMENTATION PROCEDURES

In this group of procedures the concentration changes at a series of depths within an initially homogeneous suspension are measured at a series of specified times. The region of the suspension on which measurements are made is defined as the monitoring zone or the interrogation zone. The postulated ideal monitoring zone in such systems is a thin slab of suspension of thickness Δh at a depth h from the surface of the suspension. There are several general relationships that are useful in describing the concentration changes occurring in a settling suspension. When developing these equations it is assumed that effects due to the stirring of the suspension are negligible (see Section 5.2) and that at time $t = 0$ the suspension is homogeneous so that each cubic centimeter of the suspension contains n_1, fineparticles that have a falling speed of v_1, n_2 fineparticles that have falling speeds of v_2, and so on. To develop the necessary equation, consider events occurring at a monitoring zone at depth h_1 and time t_1. A rigorous derivation of the equations would take into account the finite width of the monitoring zone since this effects the resolution of the characterization data. However, except in the case of the photosedimentometer and the x-ray sedimentometer, the exact dimensions of the effective monitoring zone of a real instrument are indeterminate. As a consequence, the effects of the finite dimensions of the monitoring zone often occur as one of many virtually inseparable ingredients in the general resolution of the characterization procedure that is determined empirically by direct comparison of data obtained by various characterization procedures. Heywood [18] carried out experimental work on the relationship between the width of an interrogation zone, when monitoring a homogeneous suspension incrementally, and the resolution of the characterization procedure and concluded that if the zone were less than one-sixth of the mean depth of the zone, error introduced into the characterization procedure would be negligible. This conclusion should not, however, be treated as a general conclusion since modern electronic equipment permits higher levels of scrutiny in characterization procedures and the effective resolution in modern methods can be very different from that in earlier instruments.

At a depth h_1 in the suspension at time t_1 in a sedimenting suspension all fineparticles with velocities $v_1 = h_1/t_1$ will have cleared the monitoring zone, but fineparticles with settling velocities smaller than v_1 leaving the zone will be replaced by an equal number of fineparticles from above the zone. Therefore, the number of fineparticles per unit volume in the monitoring zone is given by the relationship

$$(N)_{h_i t_i} = \sum_{i=s}^{i=h_1 t_1} n_i$$

where $(N)_{h_i t_i}$ = number of fineparticles per unit volume at depth h_i and time t_i

n_i = number of fineparticles per unit volume of velocity v_i

v_S = velocity of smallest fineparticle in the suspension

Consider again at at time t_2 depth h_2, where $t_2 \geqslant t_1$ and/or $h_1 \geqslant h_2$. The number of fineparticles in the zone is

$$(N)_{h_2 t_2} = \sum_{i=s}^{i=h_2 t_2} n_i$$

where

$$v_{h_2t_2} = \frac{h_2}{t_2}$$

and d_s is the smallest fineparticle present. Therefore

$$(N)_{h_1t_1} - (N)_{h_2t_2} = \sum_{i=h_1t_1}^{i=h_2t_2} n_i \tag{5.3}$$

For the situation where concentration gradients are followed by gravimetric assessment, let w_i = the weight of fineparticles of Stokes diameter. Then, assuming that the shape of the fineparticles does not vary with their size,

$$w_i = \gamma d_i^3 \rho_P$$

where γ is a shape factor. The value of γ for dense smooth spheres is $\pi/6$.

The initial weight of fineparticles per unit volume C_0 is given by

$$C_0 = \sum_{i=s}^{i=l} n_i \gamma \rho_P d_i^3 \tag{5.4}$$

where d_e is the largest fineparticle in the suspension. Let C_{ht} denote the concentration at depth h at time t, then

$$C_{ht} = \sum_{i=s}^{i(ht)} n_i \gamma \rho_P d_i^3 \tag{5.5}$$

From equations 5.5 and 5.4 R_{ht}, the percentage weight of fineparticles with Stokes diameters smaller than d_{ht} is given by

$$R_{h_1t} = \frac{C_{ht}}{C_0} \times 100 \tag{5.6}$$

It follows from this equation that the cumulative undersize distribution function of a fineparticle system can be calculated by measuring the concentration within a settling, initially homogeneous suspension at a series of specified times and/or depths and by plotting R_{ht} against d_{ht}.

The simplest procedure for monitoring the concentration changes within the suspension is to withdraw portions of suspension from a monitoring zone (or zones) using a pipette. Characterization procedures based on this strategy are termed *pipette methods*.

Andreasen pioneered the use of pipette methods for characterizing fineparticles, and accordingly, the procedures are sometimes generally referred to as Andreasen's method; however, the term "Andreasen procedure" should be restricted to the specific use of equipment designed by Andreasen. It should be noted that pipette procedures do not readily yield information on the largest and smallest fineparticle present in the suspension. All too often, Stokes diameter distributions measured by this procedure are extrapolated to what the technologists claim are reasonable upper and lower limits on the distribution. It is far better to terminate the curve linking the various data points at the first and last data points.

In the Andreasen procedure the pipette is left in the sedimenting suspension and a sequence of samples extracted through the pipette at a set of predetermined times. Each time a portion of suspension is removed from the sedimentation column, the effective depth of the pipette tip below the suspension surface changes; this is allowed for in the calculation of the Stokes diameter distribution. The Andreasen equipment is provided with several pipettes of varying stem length to facilitate the sampling of suspension at different depths; however, it is not usual to change pipettes during any particular characterization study, and a separate sedimentation vessel would normally be used for each stem length selected. Because of the low cost and the ready availability of the Andreasen equipment, data from this characterization procedure are often included in comparative data obtained by different procedures. It should be clearly understood that there is no intrinsic merit in Stokes diameter distribution measured by this procedure vis-à-vis data obtained by other procedures. Andreasen-based data are often quoted in this book for comparative purposes, but they should not be regarded as standard data to which other data should conform or approach.

An alternative to the fixed pipette procedure is to insert a pipette to the required depth at a specified time whenever it is required to examine a portion of the suspension. In this procedure, termed the *free pipette technique,* there is an unknown source of variation in the characterization data arising from the disturbance of the suspension caused by the insertion of the pipette into the sedimenting suspension. When using the free pipette procedure, it is often necessary to make a mechanical rig to aid in the exact positioning of the pipette tip prior to the withdrawal of a sample of suspension. If ample laboratory space and equipment are available, a useful procedure for avoiding uncertainties associated with the free pipette procedure is to make several equally representative suspensions of the fineparticles to be characterized, and each suspension should be used only once for a given measurement of the concentration at a given time and depth. This also has the advantage that the measurements are averaged over several samples of the fineparticle system; statistically, this is a better procedure than carrying out all measurements on one suspension.

The free pipette procedure has an advantage over the fixed pipette procedure in that there is no shadow zone effect. The shadow zone effect arises from the fact that fineparticles leaving the suspension under the column of the pipette are not replaced by fineparticles from above, thus lowering the true concentration of the monitoring zone with respect to its true value in the absence of the pipette. Furthermore, the developing zone of clear fluid immediately under the stem of the pipette is unstable with respect to the adjacent body of the suspension. As a consequence, there will be gravity convection currents in the vicinity of the pipette tip, thus causing disturbance to the trajectories of the fineparticles in the immediate vicinity of the pipette stem. When selecting a characterization strategy, the presence of the shadow zone must be traded off against the fact that the free pipette requires a ready access to the sedimentation column, thus increasing the difficulty of protecting the free surface of the sedimentation fluid from evaporative cooling. Leschonski [17] has described an equipment in which the pipette stem is continued to the base of the sedimentation vessel, this avoids problems associated with the shadow zone under the tip of the pipette. The sample of suspension is aspirated through a set of holes around the periphery of the glass stem of the sampling device. Kaye and Davies [19] have recently suggested a new design for use in pipette procedures for characterizing fineparticles in which the sedimentation vessel is a cylinder constructed out of insulated rigid foam. The glass stem of the sampling system now is able to move along the axis of the cylinder of the sedimentation vessel. This has the advantage over the Leschonski-type pipette in that it is possible to vary the height at which a portion of the suspension is extracted for examination.

The real monitoring zone of the pipette procedures differs considerably from the ideal monitoring zone postulated at the beginning of this section. It is often stated that the real monitoring zone in a pipette method is a sphere of fluid centered on the tip of the pipette. This statement is approximately correct with respect to the fluid that moves into the pipette; however, the actual fineparticles that move into the pipette depends on the flow conditions around the pipette tip. Thus, for a given volume of suspension aspirated into the pipette in a fixed time, different fineparticles will be taken into the pipette for various orifice sizes because of the different fluid flow rates that will exist for the various orifices. Normally it is desirable to extract the suspension into the pipette as fast as possible to reduce the uncertainty in t, the time at which the sample concentration is measured. If a large orifice is used on the pipette, the flow rate into the pipette will be relatively low so that larger fineparticles in the bottom hemisphere of the approximate spherical monitoring zone will not be pulled up to the orifice. All too often, experimental data do not quote the speed at which the samples of suspension are extracted from the sedimenting suspension, nor are the dimensions of the orifices used. One of the advantages of using the Anreasen pipette system, which is available commercially, is that in this equipment all dimensions are standardized. When setting up an analytical procedure using a pipette sampling procedure, it is useful to arrange for the suction on the pipette to be from a partial vacuum reservoir to ensure systematic withdrawal rates. It should be noted that the interrogation zone for the Leschonski system and the Kaye–Davies equipment is different from that of the simple pipette and consists of a series of hemispheres centered on the orifices located around the stem of the pipette.

When using a fixed pipette to take several samples out of a suspension, it should be noted that the amount of suspension between the tip of the pipette and the operating tap of the pipette remains in the pipette between successive samples. This space between the pipette tip and the tap is known as the *dead space*. The uncertainty in the characterization study due to this factor is known as the *dead-space error*. To minimize the effect of the dead-space error it is usual to specify a very narrow channel in the pipette stem. The lower limit for this channel diameter is fixed by the need to aspirate the sample in minimum time and avoid blockage of the tube in the cases where relatively coarse fineparticles exist in the suspension.

Kaye and Davies suggest that dead-space errors can be minimized with their equipment because their continuous-stem pipette permits the flushing out of the dead-space suspension between sampling operations [19].

When choosing the volume of suspension to be extracted in a pipette procedure, one should remember that the resolution of the study is increased by minimizing the volume of suspension. This requirement, however, conflicts with the need to characterize the solid contents of the extracted samples. If the solids content is to be determined chemically by dissolving the fineparticles and reacting them with appropriate reagents, one can often work with smaller volumes of suspension than in techniques in which the fineparticles must be filtered and weighed. Leschonski has suggested the use of special cups made out of light-weight plastics, with the base fabricated from a membrane filter, for assessing the solids contents of suspension samples. The fineparticles to be weighed can be filtered in the cup and washed with appropriate fluids, and the cups can be placed directly onto delicate analytical balances. Sometimes samples of suspension are evaluated by weighing the residues left after evaporation of the suspension fluid. It should be noted that in such situations the residual dispersing agent is rarely driven off by the drying process. Instead, it forms a hygroscopic ingredient in the residue that causes error in the weighing process because of the absorption of moisture from the atmosphere. For this reason the use of the membrane filters as suggested by Leschonski has the advantage that the residual dispersing agents can be washed off the fineparticles by appropriate wash fluids.

A basic disadvantage of all pipette procedures is that the need to characterize the aspirated portions of suspension usually requires relatively high solids concentrations in the initial suspension. For this reason 1–2% by volume solids concentrations are quite normal in this procedure. As explained in Section 5.4, interfineparticle effects will cause considerable disturbance to the trajectories of the fineparticles at these concentrations. The data in Table 5.2 illustrate the magnitude of the type of error that can arise from concentration effects in pipette-type characterization studies. The exact interaction effects in any given suspension will depend on the range of sizes present so that the data in Table 5.2 can be used only as a relative indication of the order of magnitude of concentration effects. The concentration effects in any particular situation would have to be investigated empirically by carrying out a series of characterization studies at a range of solids concentration.

It should be noted from the data in Table 5.2 that at the very high solids

TABLE 5.2. Stokes Diameter Distributions of a Spherical Bronze Powder, as Determined with an Andreasen Pipette at Various Suspension Concentrations[a]

Solids Concentration by Weight	0.24	0.53	0.89	2.01	2.87	3.67	4.45	13.6
Stokes Diameter, μm	Cumulative Percentage by Weight Less than Stated Stokes Diameter							
26	82.4	83.4	87.1	90.0	89.4	93.1	94.0	100.0
18.2	65.2	66.0	72.0	71.0	72.2	75.0	80.4	95.0
13.0	29.5	33.0	34.0	39.4	38.4	40.0	50.1	76.2
9.0	8.1	8.4	8.6	10.0	9.0	11.0	12.8	23.0

Source. Reference 2.

[a]Suspending medium: distilled water (637 cm^3)
viscosity 0.8937 cP
temperature 25°C
dispersing agent, Teepol (1 cm^3)—tradename for a sulfonated long-chain alcohol detergent marketed in Great Britain by Shell Oil Company

concentration there has been an apparent increase in fineness of the system. This is due to hindered settling effects. Carey and Stairmand [6] have described a pipette-type apparatus for use in the characterization of fineparticles in which the sample of suspension is driven into a pipette mounted on the side of a sedimentation column by liquid pumped into the base of the column. This sampling procedure can be carried out several times with the same suspension. The equipment suffers from dead-space error, but shadow zone effects and general disturbances to the suspension by the act of sampling are minimized. If the same fluid as that used to prepare the suspension is used to provide the fluid flowing into the base of the sedimentation column, it will be less dense than the suspension, and density-driven convection currents generated at the interface between fluid and suspension will spread throughout the sedimentation column. This source of disturbance can be avoided by using a nonmiscible denser fluid for displacing the samples of suspension.

5.7 CHARACTERIZATION PROCEDURES BASED ON THE RADIATIVE MONITORING OF THE SEDIMENTATION RATES WITHIN AN INITIALLY HOMOGENEOUS SUSPENSION

Characterization procedures for evaluating the Stokes diameter distribution of a fineparticle system based on the use of radiant energy to monitor the sedimentation kinetics of an initially homogeneous suspension include so-called photosedimentation techniques, x-ray sedimentometers, and procedures based on the used laser anemometry. In the first group of methods, changes in the energy of a beam of light passing through the monitoring zone of the suspension are used to deduce the concentration changes occurring in the suspension. In the x-ray sedimentometer, intensity changes in a beam of x-ray passing through the suspension are used in the same way. In the last group of methods the kinetics of fineparticles within a sedimenting suspension are monitored by measuring the Doppler shift in laser light reflected from a moving fineparticle.

When a very large object is placed in the path of a light beam, the light can be considered to consist of rays traveling in straight lines, and the object is said to cast a geometric shadow. In optical terms, an object is large if it has dimensions very much greater than the wavelength of light being used to inspect the objects. For a sphere of diameter d, the volume of the sphere is proportional to the diameter cubed, whereas the area of the geometric shadow is proportional to the diameter squared. Thus for a plane parallel beam of light, the power removed by a large spherical object is proportional to d^2. When considering the dimensions of a shadow of an irregular object in general, the dimensions of the shadow are related to some characteristic linear parameter of the object in the same way; that is, the volume is related to this characteristic parameter by the third power and the area of the shadow by the second power. It has been shown that if one looks at the projected shadow for an irregular object for all possible orientations in space of the object, the total sum of the shadow areas is directly proportional to the surface of that object provided there are no reentrant portions of the surface. This fact is known as *Cauchy's theorem* [49, 50]. The interaction of a fineparticle with light when the dimensions of the fineparticle are of the same order as the wavelength of the light in the beam is complex. In this situation the technologist must consider effects arising from light diffraction, light transmission, and light interference.

The theory of photosedimentometers in which the concentration of fineparticles in a settling suspension are deduced from the energy they remove from a light beam is usually treated in the following manner. Consider a beam of light used to monitor a settling suspension that passes through a tank of width L as shown in Figure 5.17. The fineparticles in

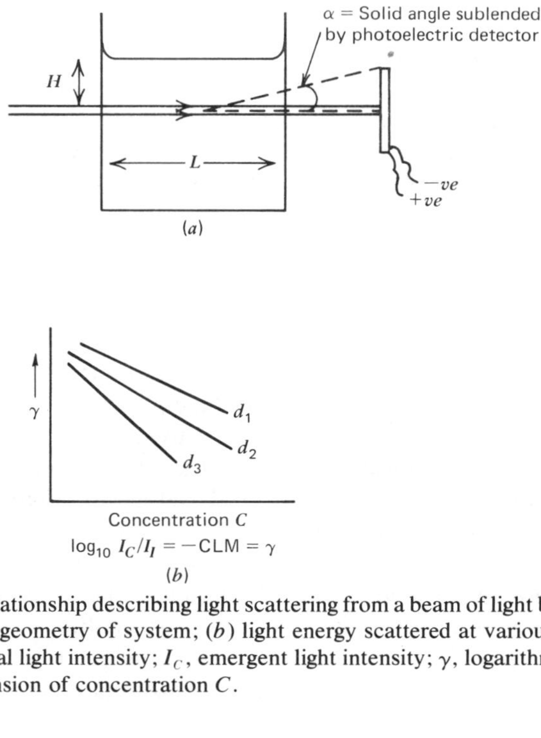

Fig. 5.17 The relationship describing light scattering from a beam of light by a suspension of fineparticles: (*a*) geometry of system; (*b*) light energy scattered at various concentrations. Symbols: I_I, initial light intensity; I_C, emergent light intensity; γ, logarithm of the transmittance of a suspension of concentration C.

suspension scatter light from this beam. The properties of the beam are then monitored by a photoelectric device placed external to the photosedimentation tank. The energy removed from the forward beam is dependent on the concentration of the fineparticles, the length of the path of the beam through the tank, and the size and optical properties of the fineparticles. Some of the light scattered by the fineparticles is received by the photocell, and the amount is governed by the solid angle subtended by the photoelectric device by the various portions of the suspension through which the light beam is passing. Normally it is assumed that some effective solid angle created with respect to the center of the sedimentation tank is useful in describing the properties of a given photosedimentometer. In the early days of the development of photosedimentometers there were many attempts to develop the relationship between the energy removed from the forward beam and the properties of the scattering fineparticles using light scattering theory. When evaluating this early work it should be appreciated that there are several detailed theories of light scattering appropriate to different fineparticle systems. When the fineparticles are of the order of 10 wavelengths of light and above, the Fraunhofer diffraction theory can be used [51]. When the fineparticles have diameters of the order of half the wavelength up to 10 wavelengths the appropriate light scattering theory has been developed by Mie; whereas for very small fineparticles less than half a wavelength of light, the appropriate light scattering theory is so-called Rayleigh theory [52].

When attempting to derive an equation linking the removal of light energy from a beam to the properties of the fineparticle in a system such as that in Figure 5.17*a* using light scattering theory, the problem is considerably simplified by considering the situation in which no scattered light is picked up by the photoelectric device; in other words, instruments in which the solid angle α is vanishingly small. To achieve such a situation physically, one has to create a large distance between the sedimentation tank and the photoelectric device. Although this approach to the design of photosedimentometers facilitates attempts at theoretical calibration, it enormously increases the dimensions of the equipment and the cost of the appropriate optical components necessary to maintain the alignment of the optical system. In addition, because of the low power being received at the photoelectric device, problems associated with signal : noise ratio in the recording of the appropriate signals causes difficulties. Early equipment designed by Rose is typical of attempts to exclude scattered light from the receiving device to facilitate subsequent data interpretation [54]. This same worker attempted to derive a so-called universal calibration curve relating the scattering power of a fineparticle to its diameter uniquely for all powders studied with photosedimentometers. This universal calibration curve has, of course, proven to be an elusive dream since the power removed from the forward-traveling beam involves interference effects as well as diffraction effects and depends on the refractive index of the fineparticle with respect to the fluid and on the wavelength of the light as well as the geometry of the fineparticles involved. The calibration curve described by Rose [54] for correcting photosedimentation data was used in some early models of photosedimentation equipment but should now be considered to be of only historic interest.

The heavy theoretical consideration of light-scattering theory and its involvement in the early discussions of the design of photosedimentometers frightened away many technologists from the photosedimentometer. This was unfortunate since the technology of electrooptics is so advanced that it is possible to build efficient low-cost monitoring equipment using the principles of photosedimentation, which are excellent for quality control and rapid generation of data in the research laboratory. Moreover, the problems of calibrating photosedimentometers are no more difficult than many of the problems of calibrating other physical procedures for characterizing fineparticles. In photosedimentation equipment one can work with very low concentration of solids in the suspension, thus circumventing all problems associated with fluid–fineparticle interaction. The photosedimentometer has the advantages that it is easily adapted for hostile environments and gives information on the largest fineparticle presence in the suspension. In the author's opinion, the number of situations in which the relation between the power removed from a beam of light and the size characteristics of the fineparticles intercepting the light beam can be calculated from theoretical considerations is very low. Therefore, one should adopt an empirical approach to the design of the photosedimentometer and thus should build the equipment to take maximum advantage of electrooptical equipment available and then calibrate the transmission signals of the light beam in terms of the solids concentration of the suspension.

The equation used in photosedimentation theory to link the opacity of a suspension to the properties of the suspension is the Lambert–Beer law. The Lambert–Beer law is an energy exponential decay law of the type that occurs in many physical situations. [Note that in some optical textbooks the law is called the *Lambert law* and in others, the *Beer law*, since different historians of science variously attribute the law to either the German scientist Johan Heinrich Lambert (1728–1777) or to the German astronomer Wilhelm Beer (1797–1850).] If I denotes the intensity of a light beam passing through a tank of suspension, then the Lambert–Beer law states that

$$I_C = I_o e^{-CLM} \tag{5.7}$$

where I_o = original intensity of the beam
I_C = intensity of the beam after passing through the suspension of concentration C
e = the exponential function
L = path length through the suspension
M = a constant that incorporates effects related to the solid angle subtended by the photoelectric detector, the spectral response of the detector, the geometry of the sedimentation tank, the nature of the walls and their thickness, and the optical characteristics of the beam and generally involves the size of the fineparticles involved in the scattering of light from the beam

The range of suspension concentrations for which the relationship is valid in any given photosedimentometer fineparticle system should always be investigated experimentally.

Equation 5.7 can be written in the form

$$\log_{10} \frac{I_C}{I_o} = -CLm \tag{5.8}$$

where m is a modified constant incorporating the essential variables present in the constant M of equation 5.7. The quantity $\log_{10} I_C/I_o$, denoted by γ, is the logarithm of the transmittance of the suspension used in standard optical terminology. It follows that

$$\gamma = -CLM = -C\mu \tag{5.9}$$

where μ is a constant for a given instrument.

Equation 5.9 predicts that the measured values of γ for a series of suspensions of a given powder plotted against concentration will generate a straight-line relationship such as that illustrated in Figure 5.17*b*. In this graph, the line marked d_g is a typical line obtained for a powder that contains fineparticles with a given range of Stokes diameters. If experiments were to be repeated with suspensions of fineparticles that all had the same Stokes diameter, a line of a different slope would be obtained. Thus for fineparticles that generally were coarser than d_g, a line such as that labeled d_1 would be obtained, whereas for fineparticles much smaller than d_g, the slope would be similar to that labeled d_3.

Equation 5.9 can be written in the form

$$C = \left| \frac{\gamma}{\mu} \right| \tag{5.10}$$

where the symbol | | denotes the positive value of the enclosed symbols. This equation is used to deduce concentrations from measured changes in transmittance γ when we know that the change is due to the light scattering of fineparticles of known magnitude.

The simplest way in which optical measurements have been used to characterize a fineparticle system is by way of turbidity (i.e., cloudiness) measurements of a stirred suspension. Although such a procedure is not, strictly speaking, a use of photosedimentometers, photosedimentometers historically developed from turbidimeters, and thus it is useful to review the use of turbidity measurements to monitor the properties of powder. Thus if one were to be engaged in the quality control of a product such as cement, one would have to follow variations in the fineness of the product. The transmittance of a series of suspensions of the same solids concentration for a series of samples of known fineness would be measured. A calibration chart of the type shown in Figure 5.18*a* would be constructed using the data. If, now, a sample of cement were taken from a process and a suspension made at the fixed concentration C_T, then the measured value of γ for this suspension would yield a measurement of the fineness of the powder directly, provided that it lay between the values established using the comparison powders.

The first publication to describe the use of measurements on the opacity of a suspension to deduce fineparticle properties is that by Wagner, who used the system shown in Figure 5.17*a* to monitor the production of Portland cement [55]. Instruments used to measure the turbidity of a suspension are also called *nepholometers,* a term derived from the Greek word *nephos,* meaning "cloud."

When monitoring the opacity changes in a sedimenting suspension, it is usual to record the opacity of the suspension on a pen recorder to produce a data display of the type illustrated in Figure 5.18. For historic reasons it has become usual to arrange such transmission records in such a way that high intensity of transmission through the clear fluid before the fineparticles are added corresponds to the bottom position of the pen recorder. Therefore, in the chart shown in Figure 5.18*b* the line indicated as I_I is the light transmitted through the sedimentation tank full of clear fluid to be used in preparing the suspension. When the fineparticles are added to the sedimentation column, there is an immediate drop in the transmission of the light beam to the level I_C as shown. At time 0, indicated on the chart, the stirring of the suspension ceases and the strength of the transmitted beam remains the same until the largest fineparticle present in suspension moves past the monitoring zone, time T_L, as indicated in the diagram. If the monitoring is being carried out at a depth H below the top of the fluid in the sedimentation column, then the Stokes diameter of the largest

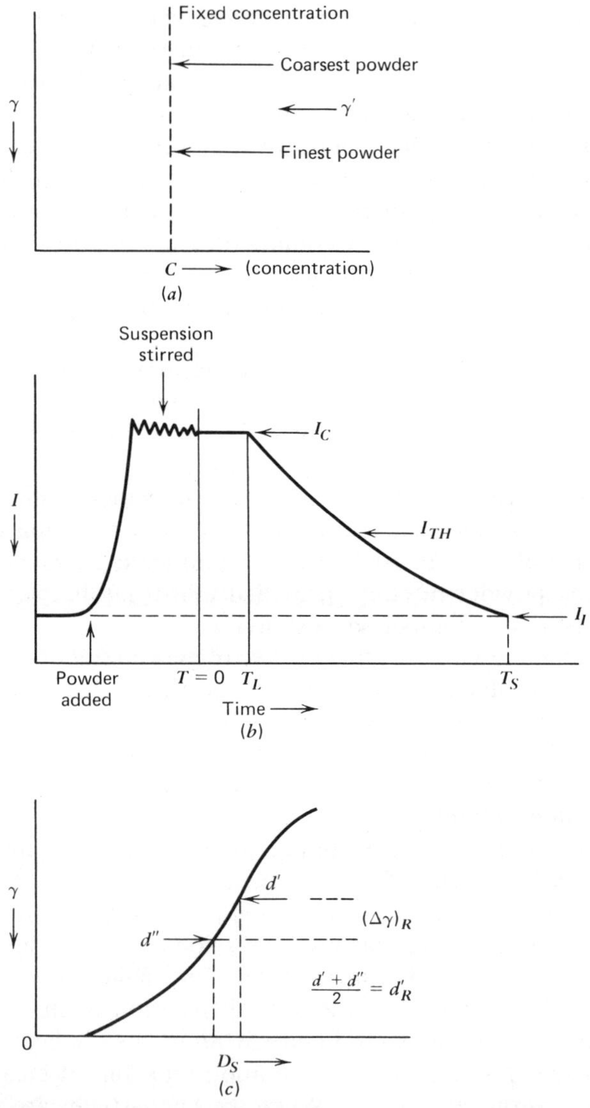

Fig. 5.18 Calculation of Stokes diameter distribution from photosedimentation data: (*a*) calibration curve for turbidometric quality control; (*b*) opacity–time curve for a settling suspension; (*c*) log of transmittance–Stokes diameter curve. Symbols: I, intensity of the light beam; γ, log of the transmittance; D_S, Stokes diameter.

fineparticle present equals a constant K time $\sqrt{H/T}$. The constant K can be calculated from the known parameters of the suspensions such as the viscosity, the density of the suspending fluid, and the density of the fineparticles. In an ideal situation the transmission of the suspension drops to the initial value I_I at T_S after the smallest fineparticle has moved through the monitoring zone. Sometimes during the adding of the powder at the sedimentation column in this type of simple instrument the exterior of the sedimentation tank is smeared, in which case there is an error introduced into the transmission of the empty cell with a corresponding failure of the signal to return to the original value of I_I at time T_S.

The first stage in the transformation of the transmission-time data is to construct a graph of the log of the transmittance against Stokes diameter using $\log_{10}(I_{TH}/I_I) = \gamma$. Consider an element of the γ/Stokes diameter curve bounded by the diameters d' and d'', it can now be assumed that the drop $(\Delta\gamma)_R$ is caused by fineparticles with Stokes diameters in the range $d' \rightarrow d''$, which can be considered to have an average diameter of d_R. To transform $(\Delta\gamma)_R$ into an estimate of the concentration of fineparticles having Stokes diameters of size d_R, one must either know or make some assumptions about its dependence on d_R. Assume that calibration studies using fineparticles of size range $d^1 \rightarrow d^{11}$ have established that μ_R and d_R are related for a range of values of d_R by the relationship shown in Figure 5.19*a*.

Data points on this graph should not be regarded as defining a general calibration curve since the values of d_R depend on the range $d^1 \rightarrow d^{11}$ used in exploring the opacity–concentration relationship. Higher resolution is defining the range $d^1 \rightarrow d^{11}$ for a given d_R would alter the value of μ_R. The weight of a powder of average Stokes diameter d_R is given by $(\Delta\gamma)_R/\mu_R$, and the total weight of the powder is

$$W_T = \sum_{d_S}^{d_L} \frac{(\Delta\gamma)_R}{\mu_R}$$

Using these relationships, the Stokes diameter distribution can be deduced.

In the absence of detailed information on the value of μ_R, some workers have applied a first-order adjustment to the transmittance data from approximate theory based on considerations of similar situations in the realm of geometric optics. Thus, if we consider several spherical objects, then for a given weight of material W the number of objects N is given by the relationship

$$N = \frac{W}{\pi/6 \cdot D^3}$$

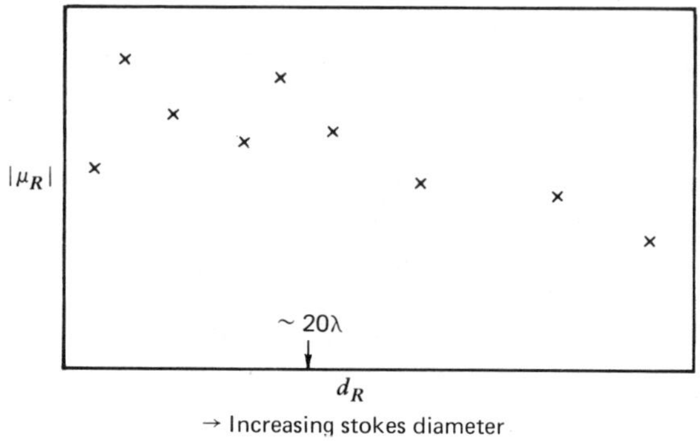

(a)

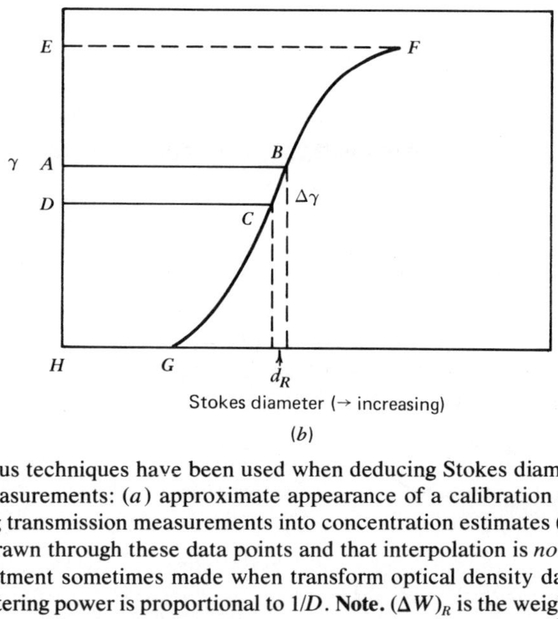

(b)

Fig. 5.19 Various techniques have been used when deducing Stokes diameter distributions from optical measurements: (*a*) approximate appearance of a calibration curve (empirical) for transforming transmission measurements into concentration estimates (note that a curve should *not* be drawn through these data points and that interpolation is *not* permitted; (*b*) a first-order adjustment sometimes made when transform optical density data are to assume that optical scattering power is proportional to $1/D$. **Note.** $(\Delta W)_R$ is the weight of fineparticles of size d_R given by $\Delta\gamma \cdot d_R \cdot k$. Symbols: γ is logarithm of the transmittance; $|\mu_R|$ is a constant dependent on $_R$, d_R Stokes diameter.

The area of the geometric shadows Sh cast by these N objects is given by

$$Sh = \frac{W}{\pi/6 \cdot D^3} \cdot \pi \frac{D^2}{4} \tag{5.11}$$

Thus

$$Sh = \frac{W}{D} \cdot \frac{3}{2} \tag{5.12}$$

Thus the total geometric shadow for a given weight of powder and, therefore, the power removed from a light beam if geometric optics could be used to interpret transmission data, is proportional to $1/D$. In general, for fineparticles larger than 10 μm, the light-scattering power of a suspension of a given weight of the fineparticles is proportional to $1/D$. For this reason, then a first-order adjustment for size when transforming data such as those in Figure 5.18*b* is to multiply $(\Delta\gamma)_R$ by d_R (see Figure 5.19*b*). This procedure for allowing for the effects of the size of the fineparticles in suspension is the recommended transformation procedure for use with the Boundbrook® photosedimentometer to be described later in this section.

Attempts to extend geometric optics thinking to a description of the opacity of suspensions of fineparticles led to the use of the extinction coefficient of a fineparticle. One widely used definition of the extinction coefficient E_R of fineparticles of Stokes diameter d_R is

$$E_R = \frac{\text{scattering power of fineparticles of diameter } d_R}{\begin{array}{c}\text{scattering power fineparticles would have if}\\ \text{geometric optics was a valid descriptive theory}\end{array}} \tag{5.13}$$

By using such a coefficient, the data transformation in Figure 5.19*b* could be modified by using tabulated values of E_R to carry out the operation

$$(\Delta W)_R = \frac{(\Delta\gamma)_R \cdot d_R}{E_R} \tag{5.14}$$

For many control procedures, it is quite adequate to assume $\mu_R = 1$ because the technologist is usually looking for differences between powders, not their absolute fineness. However, if quality control is the aim of photosedimentation characterization, then even this type of transformation is unnecessary since in such situations light transmission–time graphs can be used to "fingerprint" the powders.

The basic concepts involved in the fingerprinting approach to quality control are illustrated in Figure 5.20. Suspensions for quality control calibration are prepared by using a fixed amount of powder. The electronics of the instrument are adjusted so that for each suspension investigated the transmission drop caused by the suspension in the test tank is always of the same magnitude. (This makes it easier to carry out the direct comparison of the transmission–time curves.) Let us now consider the case of the quality control testing of a ball-milled powder. The powder emerging from the ball mill will either be too coarse, acceptable, or too fine. The test equipment is first calibrated with a powder that is just too coarse and then with one that is just too fine. These two curves between them define an acceptance zone, or an acceptance window of the type shown in the lower

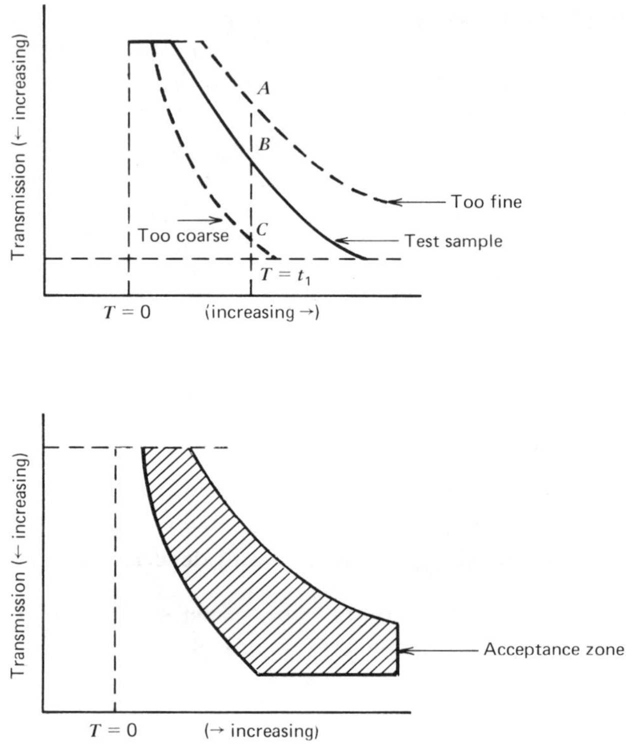

Fig. 5.20 "Fingerprinting" of test sample and comparison with the fingerprints of standard powders may be sufficient for quality control of a system.

part of Figure 5.20. Any test sample that has a transmission–time curve within this acceptance zone is acceptable, whereas one with a portion outside the acceptance window is subjected to further testing and is taken to indicate the need to adjust the process. Thus the test curve of transmission–time is a "fingerprint" of the powder used to identify its acceptability.

The most straightforward way of evaluating the value of μ_R in Figure 5.19 is to prepare a set of small-diameter-range powders of various average size, and then μ_R is measured by plotting a graph of γ against concentration for each powder. Unfortunately, it is not easy to fractionate powders accurately (see discussion of fractionation efficiency curves for elutriators in Chapter 7). In the absence of closely sized fraction of powder it is common practice to calibrate photosedimentation data by using a powder whose size distribution function has been established by another characterization procedure [56, 57].

The first photosedimentometer to be described in the fineparticle literature evolved out of the Wagner turbidimeter. It was soon discovered that because the Stokes diameter of fineparticles was related to the square root of the reciprocal of the settling time required to reach a fixed monitoring zone at a depth H below the surface of the suspension, the use of a monitoring beam at a fixed depth required very long sedimentation times for the characterization of a wide-range powder. Later models of the Wagner turbidimeter were built with the optical system mounted so that it could be moved up and down the outside of the sedimentation column. Using such a system, the characterization of a cement powder in kerosene could be completed within 30 minutes [58]. In 1958 Harner and Musgrave of the Eagle Pincher Company described a modified version of the Wagner turbidimeter in which the sedimentation tank was mounted on a platform and moved up and down through the monitoring beam [59]. Although it was mechanically easier to move the tank rather than the somewhat cumbersome optical system used in these early photosedimentometers, there was always a danger of jolting the sedimentation tank so that fluid spilled over the side of the tank, resulting in a smear down the side of the tank.

The sensitivity of the early photosedimentometers was limited by the problems of keeping the light beam at constant power level throughout the characterization of the fineparticles. Any attempt to improve the thermal insulation of the system with accompanying longer sedimentation times usually engendered problems of drift in the power levels of the light and in the response of the selenium barrier layer photocells used in the early instruments. Usually the only choice open to the scientists was to check the power level of the beam at the beginning and the end of the experiment and to replace the bulb or try again if the power level had shifted. Large constant-voltage transformers and/or power supplies were used in some of the early instruments in attempts to stabilize the power level of the lamp. For the same reason, the lamp had to be switched on for some time before the commencement of a characterization study to allow the lamp and its housing to come into thermal equilibrium with their surroundings. When using an incandescent lamp source in a photosedimentometer system, it is advisable to place an infrared filter in the path of the light beam to remove radiation that could upset the thermal equilibrium of the sedimentation column.

Heywood modified Wagner's system so that the optics were carried on a board capable of swinging in an arc. Several sedimentation tanks were then placed in an arc and monitored with the same optical system. Rose was the first worker to attempt to use small solid-angle optical systems so that attempts could be made to calculate the extinction coefficients of the fineparticles from theoretical considerations. Rose introduced the use of an outer tank with water as a thermal insulation device. Three commercial instruments evolved out of the work of Rose. One of them is known as the Evans Electroselenium (EEL) photosedimentometer (the acronym is derived from the initial letters of the company making the equipment) [60]. In the EEL photosedimentometer the optical system is placed on a small trolley so that it can be moved along a line of sedimentation tanks for sedimentation to be monitored concurrently in a series of tanks.

The second commercial system based on the work of Rose is the Boundbrook photosedimentometer developed by Morgan [61, 62]. The name of this equipment comes from the fact that Morgan was associated with Boundbrook Bearings Company. The Boundbrook photosedimentometer was the first to use matched photoelectric cells in an attempt to cope with the problem of optical drift in the equipment. Light from an incandescent electric bulb was split into two beams. One went through the sedimentation tank and the other through a dummy glass tank filled with clear sedimentation fluid to match the opacity of the two beams so that a balanced system could be employed in the measurement of the chamges in transmission of the sedimenting suspension. The Boundbrook photosedimentometer used mechanical scanning of the sedimentation tank through the light beam. The Boundbrook photosedimentometer was further developed by the Goring Kerr Company [63]. This equipment proved to be useful in the quality control of metal powders. Some comparative data on the Stokes diameter distribution for a tin powder reported by Morgan are summarized in Figure 5.21. Volume concentrations used in the characterization studies are not given; thus a detailed analysis of the discrepancies between the three methods is not possible. However, it is probable that the photosedimentation data are finer than the Andreasen data because of an incomplete compensation in the transformed data for the optical scattering power of the smaller fineparticles and also from a possible coarsening in the Andreasen data from concentration interaction effects.

Another photosedimentometer embodying the basic design philosophy of Rose was described by Muta [64]. In this equipment the sedimentation column was scanned optically by using a rotating concave mirror. The height that could be scanned was limited by the diameter of the lens used on the outside of the sedimentation tank. Another commercial photosedimentometer is available from the Micromeritics Corporation [74].

Once experimental studies started to indicate that the calibration of photosedimentometers could be facilitated by using large solid angles at the photoelectric detector with respect to the sedimentation tank, the basic design of photosedimentometers started to change [27, 57]. In the

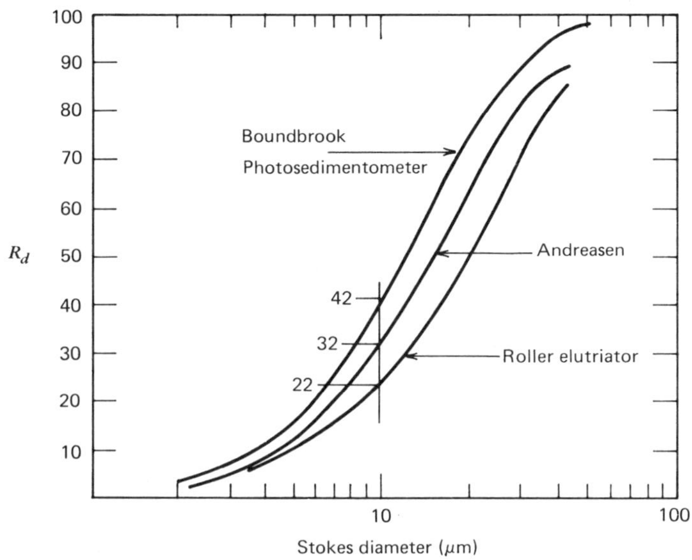

Fig. 5.21 Comparative Stokes diameter distributions reported by V. T. Morgan (no volume concentrations given): R_d, cumulative percentage less than Stokes diameter d (by weight).

wide-angle scanning photosedimentometer (WASP) described by Allen, the photocell used to monitor the transmission of light through the sedimentation column is placed as close to the column as is convenient from mechanical considerations [14]. In this equipment a reference photocell is used to monitor the light beam and correct for any optical drift that might occur over the duration of the characterization study. Again, mechanical movement of the sedimentation column is used to accelerate the collection of experimental data [65, 66].

The availability of fiber optics and cheap solid-state integrated circuits made it possible to start considering systems having any desired optical configuration in which the sedimentation column is scanned optically over great heights without any moving parts [70]. Davies and co-workers described an instrument in which fiber optics were used to take the light beam to the sedimentation tank. An arc of fiber optics was placed around the periphery of the horizontal plane through the sedimentation tank to pick up wide-angle-scattered light [71].

Naylor and Kaye have used a fiber optic-based system to make measurements on a sedimentation column of up to 1 m in height. This procedure facilitates rapid collection of information on the largest and smallest fineparticle present in an industrial suspension. The basic principles employed in the Naylor–Kaye equipment are illustrated in Figure 5.22*a*.

Light is piped to the outside of the sedimentation column in a fiber optic bundle containing many hundreds of small hairlike fibers. En route to the sedimentation tank a small subset of the fibers is twisted out of the main fiber and taken to an interface with a solid-state circuit device. Another fiber optic bundle is placed on the opposite side of the sedimentation column and used to monitor transmitted light. This second fiber optic bundle is also interfaced with the solid-state circuit device. This solid-state device is able to indicate in real time the ratio of light energy going to the tank and light energy leaving the tank. An additional integrated circuit can be used to give directly the logarithm of the transmittance of the output. Alternatively, the output can be connected directly to a minicomputer. The use of the solid-state ratio circuit means that all problems related to unstable light sources have been eliminated. This enables the use of low-cost incandescent light sources when designing photosedimentation equipment. The equipment can be plugged into any available power supply. Light beams leaving a fiber optic bundle are slightly divergent; this has to be taken into account in assessing the performance of the instrument. Fiber optic bundles can be placed at any particular height and the whole of the sedimentometer tank plus fiber optic assembly potted in a matrix of expanded plastic foam, thus achieving mechanical robustness and good thermal insulation. The whole assembly is so robust that dropping the equipment has no effect on its performance. Measurements can be rapidly taken at any height in the tank.

Harwood and co-workers [73] have described a fiber optic probe for investigating the properties of powder mixtures that can be adapted for use in quality control photosedimentation. The fiber optic probe used by Harwood consisted of a cylindrical head approximately 1 cm across and 3 mm wide. The head was carried on the end of a narrow hollow metal tube that carried two bundles of fiber optics. One bundle was split up into several subgroups of fibers that were mounted on the periphery of the cylindrical head so that they could send light out at right angles to the plane of the probe head. The other optical fibers were mounted in between the first set of fibers so that they could collect light reflected back onto the probe. The reflected signals were then carried out of the probe by the second fiber optic bundle to a measuring device. The system is shown schematically in Figure 5.22*b*. This fiber optic probe was moved along a hollow glass tube placed along the axis of a cylindrical mixer to monitor the progress of a blending procedure.

This type of fiber optic probe could be used to make a photosedimentometer as illustrated in Figure 5.22. A Dewar flask could be used to provide both the reflecting optical system and the thermally isolated sedimentation column. The Dewar flask would be mounted in a thermal

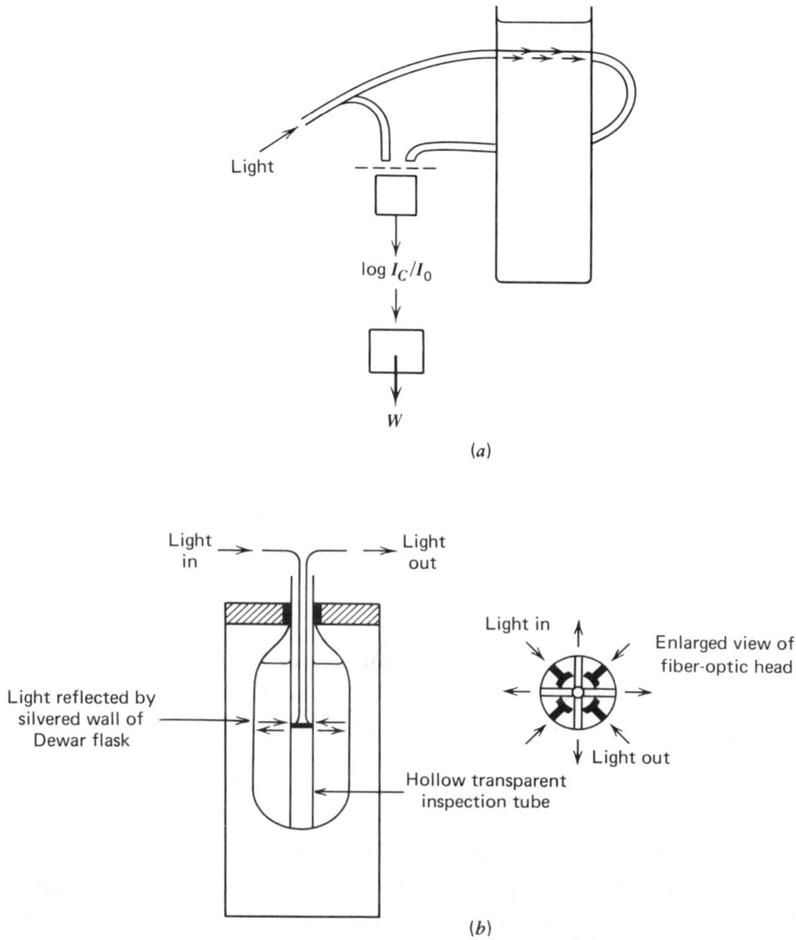

Fig. 5.22 Fiber optics and solid-state circuits are creating flexibility in the design of photosedimentometers: (*a*) optical circuit used in the photosedimentometer system described by Kaye and Naylor [72], ---- symbolic representation of fiber-optic–solid-state circuit interface; (*b*) fiber-optic–solid-state system for quality control.

protection chamber by a head-insulating lid as shown in Figure 5.22. An inspection column made of hollow glass would extend from the neck to the base of the Dewar flask. Between measurements it would be sealed to the outside. The stopper carrying the inspection tube could serve to seal the Dewar flask when the sedimenting suspension was in position. At a given time the fiber optic probe could be moved down the inspection column and the light transmission at various depths calculated by measuring the reflected signal from the walls of the flask as shown in Figure 5.22 [75].

In x-ray sedimentometers the concentration gradients within a sedimenting suspension are measured by means of x-ray absorption studies. It is claimed that the transmittance of the x-ray beam is directly related to the weight of fineparticles in the beam by a simple exponential relationship similar to the Lambert–Beer relationship used to describe the optical properties of a suspension. Thus Olivier and co-workers give the relationship as

$$\ln T = -A\phi_s \tag{5.15}$$

where T = the measured transmittance I_C/I_o
I_C = intensity of the x-ray beam passing through a suspension of solids concentration C
I = intensity of x-ray beam when passing through the sedimentation column filled with clear fluid
ϕ_s = weight fraction of solids present in the beam
A = a constant for a given equipment, powder and suspending fluid [21]

It is claimed that measurements can be made on systems containing elements with atomic numbers as low as 13. As the atomic number and/or the concentration of the heavier elements present as components of a fineparticle falls, the solids concentration of the suspension has to be increased to maintain the overall sensitivity of concentration measurements made with this type of instrument. The originators of one commercially available system state that "one may have to use volume solids concentrations of the order of 0.5 to 3.0%" [21]. At such high solids concentration, significant interfineparticle fluid dynamic interaction may distort the measured Stokes diameter distribution.

Brown and Skrebowski [76] have described equipment in which a water-cooled x-ray tube is used to generate 18 keV radiation that, after traversing the tank, was detected in an ionization chamber. They used a reference beam, from the same x-ray source, that passed through a dummy sedimentation tank to back off the zero-level signal before fineparticles were added to the sedimentation column. Allen has pointed out that because of the instability problems encountered in this type of x-ray source, the experimental design of Brown and Skrebowski [14] does not entirely compensate for drift problems if sedimentation times are relatively long.

The "Sedigraph model 5000" is a commercially available x-ray sedimentometer [77, 78]. The x-ray tube of this equipment is a small air-cooled unit with a tungsten target. The x-rays used in measuring the concentration gradients in the suspension are primarily the tungsten L lines. X-rays passing through the sedimentation column are detected by a scintillation detector connected to an electronics system. The x-ray beam is moved up and down the sedimentation column to achieve rapid collection of information. The sedimentation column is part of a fluid circuit containing a peristaltic pump, with the suspension pumped into the bottom of the column. At all times the column is completely full of fluid to prevent problems associated with a free surface in the sedimentation column. The suspension to be characterized is circulated through the column until the commencement of the sedimentation studies. The scan rate of the tank is adjusted so that the axis of the pen recorder could be calibrated directly in Stokes diameters.

Hendrix and Orr [78] carried out comparative size characterization studies using the Sedigraph model 5000 and other characterization procedures and reported good agreement between the size distribution data obtained by various methods. Recently, however, Allen and Baudet [79] have criticized the performance of the Sedigraph system for the characterization of very fine powders.

In the LADAL x-ray sedimentometer the x-ray beam is generated by a radioactive source. The source used is a promethium 147–aluminum isotope source with a half-life of 2.6 years. The x-rays passing through the system are detected with a xenon-filled proportional counter. The sedimentation tank is scanned during a characterization study by moving the x-ray system along the outside of the sedimentation column. It is claimed that the sensitivity of this instrument is such that concentrations need not exceed 1% by volume and that sedimentation characterization studies can be completed within 14 minutes for most powders [14, 80].

In the sedimentation equipment described by Talbot and Jacobs the fineparticles to be characterized move downward through a light beam, and their shadows are projected by an optical system onto a grating. Talbot and Jacobs used fineparticles ranging up to 300 μm. As the image of a fineparticle moved past the grating, it modulated the output of a photoelectric detector to generate an ac wave that can be linked directly to the velocity of the fineparticle. Thus a large fineparticle will generate a high-frequency wave, whereas a small fineparticle will generate a lower-frequency signal. When many fineparticles are moving past the grating at the same time, many fluctuating signals will be generated concurrently and will be lumped together in the output of the photoelectric device.

The output from the photoelectric device receiving the grid-encoded signal can be analyzed in many ways. Talbot [81] chose to analyze the signal with electronic circuits that could break down the gross output of the instrument into signals representing various frequencies. In this way he could detect the presence and the concentration of various-sized groups of fineparticles. Modern procedures based on Talbot's original concept would probably use a digital computer to perform a Fourier analysis of the complex waveform generated by this type of instrument. It should be noted that in the original equipment as described by Talbot and Jacobs the range of fineparticles studied was relatively coarse and the resolution requirements were not high. Thus they examined fineparticles within four groups in the ranges of 0–75 μm, 75–150 μm, 150–225 μm, and 225–300 μm Stokes diameter. For this relatively coarse set of fineparticles studied at low size resolution, the electronic filtering of the output was a sufficiently sensitive procedure.

In physics the Doppler effect is a general term applied to the observed frequency shift occurring when a wave is either generated by a moving source or is reflected from a moving object. The name comes from the fact that Christian Johan Doppler, an Austrian physicist (1803–1853), was the first to work out the mathematical theory describing the magnitude of such frequency shifts. His original work was concerned with sonic waves; however, the phenomenon is a general one and is manifest by all wave motions. The frequency shift caused by reflecting a light beam off of a moving fineparticle at the kind of velocities that occur in gravitational sedimentation situations is extremely small but can be measured by use of modern laser technology. The technique has proven to be a highly sensitive one for measuring the velocities of the fineparticles in the smoke trails used in a wind tunnel. The technique for measuring fineparticle velocities by measuring laser frequency shifts on reflection has become known in Europe as *laser Doppler anemometry*. In the United States the term used for such methods is *laser Doppler velocity* (LDV) *measurement*.

In the LDV technique for measuring the velocities of sedimenting fineparticles developed by Chabay, the light scattered by the moving fineparticle is allowed to mix with a portion of the original laser light. The two signals interfere with each other to produce a composite wave that can be regarded as being similar to a beat frequency produced when two sound waves of very similar frequency are listened to at the same time. When this process is carried out with two optical waves, the process is often known as *optical mixing spectroscopy*. In electrical engineering the process of mixing two waves of almost the same frequency to produce a beat frequency in a nonlinear detector is known as "*heterodyning*" (from Greek roots *hetero*, other, different; and *dyn*, power, with the idea that the combining of the two waves gives a combined signal of varying power). Chabay [82] described his experimental strategy as "optical heterodyne spectroscopy of droplet sizes." In Chabay's system, reflected laser light was observed at a small solid angle with respect to the horizontal. Part of the incident beam reflected off the walls and windows of the sedimentation chamber onto the photomultiplier tube surface provided the reference signal at the original frequency of the laser light. This reference signal mixes with the scattered signals from the moving fineparticles on the actual surface of the photomultiplier tube. The optical beat frequencies then occur as amplitude fluctuations in the current from the photomultiplier tube. This current is passed to a real-time spectrum analyzer that prints out the velocity data for the various fineparticles moving through the laser beam. The actual interpretation of the data in terms of the sizing of the fineparticles and their concentration is complicated, and the reader is referred to Chabay's publications for the necessary theory [82].

Wang [83] has reviewed the various experimental techniques that can be utilized in laser anemometry. Three different systems that can be employed are shown in Figure 5.23. The electrical engineering terms for the various experimental systems are given in this diagram. From a traditional physically intuitive perspective, one is tempted to interpret a system such as that shown in Figure 5.23*a* from the point of view that where the two laser light beams cross, the formation of optical interference fringes can be anticipated. Discussing this intuitive interpretation of the system, Wang points out that

> a useful but not rigorous way of analyzing the Doppler Heterodyne arrangement is the so-called fringe model. In such a model the two beams that interfere in the scattering region may be considered to form a set of parallel interference fringes. The traversing of the fineparticle across the fringe pattern gives rise to scattered light which is modulated at the fringe crossing frequency. This frequency is identical to the Doppler shifted frequency. It should be noted that no non-linear processes occur in the scattering volume, hence there are no actual fringes. Too strict an interpretation of the fringe model can therefore result in error but the model has heuristic appeal.

Again from an intuitive perspective, measuring the movement of the fineparticles down through a region such as that shown in Figure 5.23*a* is similar to the principles adopted in the Talbot–Jacobs sedimentometer, except that the grid being used to encode the movement of the fineparticles is not a physical grid but can be considered to be similar to the interference fringes produced by the interaction of two coherent monochromatic beams as in the classic Young interference experiment.

As so often happens in technology, the cautious words of the theoretical physicist are not always adhered to, and, in some of the later publications on the use of LDV for assessing the velocities of fineparticles, the

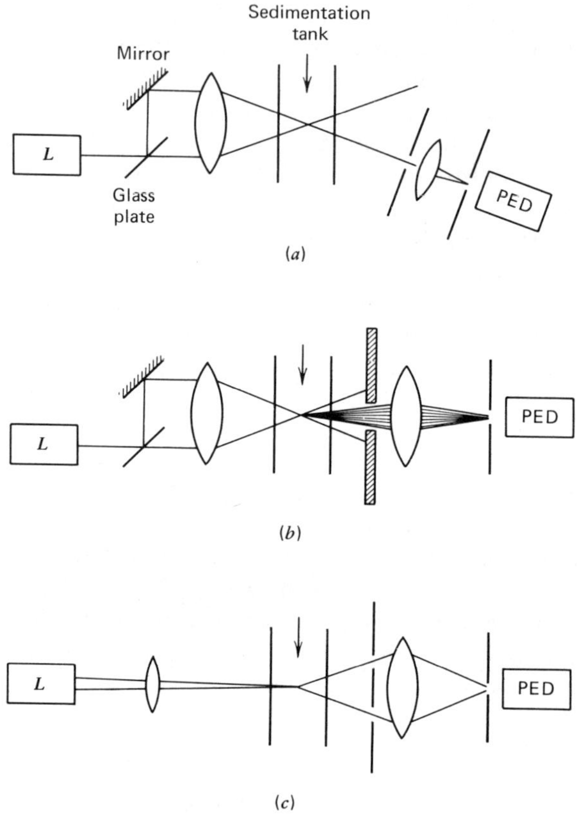

Fig. 5.23 Several experimental arrangements are possible in the design of photosedimentometers exploiting the Doppler effect: (*a*) local oscillator heterodyne setup; (*b*) differential heterodyne optical arrangement; (*c*) symmetric heterodyne arrangement. Symbols: *PED*, photoelectric detector; *L*, laser.

semantic caution expressed by Wang is abandoned and the writers describe their instruments as if the fringes exist physically in space and are physically interrupted by the passage of the fineparticles. In such research papers the explanation of the generated signal is couched in terms of "fineparticles moving across the fringes" and the connection between the experimental strategy used and the Doppler effect is obscure. Some writers even provide pictures of the fringes that are used to track the fineparticles, and mention is made of altering the fringe spacing to change the sensitivity of the measurements. Such a semantic perspective is permissible, as Wang has pointed out, as long as it is realized that from a pure physics point of view the fringes do not exist until they are photographed with a nonlinear receptor (a camera) or viewed with a similar device (like the eye!). As pointed out by Wang, there are several different possibilities for analyzing the frequency-shifted signals arriving from the moving fineparticles, and various strategies have been adopted by various workers.

The resolution and the sensitivity obtainable in a laser Doppler shift sedimentometer system are illustrated by the work of Stenhouse and coworkers [84], who were able to measure the Stokes diameter of fineparticles as small as 0.5 μm and were able to quantitatively investigate the effects of Brownian motion. The power and the potential of the methodology can be appreciated from the fact that workers have already succeeded in measuring the velocity of blood flow from optical Doppler shifts of blood cells by using a fiber optic catheter and the growth of carbon fineparticles within a flame [83, 85, 86].

Attempts have been made to use radiant energy of very short wavelength to construct sedimentometers. Thus Ross used an americium 241 γ-ray source. The strength of the source was 150 mCi, and the energy of the radiations corresponded to 60 keV with a half-life of 470 years. Such a source, encapsulated in a thermally isolated sedimentometer, could prove very advantageous for quality control situations provided adequate safety precautions are maintained [88].

5.8 SEDIMENTATION PROCEDURES BASED ON HYDROMETERS

Incremental monitoring of the settling dynamics of an initial homogeneous suspension can be carried out by measuring density gradients within the suspension. The density changes within a homogeneous suspension can be related to the Stokes diameter distribution function of the sedimenting fineparticles by using the following relationships [87]. The relationship between the solids concentration of a suspension and its density is

$$\rho_{\text{su}} = \rho_L + \alpha\,(\rho_P - \rho_L) \tag{5.16}$$

where ρ_{su} = density of suspension
ρ_L = density of dispersing liquid
ρ_P = density of powder fineparticles
α = volume fraction of solids per cubic centimeter

If ρ_I is the initial density of the suspension, it can be shown that

$$R_{ht} = \frac{(\rho_{\text{su}})_{ht} - \rho_L}{\rho_I - \rho_L} \tag{5.17}$$

where R_{ht} = percentage by weight of fineparticles of Stokes diameter less than d_{ht} that falls h cm in t seconds

$(\rho_{su})_{ht}$ = density of suspension at depth h, t seconds after settling commenced (and other symbols are as defined)

Hydrometers can be used to follow the density gradients established within a settling suspension. A hydrometer is a device normally used to measure the density of a homogeneous fluid. The idealized hydrometer shown in Figure 5.24 consists of two main parts, a symmetrical bulb and a long stem. When placed in a fluid, the bulb descends until the weight of the fluid displaced equals that of the hydrometer. When used to measure the density gradient within a suspension, the finite volume of the bulb of the hydrometer determines the resolution of the experimental measurements. The insertion of the bulb also causes a disturbance in the suspension, and it displaces the suspension in the interrogation zone, thus causing uncertainty in the interpretation of the experimental data. When interpreting data obtained by this technique, it is usual to assume that the

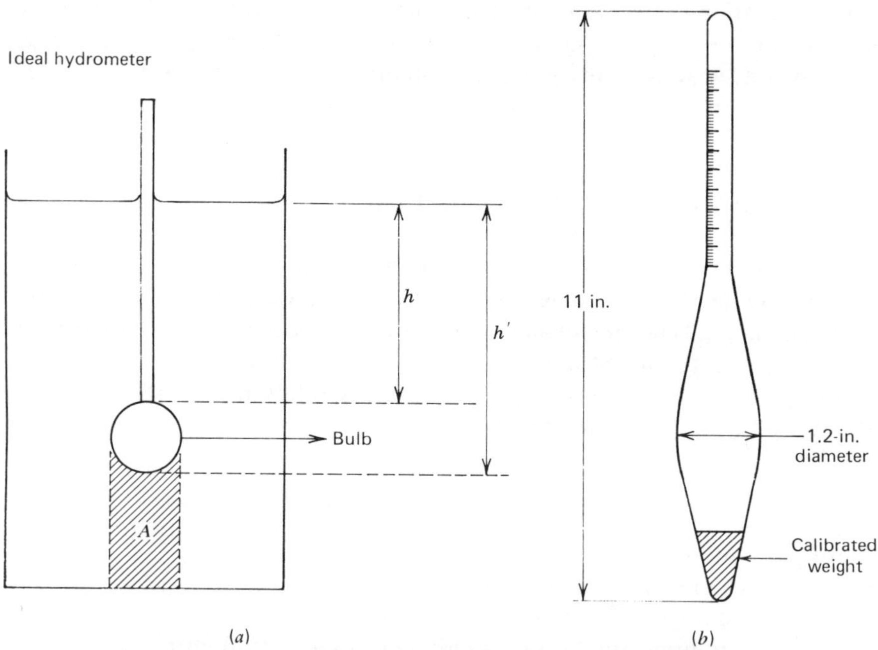

Fig. 5.24 Considerable uncertainties exist in determining the significance of the interrogation zone of the suspension evaluated by the hydrometer method: (*a*) idealized hydrometer; (*b*) streamlined hydrometer specified for use in characterization of soils [89]. Symbol A = shadow zone.

volume of the stem is negligible in comparison with that of the bulb and that the density measured is an average value over the depths h and h^1 as shown in Figure 5.24. The exact nature of this average will depend on the shape of the hydrometer bulb. In the various standard procedures established for hydrometer–sedimentation characterization techniques the shape of the hydrometer is strictly specified, and the appropriate formulas for calculating the effective value of h from the dimensions of the hydrometer and the depth to which it sinks are given [89, 90].

Two experimental procedures are possible when using the hydrometer to follow density shifts within a settling suspension: (1) the hydrometer is placed in the suspension and readings of the depth of immersion taken at a series of times or (2) the hydrometer is placed in the suspension prior to each measurement and withdrawn immediately afterward.

When the hydrometer is left in the suspension, fineparticles settle from the suspension onto the shoulders of the hydrometer. This changes the effective weight of the hydrometer, thus causing systematic error in the density readings taken from the stem of the hydrometer, which is usually calibrated to give density readings directly according to the depth to which it sinks in the suspension. The streamlined hydrometer in Figure 5.24*b* has sloping sides designed to help the hydrometer shed any fineparticles falling on it. However, the flow of fineparticles shed by the hydrometer will cause gravity-type convection currents around the bulb and further disturb the settling suspension. If the hydrometer is left in position during settlement, another source of error is the shadow zone underneath the bulb, a region denoted *A* in Figure 5.24*a*. Within this zone fineparticles settling out are not replaced. For this reason, after a period of time, the density of zone *A* is less than in the surrounding suspension. This causes turbulent mixing at the boundaries and thus disturbs the settling pattern and causes error of unknown magnitude.

The alternative experimental procedure of withdrawing the hydrometer between readings over a period of time gives rise to two types of error. When the hydrometer is inserted into the suspension, it displaces and generally disturbs the suspension as it sinks through it. The effect of this disturbance is difficult to estimate, and the cumulative result of a series of insertions and withdrawals can be measured only empirically. Factors that influence the magnitude of the error caused by this disturbance of the suspension are (1) the ratios of the bulb dimensions to those of the sedimentation vessel, (2) the height to which the hydrometer sinks before it reaches equilibrium, and (3) the rate of fall of the hydrometer. The other main source of error is caused by the fact that it is difficult to decide when the hydrometer is in equilibrium with the surrounding suspension. This problem is usually overcome by specifying arbitrarily some time period between the insertion of the hydrometer and the taking of a reading; again, errors due to this factor are difficult to estimate.

The higher the suspension concentration used in the characterization study, the easier the calibration of the hydrometers because the density changes to be measured are varying over a wider range. The advantage gained in hydrometer calibration by using higher solids concentrations is offset by the fact that dynamic clustering becomes a severe problem at higher concentrations. Usually it is not practical to use concentrations less than 2% by volume, and unavoidable concentration effects are accepted as a tolerable penalty to be traded off against low cost, portability, and ruggedness of the procedure and equipment.

Surface tension effects in the emergent hydrometer stem are another source of error. A correction factor can be calculated for this from the known dimensions of the stem and the surface tension of the fluid, plus dispersing agent.

In spite of its many sources of error, the hydrometer method is widely used in the characterization of soils and in the characterization and the quality control of large-tonnage–low-cost material such as clay and crushed calcium carbonate because of the low cost and the simplicity of the equipment. Provided all experimental conditions and equipment dimensions are carefully specified, the techniques give reproducible results, but it is difficult to relate data obtained by this procedure to those generated by other sedimentation methods.

Berg developed the use of miniature hydrometers that can be completely immersed within the suspension. These miniature hydrometers are called *divers*. The major dimension of these divers is of the order of 1 cm. Because of their small volume, their disturbance of the settling behavior of a suspension is much less than that of a traditional hydrometer [91]. The sampling zone of a diver is clearly defined, and it is easier to determine the effective value of h, the depth at which the suspension density equals that of the divers. Corrections for surface tension effects have obviously been eliminated since no part of the diver is above the surface when measurements are taken. A diver, since it is completely immersed in the suspension, can measure only one zone of constant specific gravity within the suspension that moves down through the suspension at constant velocity. Therefore, a set of divers is required to obtain several points on the size distribution function of the settling fineparticles. The cost of producing a series of divers within a closely defined range of specific gravity has limited the use of the technique. Berg has personally made many sets of divers for scientists to use, but since they are constructed of mercury, glass, and steel, they are somewhat fragile and have never become available at low cost. Because of Berg's efforts (he carried out his work at the Institute for Ceramic Technology in Denmark), the diver technique has been relatively widely used in Europe but has not been used extensively in England and North America. In North America, divers are commercially available from Cargile [92].

Since Berg's divers contained a small piece of magnetic material, they would be located within the suspension by drawing them to the side of the sedimentation tank with a magnet. This method of location has the disadvantage that, since the divers cannot be seen, the magnet in drawing the divers to the side may also drag it upward or downward [91]. This would introduce error into the measured value of h. Johnson reports that divers used at the British Ceramic Research Association were also located magnetically, although in the description of the method it is stated that the diver is introduced at the side of the sedimentation vessel [93].

Jarrett and Heywood [9] developed the use of roof divers. The structure of a roof diver is shown in Figure 5.25*a*. The density of the diver was adjusted by filling the brass peg. An ac inductance bridge was used to

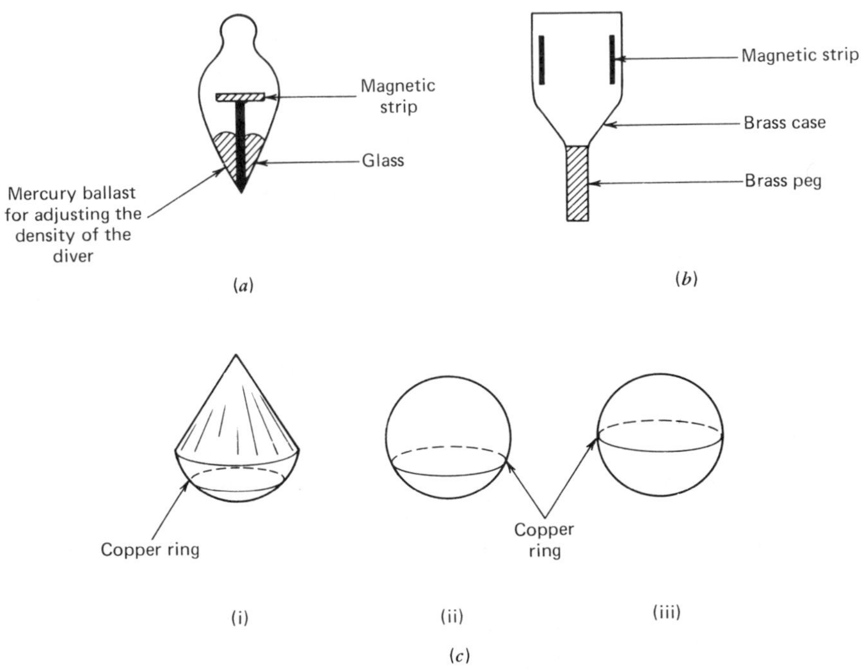

Fig. 5.25 Special small hydrometers, called *divers*, can be used to follow density gradients in a settling suspension: (*a*) Berg diver; (*b*) roof diver after Jarrett and Heywood [9]; (*c*) experimental divers, used by Kaye and James [87].

locate the divers in a suspension. In the roof diver technique the diver is placed under a column protruding vertically into the suspension. The coil to locate the divers is placed around the sedimentation tank just below the diver. When the density of the surrounding suspension equals that of the diver, the diver moves away from the end of the column and is detected by the coil. Jarrett and Heywood found it necessary to vibrate the central column to ensure that the divers fell away at the right time. The suspension space under the roof divers becomes progressively deficient in fineparticles with respect to adjacent suspensions as sedimentation proceeds. This dead-space or shadow-zone effect causes errors of unknown magnitude.

Theoretically, it can be shown that if there was no such thing as dynamic clustering, no fineparticles above the diver would be moving faster than the diver. Therefore, a demonstration of the reality of fineparticle deposition on a diver is one more piece of evidence demonstrating the reality of dynamic clustering. Kaye and James carried out an investigation into the effect of diver design on the behavior of the diver in a sedimenting suspension. Their work is reviewed in some detail, since it suggests improvements in diver design and demonstrates clearly the reality of dynamic clusters in suspension concentrations higher than 2% by volume that are usually opaque so that interaction of fineparticles cannot be seen easily in the body of the suspension. In Kaye and James's experimental work, the divers were located by placement of a copper ring in a diver molded from plastic. The diver density was adjusted by varying the proportions of copper to plastic in the overall construction of the diver. The divers were located in the suspension by means of a coil placed around the sedimentation vessel. This coil was fed with high-frequency electric current and formed a part of a detection circuit. When the copper coil of the diver lay in the plane of the detection coil, it disturbed the detection circuits so that its presence was registered on a chart recorder. An advantage of this detection mechanism was that rotation of the diver could be detected since, if the copper coil in the diver was vertical, it was not registered by the detection circuit. Therefore, the diver could be located and tracked down through the suspension and any rotation of the diver could be recorded. The divers used by Kaye and James were inserted into the suspension after the suspension had been allowed to settle for a fixed period of time. In Figure 5.26 possible paths of an inserted diver, depending on the magnitude of the various effects on the trajectory on the diver, are shown. Thus in the trajectory in Figure 5.26*a* the diver drops to the zone of equal density. It then moves down the column at the speed of that density gradient. When the diver reaches the zone of equal density, it has a finite velocity and hence kinetic energy. Therefore, it is reasonable

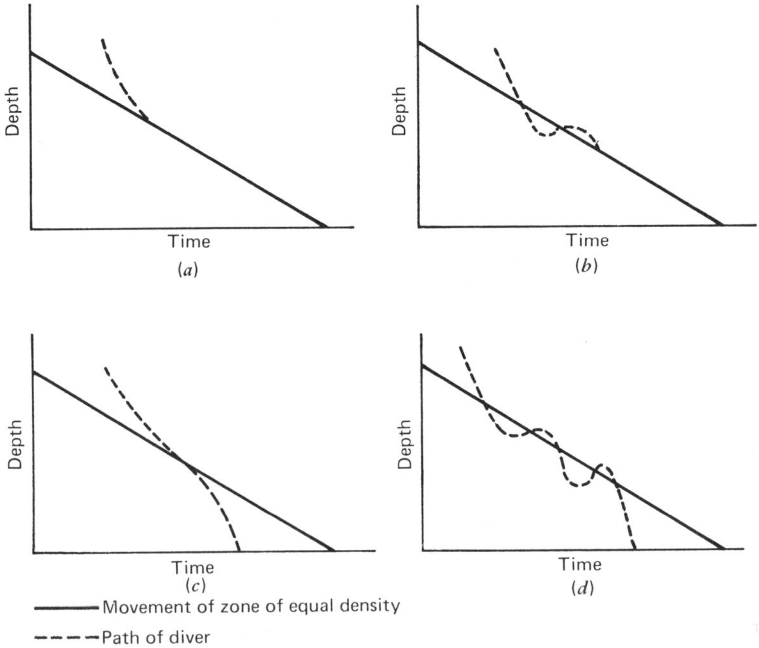

Fig. 5.26 Possible diver paths through the suspension: (*a*) ideal path; (*b*) oscillations, no settlement; (*c*) no oscillations, powder settlement; (*d*) oscillations, powder settlement.

to anticipate that it may oscillate about the zone of equal density until this energy is damped out by the fluid of the suspension and that this would generate the trajectory shown in Figure 5.26*b*. The effect of powder deposition onto the diver will be to accelerate it out of the zone of equal density. Thus deposition effects without oscillation would give rise to the trajectory shown in Figure 5.26*c*. Oscillations and powder settlement would generate the trajectory in Figure 5.26*d*.

The first experimental design of diver used by Kaye and James is shown in Figure 5.25*c*. Its pear-shaped plastic body incorporated a copper coil, the plane of which was well below the center of gravity of the whole diver. The shoulders of the pear-shaped diver were of low angle to discourage powder shedding once it was deposited on the diver so that any deposition effects would be enhanced. When this diver was inserted into a suspension of tungsten powder, it descended very rapidly to the bottom of the suspension, apparently accelerated by powder deposited on its shoulders. A typical trajectory of the pear-shaped diver is shown in Figure 5.27*a*.

The second diver design used by Kaye and James consisted of a copper ring embedded in a plastic sphere, with the copper ring situated in a plane

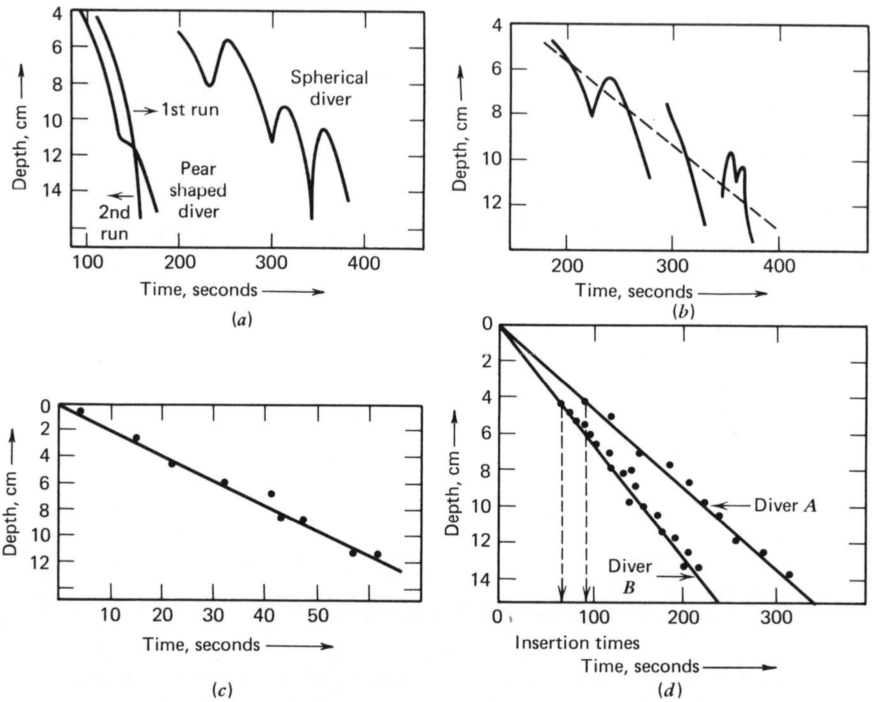

Fig. 5.27 The trajectories of various types of diver [87]: (*a*) trajectories of a pear-shaped diver and a spherical diver with off-center copper band of the same density; (*b*) trajectory illustrating tipping action of spherical diver with low center of gravity; (*c*) trajectory of duraluminum hollow-sphere diver; (*d*) trajectories of plated plastic divers.

below the center of gravity of the sphere. The density of this spherical diver was the same as that of the pear-shaped diver. When this spherical diver was placed in the suspension, it seemed to move down to the density zone matching its own density. However, it oscillated in an unanticipated manner. A typical trajectory is illustrated in Figure 5.27*a*, where it can be seen that the first oscillation has a rounded peak to it, indicating that this is probably an overshoot oscillation of the type predicted in Figure 5.26*b*. However, the next oscillation has a very sharp lower peak. Kaye and James explained this sharp reversal of direction as being due to powder shedding. According to their explanation, powder accumulates on the diver, driving it lower than the zone of equal density until the moment of the deposited powder about the center of gravity is greater than that of the copper coil in the lower portion of the diver. When this situation is reached, the diver is in a position of unstable equilibrium and any minor inequality of deposition initiates rotation with the dumping of the deposited powder. Buoyancy thrust then causes the diver to rise above the zone of equal density. The diver will then slowly move back into position until further deposition causes it to undergo another sharp reversal of direction. The powder-dumping reversals in the trajectory of the spherical diver in Figure 5.27*a* were relatively sharp, and the equipment failed to record a change in the orientation of the diver as it tipped off its powder load. However, another experiment with the same diver definitely demonstrated that in the upward-moving regions of the trajectory, the diver's orientation had changed. A typical trajectory showing these features is illustrated in Figure 5.27*b*. For two portions of this trajectory, the diver apparently disappeared in the suspension, demonstrating that its internal coil was not aligned with the detection coil for this region of its path. Trajectories such as those illustrated in Figure 5.27*b* suggested to Kaye and James that a diver with a detection coil on the equator of the diver would continually shed any small amounts of powder deposited on its surface by dynamic clustering. Therefore, this type of diver would continually spin in the zone of equal density. Although this spinning would cause a small disturbance to the suspension, it would also give the diver a dynamic stability within the zone of equal density. Experiments were carried out with divers using equatorial coils, but the constant spinning of the diver meant that it was difficult to detect with the equipment being used. Kaye and James then designed two other spherical symmetrical divers that would spin and constantly shed powder. In the first design they took a plastic sphere and plated it with copper to the required density. The copper coating made it possible to locate the diver at all times. The trajectories of two plated divers are shown in Figure 5.27*d*. They were inserted into the suspension more than a minute after settling had commenced. They appeared to take up their positions in zones of equal density very rapidly. They then followed a trajectory that extrapolated back to zero depth at initial time. This would seem to indicate that such a diver design was very effective in ridding it of powder and that such divers were following zones of equal density down through the suspension in accordance with the theoretical path predicted for ideal divers. Thus with this type of diver, overall disturbances to the dynamics of the suspension were minimal. It is even possible that the spinning transfers fineparticles around the diver into the shadow zone, thus minimizing effects due to this zone. However, further experiments would be necessary to establish this possibility. Kaye and James report that their plated divers suffered from a tendency to flake. Accordingly, they investigated a final design of diver in which the diver was fabricated from two hollow hemispheres. The trajectory of this type of diver is illustrated in Figure 5.27*c*. It can be seen from this diagram that this type of diver also followed a linear trajectory of the type predicted for an ideal diver. The aluminum alloy diver had a diameter of 1 cm and was very robust. Kaye and James suggested that this type of diver would be loaded symmetrically within the hollow sphere to give any required specific gravity. Thus, if small ball bearings were to be glued to the bottom and top hemisphere the center of gravity of the diver could be maintained coincident with the geometric center of the sphere. Divers made in this way are so robust that they can be dropped on the floor without damage. They can also be fabricated in mass quantities at low cost. It would seem that this type of diver may eventually come to be used widely in quality control. The electronics revolution of the last decade would also seem to indicate that cheap detection systems will be available in the not-too-distant future. This type of diver will probably be more useful in studying the internal dynamics of high-concentration suspensions than in accurate characterization of powders for research purposes, because of the high solids content required in a test suspension to create density changes that can be measured with relative ease.

Dotts [94] has described an instrument in which a sloping side arm monometer has been used to monitor sedimentation rates within the suspension. A commercial version of Dott's equipment is often referred to as the Fisher–Dotts apparatus for measuring Stokes diameter distributions [94].

5.9 SEDIMENTATION CHARACTERIZATION BASED ON CUMULATIVE MONITORING OF SEDIMENTS FROM AN INITIALLY HOMOGENEOUS SUSPENSION

In these procedures a record of the weight of sediment W_t at a time t is used to deduce the Stokes diameter distribution of the suspended fineparticles. At time t the sediment contains all fineparticles greater than d_{Ht} (where d_{Ht} is the Stokes diameter calculated from H, the height of the settling chamber, and t and the appropriate values of the other parameters of the suspension) and a portion of the fineparticles less than d_{Ht}. If $(dW/dt)_t$ is the rate at which fineparticles are arriving at the base of the column at time t, then the amount of fineparticles already present in the slurry smaller than d_{Ht} is given by the relationship

$$\left(\frac{dW}{dt}\right) \cdot t$$

Therefore, the weight of fineparticles greater than d_{Ht} is given by

$$W_{Ht} = W_t - \left(\frac{dW}{dt}\right)_t \cdot t$$

If, now, W_t is the total weight of fineparticles in suspension, then

$$R_{Ht} = \left(\frac{W_t}{W_T}\right) \times 100 - \left(\frac{dW}{dt}\right)_t \cdot \frac{t}{W_t} \cdot 100$$

where R_{Ht} is the weight percentage of fineparticles greater than d_{Ht}.

Most of the publications dealing with the basics of this group of methods date from before the availability of cheap computers and discuss the evaluation of $(dW/dt)_t$ by graphical procedures. These graphical methods will be reviewed in this section since they give a physical appreciation of the methodology; however, modern experimental strategy would process the data with microprocessors.

One technique for measuring the weight of sediment accumulating at the base of a sedimentation column is to place a balance pan within the sedimentation column. Characterization procedures based on this concept are termed *sedimentation balance procedures*. A balance pan hung directly in the column is subject to many sources of uncertainty affecting the characterization data. The most significant of these sources of uncertainty is the dead-zone error. The dead-zone under the balance pan is created by fineparticles settling out below the balance pan that are not replaced by fineparticles from above. This creates a density difference between the suspension above and at the side of the pan and immediately under the pan. This difference in density gives rise to turbulence around the edges of the balance pan, which leads to error in the measured weight of suspension deposited on the balance pan. Dead-zone effects can be overcome by suspending the balance pan in a reservoir of clear fluid underneath the sedimentation column and allowing fluid pressure to take care of the shifting hydrostatic pressure of the suspension column. The balance pan is also given a tall rim so that any residual motion of the pan does not result in fineparticles moving around the rim (see discussion of Leschonski's equipment later in this section).

The presence of a rod linkage between the balance pan arm and the pan creates interference with the trajectories of the fineparticles. Surface tension forces on an emergent rod linkage have to be taken into account when interpreting the data.

When weighing the fineparticles, one can either allow the balance pan to sink gradually over a period of time with the weight being registered on a torsion mechanism or a spring or use compensating forces to keep the pan at a fixed depth. The first strategy leads to gross displacement of fluid from under the balance pan, thus interfering again with the trajectories of fineparticles in the vicinity of the edges of the pan. On the other hand, minimizing the movement of the pan by using servo controlled weighing mechanisms makes the pan oscillate, which has an unknown effect on the measured data. The various instrumentation procedures that have been employed by different scientific investigators are basically strategies aimed at avoiding these several difficulties associated with the presence of the balance pan in the suspension.

Many workers who have used sedimentation balance characterization procedures report the loss of fineparticles from the sedimenting fluid onto the walls of the sedimentation column. Thus Davies and Kaye [16] report that, when using glass powders in water with the Cahn sedimentation balance, losses as high as 30% by weight were incurred when comparing the total weight of powder collected on the balance pan after prolonged sedimentation to the theoretical predictions of the ultimate weight expected on the pan from the known concentration of the suspension.

Many sedimentation balances employ an open surface sedimentation column, and at least some of the reports of apparent powder losses from the balance pan may not be entirely due to losses of fineparticles to the wall of the sedimentation column but could be the result of evaporation-driven convection currents resuspending the fineparticles and preventing their sedimentation onto the balance pan.

If the powder lost to the side of the sedimentation column is representative of the fineparticles in suspension, the weight discrepancy incurred experimentally as compared to the values predicted may not be a source of serious uncertainty in the deduced Stokes diameter distribution. Davies and Kaye compared the powder lost to the walls of the vessel to the powder that arrived on the balance pan and showed that they had similar distribution functions. In any particular situation with any particular powder the size distribution function of the powder on the balance pan should be compared to any lost to the walls of the sedimentation chamber.

Because of the loss of powder to the walls of the sedimentation vessel, any actual weight record from a sedimentation balance will usually have an experimental limit differing from the theoretical weight limit, as indicated in Figure 5.28. In such a situation one can either continue the experiment until the recorded weight no longer changes with time or make a best estimate of the ultimate weight by extrapolation of the graphical record to infinite time, as illustrated in Figure 5.28.

A graphical method for interpreting equation 5.18 is illustrated in Figure 5.28 [95]. At a point such as E at a time t, a tangent is drawn to the weight record curve. This tangent is extended until it cuts the line erected at $t = 0$. Considering the terms in equation 5.18 with respect to the intercepts shown in Figure 5.28, it can be stated that

$$\frac{W_{t'}}{W_T} = \frac{C}{L} \quad \text{and} \quad \frac{dW \cdot t}{dt \cdot W_T} = \frac{b}{L}$$

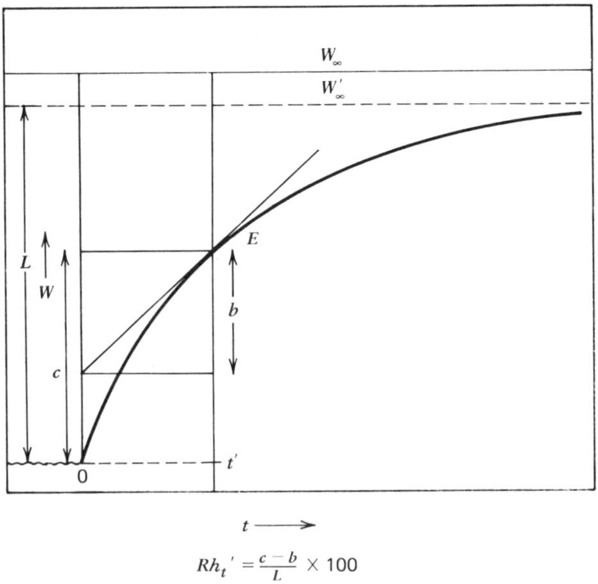

Fig. 5.28 The graphical calculation of Stokes diameter distribution from a sedimentation balance study. Symbols: t, time; W, weight deposited; R_{ht}', percentage by weight of fineparticles of Stokes diameter less than d_{ht}'; $W\infty$, theoretical sediment weight after infinite sedimentation time; W'_{∞}, estimated sediment weight after infinite sedimentation time.

Therefore

$$R_{ht'} = \frac{C - b}{L} \tag{5.19}$$

The accuracy of the evaluation of $R_{ht'}$ depends on the precision with which the tangent can be drawn at E and other similar points.

It has been pointed out by several workers that the second term of equation 5.18 can be written as the differential of the reciprocal of natural log of t. For this reason it is suggested that graphical transformation of the data is made easier by plotting the weight accumulated at time t against the logarithm of t. This really does not facilitate the transformation of the data, since although in one sense the tangents are easier to draw to the curve, it now becomes much more important to locate them precisely on the logarithmic axis; otherwise, small movements of the line can create large differences in the numerical values inserted in the calculated gradients. The choice of format in presentation and transformation of the data is simply a matter of convenience.

The reproducibility of characterization studies undertaken by using sedimentation balances and the effect of solids concentration on the data

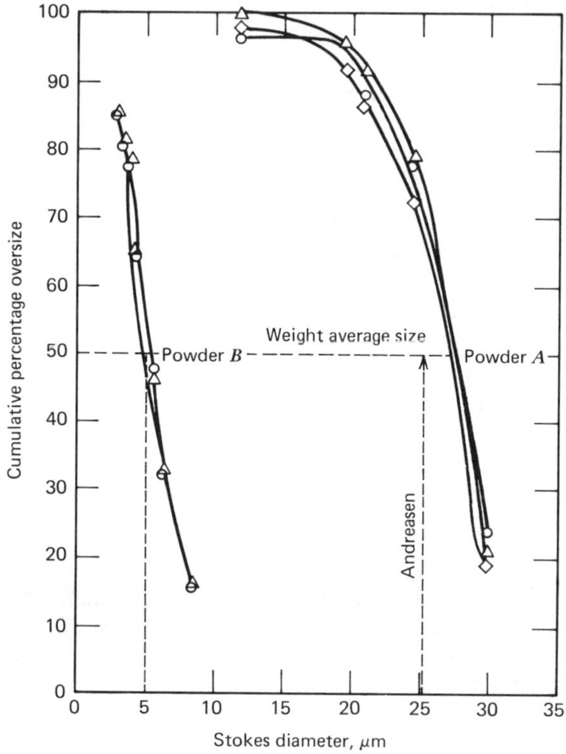

Fig. 5.29 Reproducibility of distributions measured by sedimentation balance procedures [25]: Δ, run 1; ○, run 2; ◇, run 3. Volume concentration of all measurements 0.207%.

are illustrated by data reported by Davies and Kaye [16]. Thus Figure 5.29 shows the data for three determinations of test powders *A* and *B* carried out by using a Cahn sedimentation balance at a volume concentration of 0.207% using an initially homogeneous suspension technique. For comparative purposes, the mean size of the Stokes diameter distributions as measured by the Andreasen pipette method at 0.37% by volume concentration is shown.

Davies and Kaye report that for test powder B, which is the wider-range distribution powder and hence less sensitive to fineparticle–fluid interaction, the Stokes diameter distributions measured at suspension concentrations of 0.057 and 0.414 by volume were indistinguishable. For test powder *A*, a relatively short-size-range powder, there was a gradual coarsening of the data detectable as the volume concentration of the suspension used was increased above 1% by volume; however, as shown by the data in Figure 5.30, the observed coarsening was small.

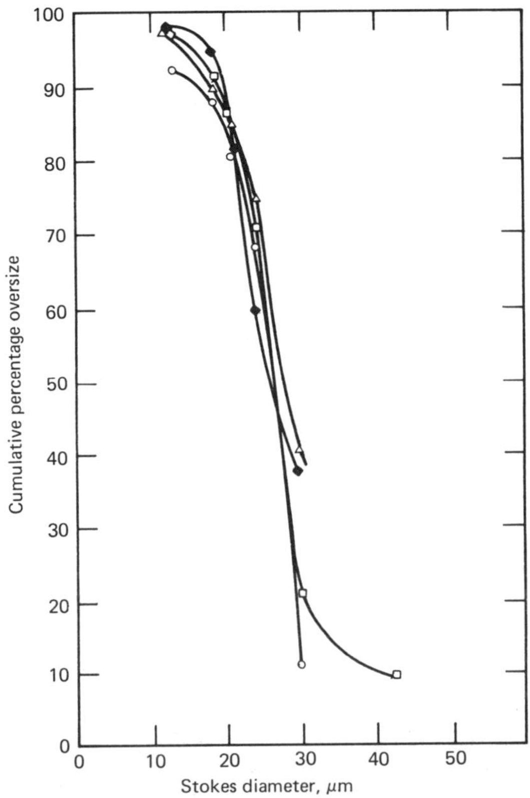

Fig. 5.30 Concentration generated errors for a narrow-size-range powder [16]. Volume concentration of suspension used: ○, 0.046; Δ, 0.103; □, 0.207; ◆, 0.322.

The data in Figure 5.30 illustrate that the modern sedimentation balance is able to operate with very low solids concentrations and that for many powders, a volume concentration of 0.1% will result in a system free from fluid–fineparticle interaction. For any particular situation, it is advisable to carry out a series of investigations with various solids concentrations to determine the upper solids concentrations that should be used with any particular powder.

The sedimentation balance of Rabatin and Gale is shown in Figure 5.31*a* [96]. The use of a sensitive spring-balance and an optical method of recording its movement make it possible to use low solids concentrations and to keep the total movement of the pan through the suspension small. As the spring balance is extended by the solids settling on the pan, an opaque shutter moves across a beam of light, and the signal received by a

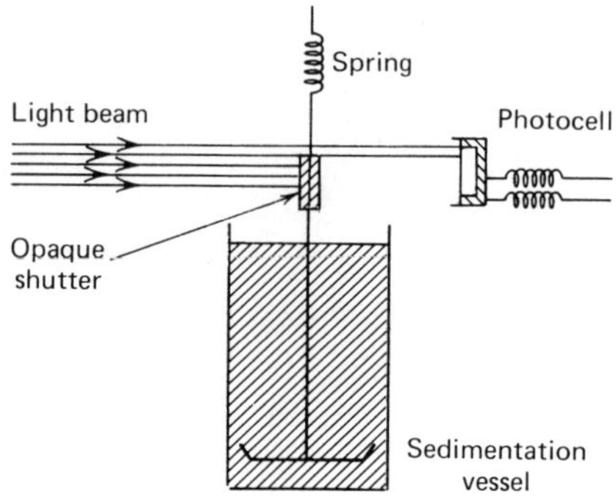

(a)

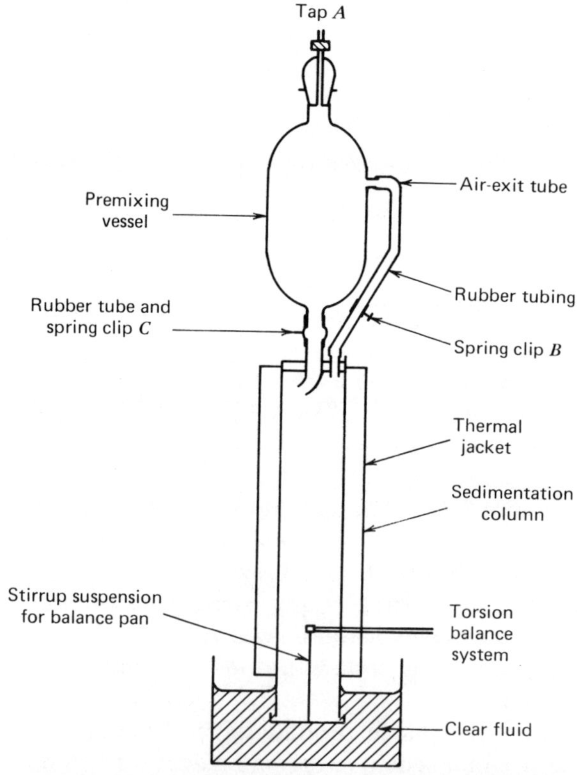

(c)

Fig. 5.31 Sedimentation balance equipment: (*a*) sedimentation balance of Rabatin and Gale [95]; (*b*) Shimadzu sedimentograph [98]; (*c*) basic system of Bostock (Gallenkamp) [99]; (*d*) Sartorius sedimentation balance described by Leschonski [17].

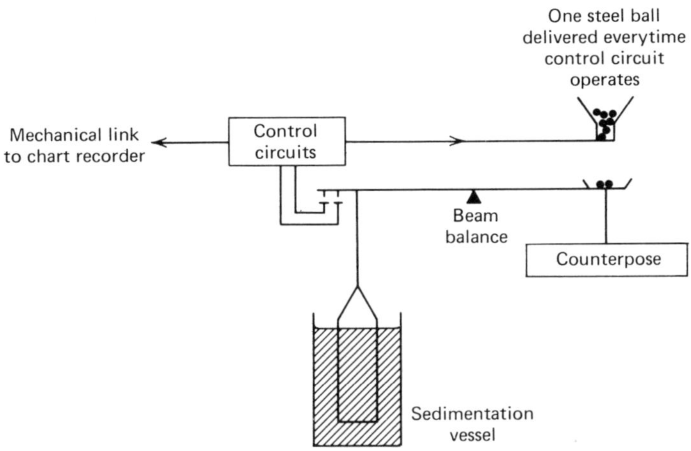

(b)

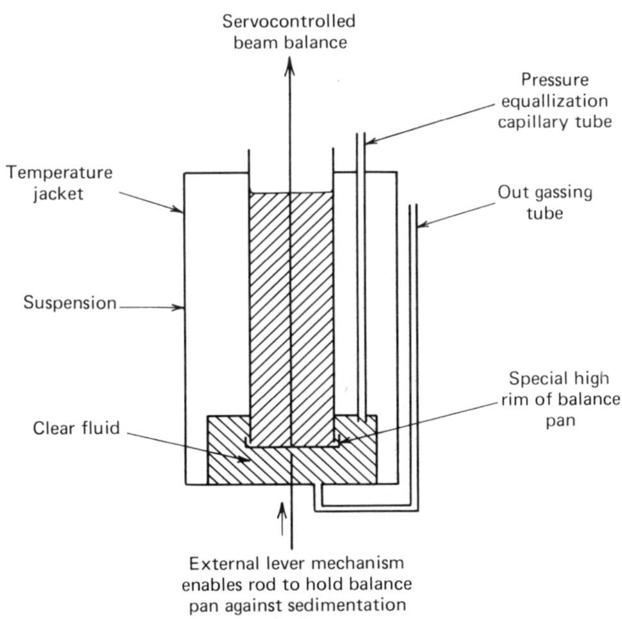

(d)

Fig. 5.31 (*Continued*)

photoelectric cell is recorded. The system can either be calibrated with known weights, or, if the response of the photocell is linear with energy received, the signal output can be taken as being directly proportional to weight. Palik [97] has described a modification of the Rabatin–Gale balance in which a torsion balance mechanism is substituted for the spring balance system. The equipment is suitable for quality control investigations; however, errors due to the dead space and the fluid movement as the balance pan sinks are not allowed for, and empirical calibration of the equipment for research purposes may be necessary.

Another instrument suitable for quality control purposes but unsuitable for research purposes because of the large solids volume concentration that have to be used, usually of the order of 2% by volume, is the Shimadzu sedimentograph. The basic principles employed in this instrument are illustrated in Figure 5.31*b* [98]. The balance pan is hung from one arm of a beam balance. The sedimentation cylinder and the balance assembly are placed in a draught-proof box, but there does not appear to have been any attempt to thermally isolate the sedimentation vessel from the rest of the equipment. The powder settling onto the suspension depresses the arm of the beam balance. When a certain displacement is reached, contacts mounted on the end of the beam complete control circuits. When the control circuits are activated, a small steel ball is released onto the opposing balance pan. At the same time as the ball is released, a mechanical link to a simple pen recorder operates and the addition of the ball to the balance pan is recorded. Steel balls of 135 or 50 mg can be used.

The system used in the Bostock–Gallenkamp sedimentation balance is shown in Figure 5.31*c* [99, 100]. The use of this instrument is described in the British Standards publications concerned with fineparticle characterization [102]. At the beginning of an experiment the pan is clamped to the base of the sedimentation column and the clear fluid placed in the reservoir. The suspension of fineparticles to be characterized is placed in the premixing vessel. It is then introduced into the sedimentation column by closing tap *A* and opening spring clips *B* and *C*. As the suspension flows into the column, the displaced air flows up the exit tube into the premixing vessel. The turbulent condition of the suspension as it flows into the column ensures good dispersion of the powder. As soon as the suspension is in the column, the balance pan is released and swings freely. The suspension falls slightly until the drop in pressure of the air in the closed system above the surface of the suspension is sufficient to balance the hydrostatic pressure of the suspension. The errors due to the balance suspension system passing through the column have been eliminated, and the fineparticle suspension is thermally insulated from the surroundings. As the density of the column changes due to fineparticles settling out, there are movements in fluid around the edges of the balance pan as the pressure system readjusts itself.

In the version of the Bostock balance described by Cohen, the weight deposited has to be noted by an operator from the movement of a pointer attached to the torsion mechanism of the balance [101]. Ames et al. [105] have described a modified version of the commercial equipment in which the weight of particles on the balance pan is recorded continuously.

The Sartorius Company provides equipment for using their analytical balances as sedimentation balances [106]. Leschonski [17] has studied the performance of this equipment and suggested several modifications leading to improved operation of equipment. The system evolved by Leschonski is shown in Figure 5.31*d*. The balance pan is provided with a high rim to minimize suspension movement around the edge of the pan and is hung in a reservoir of clear fluid. At the start of the experiment the balance pan is clamped against the base of the sedimentation column. The fluid below the pan is outgassed through the tube shown. This prevents bubbles forming and attaching themselves to the pan during the course of an experiment. (The buoyant effect of the bubbles could cause error in weight determination and could be the cause of some apparent "lost powder"). The suspension is placed in the sedimentation vessel and stirred. To begin the experiment, the locking mechanism on the balance is released. The clear fluid reservoir is fitted with a pressure-equilization capiliary tube. A slight flow of fluid into the capillary is sufficient to balance the hydrostatic pressure of the column of suspension. As the pressure from the sedimentation column varies, the flow of fluid required to maintain equilibrium at the balance pan is very small. Leschonski reports that very little powder is lost with this design of equipment.

Although the Leschonski system probably represents the best equipment configuration as yet published, there are still two features of this design that appear to be unsatisfactory: (1) the rod linking the balance to the pan disturbs the ideal sedimentation of the particles; and (2) although the sedimentation vessel is protected thermally, the surface of the suspension is open to the atmosphere, making it vulnerable to convection currents arising from surface cooling. A design of sedimentation balance that would appear to overcome these unsatisfactory features is sketched in Figure 5.32.

As can be seen from Figure 5.32, the experimental system is essentially the same as Leschonski's except that now the balance pan is supported on spring leaves. A solenoid operated piston when activated can press the balance pan to the base of the sedimentation column. Stirring is achieved by using a stirrer with a perforated disk, and the handle of the stirrer and the disk are made of thermally nonconducting materials. At the com-

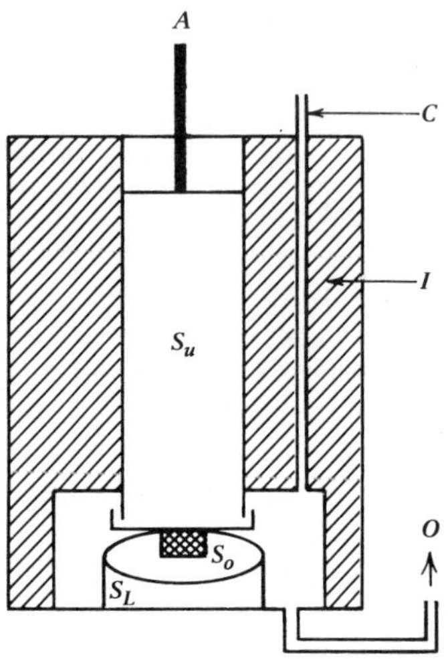

Fig. 5.32 Strain gauge–balance pan design for a sedimentation balance. Symbols: I, thermal insulation; O, out gassing tube; S_L, spring leaves; S_o, solenoid operated piston; S_u, suspension; C, clear fluid; A, stirrer with perforated disk.

mencement of the characterization study the balance pan is pressed against the base of the sedimentation column, which is then filled with suspension. Then the top of the column is sealed and the suspension stirred by vertical movement of the stirrer. At the termination of stirring the stirrer is lodged at the top of the sedimentation column. The solenoid operated piston is then moved to let the balance pan be suspended freely on the spring leaves, as indicated in Figure 5.32. Strain gauges are mounted on the spring leaves so that very small amounts of fineparticles deposited on the balance pan are registered automatically. Strain gauges of sufficient sensitivity are becoming available, so that when placed in a Wheatstone bridge arrangement on the spring leaves, they can measure very small quantities of material without very large physical movement of the system.

Radiometric techniques for evaluating accumulated solids at the base of a sedimentation column have been reported by Connor et al. [103, 104]. Another procedure for evaluating the weight of sediment accumulated at the base of the sedimentation column is to syphon off the suspension above a given depth after a certain time and then filter and weigh the fineparticles at the bottom of the column. Alternatively, chemical methods can be used to assess the concentration of fineparticles at the base of the column. This procedure is obviously time consuming, and it is not easy to syphon off a top layer of suspension without disturbing to some extent below the layer at which the syphon is placed. The method has been mentioned here because it is the standard procedure used in the pesticide industry for assessing the fineness of pesticides [107].

The technique of syphoning off the top layer of suspension is often referred to as *fractional decantation*. Fractional decantation is obviously one method of preparing fractions of powder that has a given range of Stokes diameters. However, the slurry contains not only all fineparticles with Stokes diameter greater than d_{Ht}, but a proportion of smaller fineparticles that started off their journeys lower down the sedimentation column. Thus, to achieve complete fractionation, one would have to resuspend the slurry with fresh fluid and let the fineparticle sediment for the same time before syphoning off at the same depth again to reduce the concentration of smaller fineparticles in the slurry. In theory, this would have to be carried out an infinite number of times to completely free the base slurry of the smaller fineparticles. Heywood [108] has shown that for many purposes good fractionation of fineparticles can be achieved in this method by 11 successive resuspension and decantation stages. Allen [14] has provided tables for predicting the purity of fractionation achieved after a given number of resuspensions. Horsfall and Jowett [109] have described automated equipment for fractional decantation.

In column extraction sedimentation methods the accumulated sediment from an initial homogeneous suspension is removed at periodic intervals. The weight of fineparticles in the fractions taken from the base of the column can be determined by any convenient method. From these weights and a knowledge of the settling heights and times, the Stokes diameter distribution can be calculated by using the equations developed earlier for sedimentation balances. The column extraction characterization procedures have the advantage that the equipment is usually inexpensive to build and operate. An attractive feature of the technique is that in the course of the analysis the prime suspension is split into convenient fractions. These fractions can then be used in subsequent experimental investigations. A major disadvantage is that it is usually necessary to constrict the base of the sedimentation column to facilitate the removal of the accumulated sediment. This constriction disturbs the falling pattern of the suspended fineparticles. It is difficult to estimate the effect of this disturbance on the settling velocity of the fineparticles. Another disadvantage is that fineparticles arriving at the constriction sometimes settle on the sides of the conical section of the tube instead of continuing on to the base of the column. These fineparticles are not always easily removed from the tube.

The column extraction equipment designed by Stairmand is shown in Figure 5.33*a*. The accumulated suspension is flushed from the bottom of the sedimentation column by fluid from the clear fluid reservoir [110]. Koback made extensive studies of the precision and the accuracy of the Stairmand column extraction equipment. He reported that the density

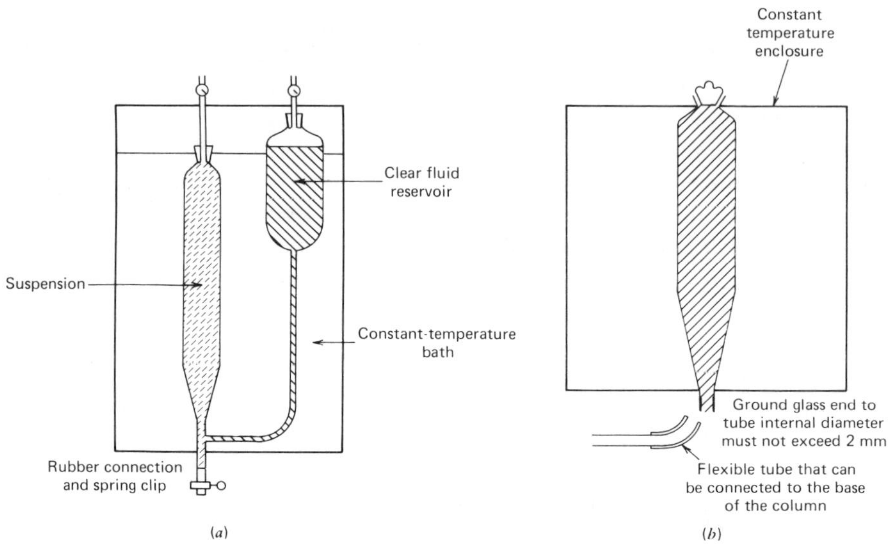

Fig. 5.33 Column extraction instruments: (*a*) suspension extraction equipment described by Stairmand [6]; (*b*) basic system of column extraction equipment of Koback and Loveridge [112].

difference between the accumulated slurry at the base of the column and the clear fluid of the liquid in the side arm leading to the reservoir causes turbulence that leads to the deposition of fineparticles on the base of the side arm tube. It proved difficult to dislodge these fineparticles with the low flow rates of fluid used in the equipment.

As in the case of powder loss during a sedimentation balance characterization study, it is usually assumed that powder loss to the walls of the equipment during a column extraction characterization study is representative of the powder as a whole and can, therefore, be discounted in the characterization of the Stokes diameter distribution. Stairmand recommended the stirring of the suspension by using compressed air to create bubbles in the column. As noted earlier, this stirring technique can introduce buoyancy errors into the sedimentation dynamics because small bubbles become attached to individual fineparticles.

Koback reported that the performance of the Stairmand equipment can be considerably improved by reducing the diameter of the side arm tube and mounting it tangentially so that the fluid entering the base of the sedimentation column comes in at a higher velocity and swirls around the walls flushing the fineparticles into the sump [111]. Koback reported that with this improved design he recovered 99% of the solids placed in the original suspension.

Koback and Loveridge [112] described the column extraction equipment shown in Figure 5.33*b*. The sedimentation column has a ground-glass tapered end. The largest permitted internal diameter for the end of the column is 2 mm. The tube is filled with a well-stirred suspension with the base of the column closed mechanically with a flap valve. When the closure stopper is in place at the top of the column, this valve is removed and the suspension is held in the tube by a combination of surface tension forces and atmospheric pressure. When it is desired to remove fineparticles from the base of the column, a small glass tube filled with clear fluid is placed at the base of the column. As the liquid surfaces in the tube and the base of the column are fused, the settled fineparticles fall into the collection tube. When the tube is removed after an appropriate time interval, the meniscus reforms immediately resealing the column. The process is repeated for any number of samples required. The main advantage of this type of equipment is its simplicity of construction and low cost.

Connor et al. [53] have measured the accumulation of sediment at the base of a sedimentation column by using β-ray adsorption. β-rays from a 1 mCi strontium-90 radioactive source are reflected from the base of the sedimentation column into a scintillation counter or an ionization chamber. As the mass of sediment at the bottom of the tube increases, the amount of radiation back scattered into the counter increases. The equipment has to be calibrated empirically [53].

NOTES AND REFERENCES

1. C. N. Davies, *Symposium on Particle Size Analysis*, The Institution of Chemical Engineers, London, 1947.
2. R. P. Boardman, *The Terminal Velocity of a Particle Falling Through a Viscous Fluid in the Presence of Other Particles*, M.Sc. thesis, London University, England, 1961.
3. B. H. Kaye and R. P. Boardman, "Cluster Formation in Dilute Suspension," *Proceedings of the Symposium on Interaction between Fluids and Particles*; published by Institute of Chemical Engineering, London, 1962, pp. 17–22.
4. S. W. Martin, "The Determination of Sub Sieve Size Distributions by Sedimentation Methods," ASTM, Special Technical Publication No. 51 (1941), p. 66.
5. W. F. Carey and C. J. Stairmand, *Engineering*, 154 (1941), p. 181.
6. C. J. Stairmand, "Some Practical Aspects of Particle Size Analysis in Industry," *Symposium on Particle Size Analysis*, The Institution of Chemical Engineers, London, 1947, p. 58.

7. E. H. Amstein and B. A. Scott, *J. Appl. Chem.*, Suppl. No. 1 (1951), p. 510.
8. P. G. W. Hawksley, "The Physics of Particle Size Measurement," Part 1, Fluid Dynamics and the Stokes Diameter, *Mon. Bull. Br. Coal Util. Res. Assoc.*, 20 (4) (1951).
9. B. A. Jarrett and H. Heywood, "A Comparison of Methods for Particle Size Analysis," *Br. J. Appl. Phys.*, Suppl. No. 3 (1954), p. 21.
10. C. Orr and J. M. Dallavalle, *Fine Particle Measurement*, Macmillan, New York, 1959.
11. R. R. Irani and C. F. Callis, *Particle Size: Measurement, Interpretation and Application*, Wiley, New York, 1963.
12. G. A. H. Elton, "Electroviscosity in Dilute Heterodisperse Suspensions," *Br. J. Appl. Phys.*, Suppl. No. 3 (1954), p. 60.
13. G. D. Parfit, Ed., *Dispersion of Powders in Liquid*, Elsevier Sequoia, Lausanne, Switzerland, 1977.
14. T. Allen, *Particle Size Measurement*, 2nd ed., 1976, Chapman and Hall, London, 1976.
15. B. H. Kaye and C. R. G. Treasure, *The Thermal Stability of Sedimenting Suspensions Used to Measure Particle Size Distributions*, Materials Research and Standards, Vol. 5, No. 11, 1965.
16. R. Davies and B. H. Kaye, "The Effect of Cluster Instability in Suspensions and Particle Size Analysis Measurements with the Cahn Balance," *Particle Size Analysis Conference* 1970, Bradford University, British Society for Analytical Chemistry, London.
17. K. Leschonski, "Vergleichende Untersuchungen der Sedimentations Analyse," *Staub*, 22 (11) (1962), 475–486.
18. H. Heywood, *Symposium on Particle Size Analysis*, Institute for Chemical Engineers, Great Britain, 1947, p. 114.
19. B. H. Kaye and R. Davies, "A Continuous Stem Pipette Equipment for Evaluating Stokes' Diameter of a Fineparticle System," *Proceedings of the International Powder and Bulk Solids Handling and Processing,* Rosemont, Illinois, May 1978; published by Industrial and Scientific Conference Management Inc., 222 West Adams Street, Chicago, Illinois, 60606, pp. 351–352.
20. E. W. Hall, Ph.D. thesis, 1956, University of Birmingham.
21. J. P. Olivier, G. K. Hickin, and C. Orr, Jr., "Rapid Automatic Particle Size Analysis in the Sub Sieve Range," *Powder Technol.*, 4 (1970/71), 257–263.
22. Andreasen pipette equipment is available from Gallenkamp, Portrait Lane, Stockton-on-Tees, Durham County, England.
23. B. E. Worsnop and H. T. Flint, *Advanced Practical Physics for Students*, 9th ed., Methuen Educational Ltd., London, 1969.

24. V. Fidleris and R. C. Whitmore, *Br. J. Appl. Phys.*, 12 (1961), 490–494.

25. B. H. Kaye and R. Davies, *Investigations into the Hindered Settling Method of Particle Size Analysis Symposium on the Interaction between Fluids and Particles*; published by the Institution of Chemical Engineers, London, 1962.

26. K. W. Pearce, "Settling in the Presence of Downward Facing Surfaces," in *Proceedings of the Symposium on the Interaction Between Fluids and Particles*; published by the London Institute of Chemical Engineers, 1962.

27. B. H. Kaye, Ph.D. thesis, "Physical Problems in Particle Size Analysis," London University, 1961.

28. R. Johne, *Einfluss der Konzentration einer monodispersen Suspension auf die Sinkgeschwindigkeit ihrer Teilchen*, Dr.-Ing. thesis, Karlsruhe, Germany, 1965.

29. B. Koglin, "Settling Rate of Individual Particles in Suspension," *Loughborough Conference on Particle Size Analysis, 1970 Proceedings*; published by the British Society for Analytical Chemistry.

30. B. Scarlett and S. K. Cowlam, "Concentration Effects in Particle Size Analysis by Sedimentation," *Conference on Particle Size Analysis, September 1977, Salford, England*; proceedings published by the British Society for Analytical Chemistry.

31. S. K. Cowlam, "Particle Interaction in Dilute Slowly Sedimenting Systems," Ph.D. thesis, Loughborough University, England, 1976.

32. B. H. Kaye, M. R. Jackson, and R. Kahrun, "The Stability of Line Start Suspension Systems in the Centrifugal Disc Photosedimentometer," *Powder Technol.*, 2 (1968/69), 290–300.

33. B. H. Kaye and M. R. Jackson, Size Analysis of Latex Emulsions Using a "Centrifugal Disc Photosedimentometer," *Powder Technol.*, 1 (1967), 81–88.

34. J. Happel and H. Brenner, *Low Reynolds Number Hydrodynamics with Special Application to Particulate Media*, Prentice-Hall, Englewood Cliffs, New Jersey, 1965.

35. J. S. McNown and P. N. Lin, *Proceedings of the Second Midwestern Conference on Fluid Mechanics*, English reprint No. 109, Iowa State University, Ames, Iowa, 1951.

36. D. Werner, "A Simple Method of Obtaining the Size Distribution of Particles in Soils and Precipitates," *1925 Transactions of the Faraday Society*, Vol. 21, p. 381. Commercial information is available from Palo-Meyers, New York.

37. See commercial literature of the Cahn Division of Ventron Instruments Corporation, 7500 Jefferson Street, Paramount, California 90723.

38. K. T. Whitby, "A Rapid General Purpose Centrifuge Sedimentation Method for Measurement of Size Distribution of Small Particles," *Heat., Piping Air Cond.*, Part 1 (January 1955), 231–237; Part 2 (June 1955), 139.

39. Technical Literature on Micromerograph, from Franklin Electronics, Inc., East Fourth Street, Bridgeport, Mont. W., Pennsylvania.

40. F. S. Eadie and R. E. Payne, "New Instrument for Analysing Particle Size Distribution," *Iron Age*, 174 (1954), 99; *Br. Chem. Eng.*, 1 (1956), 306.

41. J. M. Dallavalle, *Micromeritics, the Technology of Fineparticles*, Pitman, New York, 1948.

42. S. M. Kaye, D. E. Middlebrooks, and G. Weingarten, *Evaluation of the Sharples Micromerograph for Particle Size Distribution Analysis*, Technical Report, FRL-TR-54, Picatinny Arsenal, Dover, New Jersey, February 1962.

43. G. L. Duchesne, *Cardre Technical Note 1622*, Canadian Armament Research & Development Establishment, Valcartier, Quebec, December 1964.

44. W. J. Ullrich [Metals Disintegrating Corporation, Box 290, Elizabeth B., N. J.], "Rapid Particle Size Analysis of Metal Powders with an Electronic Device," preprint No. 6, *International Powder Metallurgy Conference*, New York, June 14–17, 1965.

45. K. T. Whitby, A. B. Algren, and J. C. Annis, *1958 Special Publication of ASTM*, No. 234, p. 117.

46. Mines Safety Appliance Company, 201 Braddick Avenue, Pittsburgh 8, Pennsylvania.

47. CILAS, marketed by "CBL," Compagnie Belge des Lasers SA/Belgische Vennootschap voor Lasers NV, Meersstraat, 130-B 9000 GENT, Belgium.

48. Leeds & Northrup Company, MD 353, North Wales, Pennsylvania 19454.

49. For a discussion of Cauchy's Theorem, see M. J. Kendall and P. A. P. Moran, *Geometrical Probability*, Griffin's Statistical Monographs No. 10, 1963, Hafner, New York.

50. J. Bodziony, "Determination of the Surface Area of the Convex Solid Bodies by Means of Measuring the Surface Area of Their Shadows," *Archiwun Gornictwa*, 20 (4) (1975). In this paper Dr. Bodziony discusses the application of Cauchy theory to the average value of the projected shadow. He also uses the principle of antithetic variates to increase the confidence in a measured value of the shadow.

51. T. Allen, 1967, Ph.D. thesis, University of Bradford, England.

52. For a general discussion of all the various theories of light scattering, see the book by H. C. Van de Hulst, *Light Scattering by Small Particles*, Wiley, New York, 1957.

53. P. Connor, W. H. Hardwick, and B. J. Laundy, *J. Appl. Chem.*, 9 (1959), 525.

54. H. E. Rose, *The Measurement of Particle Size in Very Fine Powders*, Constable and Company, London, 1953.

55. L. A. Wagner, "A Rapid Method for the Determination of the Specific Surface of Portland Cement," *Proc. ASTM*, 33, Part 11 (1933).

56. K. Leschonski, K. L. Metzger, and U. Schindler, "A New On-Line Parti-

cle Size Analyzer," *Salford Conference on Particle Size Analysis*, September 1977, Proceedings, M. J. Groves, Ed., Heyden and Son, London, 1978.

57. B. H. Kaye and T. Allen, "Optical Scattering Cross-section of Small Particles and the Design of Photosedimentometers," *Analyst*, 90 (1965), 147.
58. Inquiries as to the commercial availability of the Wagner Turbidimeter should be directed to the Lapine Scientific Company, Chicago and New York.
59. H. R. Harner and J. R. Musgrave, "Particle Size Measurements," ASTM Special Technical Publication No. 234 (1958), 172–179.
60. Commercial information from Evans Electroselenium Limited, Halstead, Essex, England.
61. V. T. Morgan, *Iron & Steel Institute Special Report No. 58*, 1954, p. 81.
62. V. T. Morgan, Contribution to the symposium discussion in the Physics of Particle Size Analysis Supplement No. 3, *Br. J. Appl. Phys.* (1954), S207, S208.
63. Goring Kerr Ltd., Station Road, Gerard's Cross, Buckinghamshire, England.
64. A. Muta, "A New Particle Size Analyzer of the Photo Scanning Type," *Proceedings of the Conference on Particle Size Analysis*, Loughborough, 1966; published by the British Society of Analytical Chemistry, London, pp. 212–221. This instrument was made available commercially by the Hitachi Company of Tokyo, Japan.
65. Commercial equipment using wide-angle optics available from Microscal Ltd., 20 Mattock Lane, Ealing, London W5.
66. Commercial equipment using wide-angle optics available from Ladal Scientific Equipment Ltd., Warlings, Warley Edge, Warley, Halifax, Yorkshire HX2 7RL, England.
67. J. Brezina, "Particle Size Analysis," *Soc. Anal. Chem.* (1970), 225–266.
68. J. Schlee, "A Modified Woods Hole Rapid Sediment Analyzer," *J. Sed. Petrol.*, 36 (2) (1966), 403–413. Commercial information on the system available from Benthos, Inc., Edgerton Drive, Falmouth, Massachusetts.
69. For additional information on Werner Sedimentation System, see P. M. Travis, "Measurement of Average Particle Size by Sedimentation and Other Physical Means," *ASTM Bull.* 29 (102) (1940).
70. B. H. Kaye, "Automatic Decision Taking in Fineparticle Science," *Powder Technol.* 8 (1973), 293–306.
71. R. Davies, private communication.
72. A. G. Naylor, private communication.
73. C. Harwood, R. Davies, and J. Stockham, "Review of Optical Devices for Measurement of Mixture Composition," in *Proceedings of Powder Technology*, Harrogate, May 1971; published by the Powder Advisory Centre, London, 1971.

74. Commercial information on this photosedimentometer, Sedigraph 5500L, available from Micromeritics Corporation, 5680 Goshen Springs Road, Norcross, Georgia 30093.

75. B. H. Kaye, "Feedback Loops for Fineparticle Systems: Past, Present and Future," in *Proceedings of the In Stream Measurements of Particulate Solid Properties*, Bergen, Norway, 1978; published by the Chris Michelson Institute, Bergen, Norway.

76. K. F. Brown and J. N. Skrebowski, "Contribution to the discussion at the Symposium on Particle Size Analysis," *Br. J. Appl. Phys.*, Suppl. No. 3 (1954), S27.

77. Commercial information on the Sedigraph 5000 is available from the Micromeritics Corporation; see reference 74.

78. W. B. Hendrix and C. Orr, *Automatic Sedimentation Size Analysis Instruments Particle Size Analysis Conference*, Bradford, England, 1970; published by the Society for Analytical Chemistry, 9–10 Saville Row, London, W1.

79. T. Allen and M. G. Baudet, "The limits of Gravitational Sedimentation," *Powder Technol.*, 18 (2) (1977), 131–138.

80. Commercial information on LADAL x-ray sedimentometer available from the company listed in reference 66.

81. J. H. Talbot and D. J. Jacobs, "An Onstream Particle Size Analyzer for a Mill Pulp," *Automatic Sedimentation Size Analysis Instruments Particle Size Analysis Conference, Bradford, England, 1970*, Society for Analytical Chemistry, 9–10 Saville Row, London, W1.

82. I. Chabay, "Rapid Measurement of Droplet Size Distributions by Optical Heterodyne Spectroscopy Symposium on Aerosol Measurements," *NBS Special Publication 412*, p. 65, obtainable from the U.S. Department of Commerce.

83. C. P. Wang, "Laser Anemometry," *Am. Sci.*, 65 (3) (May/June 1977), 289–293.

84. J. I. T. Stenhouse, B. Scarlett, and I. Sinclair, "Laser Doppler Techniques in Aerosol Size Analysis," paper presented at the Salford Conference on Fineparticle Characterization, England, September 1977. Proceedings to be published by the British Society for Analytical Chemistry, London.

85. C. P. Wang, "Application of Lasers in Atmospheric Probing," *Acta Astronaut.*, 1 (1974), 105–123.

86. J. M. Bernard and S. S. Penner, "Determination of Particle Sizes in Flames from Scattered Laser Power Spectra," paper presented at the AIAA 14th Aerospace Sciences Meeting, January 1976, Washington, D.C. Information from Dr. J. M. Bernard, Director of the Energy Center, University of California, San Diego, La Jolla, California.

87. B. H. Kaye and G. W. James, "Investigations into the Diver Method of Particle Size Analysis," *Br. J. Appl. Phys.*, 13 (August 1962), 415–419.

88. C. F. Ross, *Anal. Chem.*, 31 (1958), 337.

89. See, for example, British Standard 1377, 1948, *Methods of Test for Soil Classification and Compaction.* (This standard was revised in 1961.)

90. ASTM, *Procedure for Soil Testing*, Part IV, 1961, p. 1272.

91. S. Berg, "Determination of Particle Size Distribution by Examining Gravitational and Centrifugal Sedimentation According to the Pipette Method and with Divers," *ASTM*, Special Technical Publication No. 234, 1958, pp. 143–171.

92. R. P. Cargille Laboratories, Inc., Cedar Grove, New Jersey 07009; telephone (201) 239-6633.

93. R. Johnson, 1955, *Transact. Br. Ceramics Soc.*, 55 (1955), 237.

94. W. M. Dotts, *Ind. Eng. Chem.*, 18 (1946), 326. Commercial information is available from Fisher Scientific Corporation.

95. S. C. Kim, G. Schlotzer, and E. C. Palik, "A Method of Calculating Particle Size Distribution from Sedimentation Balance Data," *Powder Technol.*, 1 (1967), 54–57.

96. J. G. Rabatin and R. H. Gale, "Determination of Particle Size with a Simple Simple Recording Balance," *Anal. Chem.*, 28 (1956), 1314.

97. E. S. Palik, "Particle Size Measurements with an Improved Continuously Recording Sedimentation Apparatus," *Anal. Chem.*, 33 (1961), 956.

98. E. Suito and M. Arakawa, *Bull. Inst. Chem. Res.* (*Kyoto Univ., Jap.*) 23 (7) (1950), 17. Technical information is available from foreign trade department of Shimadzu Seisakusho Ltd., Kanda-Mitoshiro-cho, Chiyoda-ku, Tokyo, Japan.

99. W. Bostock, "A Sedimentation Balance for Particle Size Analysis in the Subsieve Range," *J. Sci. Instrum.*, 29 (1952), 209.

100. A. Gallenkamp Company, London. U.S. agents: A. S. La Pine and Company, 6001 South Knox Avenue, Chicago, Illinois.

101. L. Cohen, "A Sedimentation Balance for Particle Size Analysis," *Instrum. Pract.* (October 1959).

102. British Standard 3406, Part 2, 1962.

103. P. Connor, W. H. Hardwick, and B. J. Laundy, *J. Appl. Chem.*, 8 (1958), 716.

104. P. Connor and W. H. Hardwick, *Ind. Chem.* (September 1960).

105. D. P. Ames, R. R. Irani, and C. F. Callis, "The Agreement between Independent Methods for Particle Size Distribution Measurements on Finely Divided Powders including Phosphates," *J. Phys. Chem.*, 63 (1959), 531.

106. Manufactured by Sartorius Werke, AG, Gottingen, Germany. U.S. agents: C. A. Brinkman and Company, 115 Cutter Mill Road, Great Neck, New York.

107. Collaborative Pesticides Analysis Committee of Europe, recommended methods of analysis for pesticides, No. 11, "Determination of Suspensibil-

ity of Water Dispersible Powders.'' See also classification of methods for determining particle size, a review prepared by a committee of the Society for Analytical Chemistry, *The Analyst*, 88 (1044) (March 1963), 156–187.

108. H. Heywood, ''Recent Developments in Mineral Dressing,'' *Inst. Min. Met.* (1952), 31.
109. F. Horsfall and A. Jowett, *J. Sci. Instrum.*, 37 (4) (1960), 120.
110. C. J. Stairmand, ''A New Sedimentation Apparatus for Particle Size Analysis in the Sub-Sieve Range,'' *Symposium on Particle Size Analysis*, London Institute of Chemical Engineers and Society of Chemical Industry, 1947, vol. 25, p. 128.
111. J. Koback, ''Modification to Sedimentation Apparatus,'' *J. Sci. Instrum.*, 32 (1955), 153.
112. J. Koback and D. J. Loveridge, ''Single Tube Sedimentation Apparatus for the Measurement of Particle Size Distribution,'' *J. Sci. Instrum.*, 37 (1960), 266.

CHAPTER

6

CHARACTERIZATION OF FINEPARTICLES BY CENTRIFUGAL SEDIMENTATION

6.1 INTRODUCTION

A laboratory centrifuge in which test tubes are hung at the end of radial arms able to spin about a central axis is termed a *tube centrifuge*. When the centrifuge is switched on, these tubes assume a horizontal position as they rotate rapidly. The earliest attempts to characterize the Stokes diameters of fineparticles by centrifugal sedimentation used tube centrifuges. In a characterization study, usually the centrifuge has to be taken up to speed, run for a set time at top speed, and then stopped to enable determination of the movement of fineparticles in the tube. The starting and stopping of the centrifuge results in a varied centrifugal acceleration experienced by the fineparticles during the time that the centrifuge is operating. Although one can calculate the force at any given time at any given position in the centrifuge and integrate the overall effect on the trajectory of the fineparticle, this is not a sufficient correction for the fact that during the acceleration and deceleration the fluid in the tube is made to spin by the force system exerted in the centrifuge. This fluid spinning can be demonstrated by viewing with a stroboscopic light an open centrifuge equipped with transparent tubes. With the stroboscope adjusted to the speed of the centrifuge, one can note the time at which the fineparticles are all completely sedimented onto the base of the tube. If the centrifuge is switched off at this time, the swirling currents caused by the deceleration of the centrifuge will resuspend fineparticles lodged on the bottom of the tube.

The second disadvantage of a tube centrifuge is that the fineparticles follow radial trajectories. Therefore, sedimenting fineparticles impinge on the side walls of the centrifuge tube and either adhere to the sides or initiate currents down the sides and up the middle of the tube as accumulated fineparticles on the walls slip down into the base of the centrifuge tube.

Tube centrifuge characterization systems have been used for quality control of fineparticle systems; however, any variation in the experimental technique usually leads to shifts in the measured Stokes diameter distributions.

To avoid the problems associated with tube centrifuges, many investigators have used shallow bowl, or disk centrifuges, in which the fluid supporting the fineparticles is centrifuged in a hollow disk rotating about its major axis. Problems associated with radial wall impingement are eliminated by such centrifuge design. To avoid the problems associated with the stopping and starting of the centrifuge, suspension to be characterized can be injected into the centrifuge already running at a steady speed. The many experimental difficulties associated with such injection procedures are a major area of controversy in the methodology of centrifugal sedimentation procedures.

All the cautionary statements with regard to concentration effects and fluid fineparticle dynamic interaction discussed in Section 5.4 apply to fineparticle dynamics in centrifugal fields. Unless otherwise stated, it is assumed that flow conditions in the characterization system are laminar.

6.2 FINEPARTICLE MOVEMENT IN A CENTRIFUGAL FIELD

The instantaneous velocity of a fineparticle moving along the radius of the centrifuge is given by the relationship

$$\frac{dx}{dt} = V_c = \frac{(\rho_P - \rho_L)}{18\eta} \cdot D_s^2 \cdot \omega^2 x \tag{6.1}$$

where V_c = velocity of fineparticle along radius of centrifuge
x = distance of fineparticle from axis of revolution
D_S = Stokes diameter of fineparticle
ω = angular velocity of centrifuge (radians per second)
η = viscosity of fluid

Integration of equation 6.1 for a given array of fineparticles in a given centrifuge forms the basis of the data transformation in Stokes diameter characterization studies using centrifugal forces. Comparing this to the velocity of the fineparticle sedimenting under gravity, we have

$$V_c = \frac{\omega^2 x}{g} \cdot V_g \tag{6.2}$$

where V_g is the terminal velocity under gravitational forces in the same fluid. The dependence of the instantaneous velocity of a fineparticle sedimenting under centrifugal force on its distance from the center of rotation makes the interpretation of the dynamics of an initially homogeneous suspension complex. To simplify data interpretation in centrifugal sedimentation studies, long-arm centrifuges have been developed so that the total movement of the fineparticles from x_1 to x_2 the initial and final position of the fineparticle is such that the difference between $\omega^2 x_1$ and $\omega^2 x_2$ can be neglected. Long-arm centrifuges have not gained wide acceptance, probably because of the mechanical problems of building stable centrifuges of large dimensions within acceptable cost limitations.

6.3 LINE START SEDIMENTATION PROCEDURES USING CUMULATIVE MONITORING OF ACCUMULATED SEDIMENT

In the line start procedure since all the fineparticles start at essentially the same distance from the center of rotation of the centrifuge, the location of the fineparticles at any time t after the commencement of centrifugation can be calculated by using equation 6.1. Let all the fineparticles start at a distance H_1. By integrating equation 6.1 over the distance $H_1 \rightarrow H_2$, it can be shown that all fineparticles greater than

$$D_{Ht} = \sqrt{\frac{18\eta \ln H_2/H_1}{(\rho_P - \rho_L)\omega^2 t}} \tag{6.3}$$

will have reached the position H_2 at time t.

When H_2 represents the base of the centrifuge, then D_{Ht} is the Stokes diameter of the fineparticles that will just have reached the base by time t. Let W_t be the weight deposited on the base at time t; then

$$R_{H_2t} = \frac{W_t}{W_T} \times 100 \tag{6.4}$$

where W_T is the total weight of powder in suspension and R_{H_2t} is the percentage of fineparticles greater than D_{Ht}. By studying the rate of accumulation of sediment at the base of the centrifuge, equations 6.3 and 6.4 can be used to calculate the Stokes diameter distribution of the sedimenting fineparticles.

The first line start sedimentation procedure based on centrifugal sedimentation was described by Marshall [2], who was aware of the problems involved in floating a layer of suspension onto a column of clear fluid because of reported difficulties with the Werner sedimentation tube (see Chapter 5) and attempted to achieve a stable suspension system by using a suspension of lower density than the liquid column through which the fineparticles were to be sedimented. He used a sucrose solution for the column fluid and a water-based suspension of fineparticles. This strategy only eliminates gross inversion of the floating layer of suspension. As demonstrated by the data of Davies and Kaye discussed in Chapter 5, the use of different fluids in the suspension layer and the sedimentation column can result in a general coarsening of the measured Stokes diameter distribution. At the end of a specified centrifugation, Marshall decanted the suspension from above the sediment deposited at the base of the tube and weighed the deposited fineparticles. This is time consuming, and the method has not been widely used in recent times. However, the equipment costs are low and the procedure is useful when the work load demands are low and accuracy demands are not high.

The MSA centrifuge procedure, based on the work of Whitby and coworkers [3], is a centrifugal adaptation of the Werner sedimentation equipment described in the previous chapter. The dimensions of the sedimentation tubes used are illustrated in Figure 6.1. In an attempt to stabilize the line start procedure, Whitby used Marshall's technique of using a lower-density fluid to prepare the suspension to be floated onto the column of clear fluid. In the MSA commercial equipment a special mesh-bottomed filling chamber is used to help prevent immediate penetration of the layer suspension into the column of fluid. The mass of fineparticles accumulating at the base of the centrifuge is taken to be directly proportional to the height of the sediment in the narrow constriction at the base of the tube. The constricted base of the tube interferes with the sedimentation of the fineparticles, and it is difficult to calculate the height of free

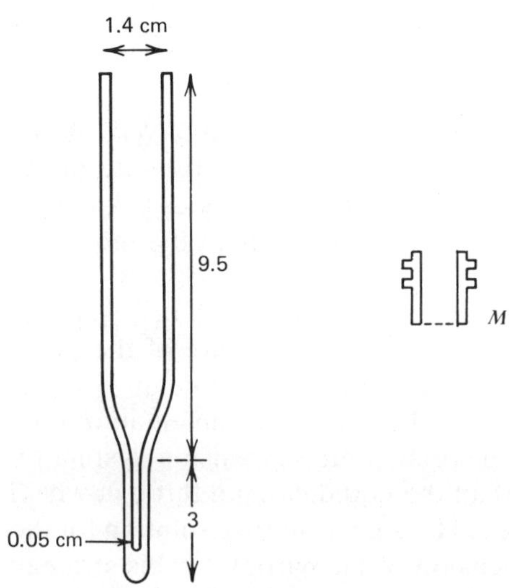

Fig. 6.1 The MSA tube used in centrifugal sedimentation studies of fineparticles (*M*, mesh-bottomed filling chamber).

fall through which a fineparticle moves under the influence of centrifugal force.

The MSA equipment suffers from the basic problems of the Werner sedimentation procedure as discussed in Chapter 5 and also from problems arising from collapse of sediment in the base of the tube.

Thus in a characterization procedure the suspension under test is floated onto the top of the clear column and allowed to settle under gravity until all fineparticles greater than 8 μm Stokes diameter have reached the base of the column. The tube is then placed in a centrifuge and subjected to centrifugal forces for a given period. The MSA equipment procedures manual gives formulas for allowing for the acceleration and deceleration stages of the centrifugal action. To evaluate the sediment height, the centrifuge is stopped and the tube inspected. No attempt is made in the recommended data transformation procedures to allow for the fact that the swirling action initiated by the deceleration of the centrifuge can cause resuspension of fineparticles precipitated onto the walls because of the radial trajectories of the fineparticles. To achieve the characterization of a wide-range fineparticle system, it is necessary to repeat the centrifuging of the column for various periods of time or at various centrifuge speeds. Davies has reported that the use of various speeds to achieve rapid characterization of very small fineparticles can result in discontinuities in the measured distribution curve [4]. Davies attributes these discontinuities to the fact that centrifugation at the higher speeds can cause both a consolidation of the accumulated sediment and a slumping of sediment collected on the conical portion of the sedimentation column that can maintain its position at low centrifuge speeds but that is sloughed off at the higher speeds into the narrow base section. Davies has pointed out that it is very difficult to allow for these effects from theoretical considerations and that when using the MSA centrifuge procedure to characterize the powder, one should rigorously standardize time and speeds centrifugation used in characterizing the fineparticles. In certain situations size distributions measured with the MSA centrifuge can agree quite well with data from other methods probably because of compensating errors. Thus clustering dynamics will coarsen the measured Stokes diameter, whereas the swirling action from deceleration will give an apparent increase in the fineness of the fineparticle system. There is no doubt that the MSA centrifuge has proved to be a useful workhorse for quality control in many industries; however, its use in the characterization of powders that are subsequently used in fundamental studies into such subjects as dust hazards in industrial environments is open to many questions unless performance behavior is empirically correlated with other more fundamentally accurate size information.

6.4 CENTRIFUGAL LINE START SYSTEMS WITH INCREMENTAL MONITORING OF THE FINEPARTICLES

Many systems based on a line start suspension system with incremental monitoring of the sedimenting fineparticles by use of a light beam have been described. All the problems associated with the extinction coefficient, the solid angle subtended at the photoelectric detector, and the wavelength of the scattered radiation discussed in connection with gravity photosedimentometers are encountered in centrifugal photosedimentation techniques. Again, as in the case of gravity photosedimentometers, there are two approaches to the problem of interpreting the optical signals in terms of fineparticle concentrations. One relies on empirical calibration of the transmission signal in terms of true concentrations of fineparticles, and the other attempts to calculate the scattering power of the fineparticles from the light-scattering theory. In the case of centrifugal sedimentation, many of the fineparticles are of the order of the wavelength of light or less and the theoretical calculations become very difficult. In particular, as one considers fineparticles down to sizes smaller than 0.1 μm Stokes diameter, the scattering power of the fineparticles is very small. This factor tends to create a lower size limit below which the characterization procedure is not useful.

The basic system of a centrifugal disk photosedimentometer is illustrated in Figure 6.2. Movement of fineparticles in the spinning disk is monitored by the light beam as shown. In a line start procedure the disk is run up to speed partially filled with clear fluid termed the *spin fluid*. Under centrifugal forces the free surface of the fluid becomes a cylindrical surface coaxial with the axis of the rotation. The suspension to be characterized is injected into the centrifuge through the axial entry port shown in Figure 6.2. The cylindrical free surface of the spin fluid is moving with high energy so that the injected suspension seldom penetrates the surface before it takes up the speed of the centrifuge. In this short time period it spreads out into a thin-layer configuration, as illustrated in Figure 6.2. Because the volume of any particular monitoring zone is larger than a comparable width of suspension at the line start position, the dilution of the fineparticles as they move out from the center has to be accounted for in the transformation of the experimental data if more than one interrogation zone is used during a characterization study. Also, the effect of the width of the beam of light used to monitor the movement of the fineparticles has to be taken into account.

Treasure [5] has studied the effect of the finite width of the monitoring beam. He has shown that the range of fineparticles present in the beam can be taken to be a linear function of D_m, the mean size of the fineparti-

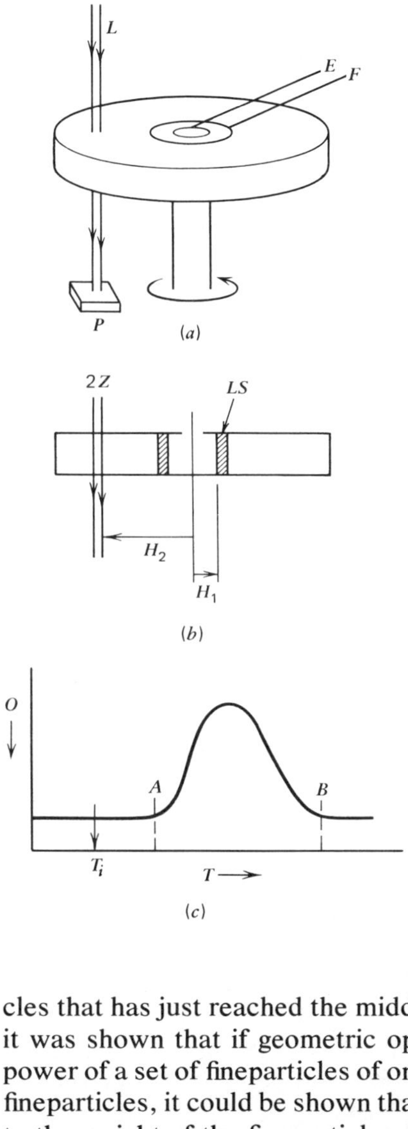

Fig. 6.2 Basic system used in photocentridisk when following a line start procedure: (*a*) perspective view of photocentridisk; (*b*) line start system; (*c*) typical signal trace. Symbols: *E*, axial entry port; *F*, free surface of fluid; *L*, light beam; *P*, photoelectric cell; *B*, smallest fineparticle leaving the beam; *LS*, layer of suspension; *Ti*, injection time; *T*, time; *O*, optical transmission; *A*, largest fineparticle entering the beam.

cles that has just reached the middle of the monitoring zone. In Chapter 5 it was shown that if geometric optics were used to relate the obscuring power of a set of fineparticles of one size in a beam of light to the weight of fineparticles, it could be shown that the scattering power was proportional to the weight of the fineparticles divided by the diameter of the fineparticles. It was pointed out that because of this fact, a widely used data transformation in photosedimentation techniques that made a first-order allowance for the scattering power of fineparticles of different sizes was to multiply the logarithm of the transmittance (i.e., $\log I_t/I_o$), caused by a set of fineparticles, by D the diameter of those fineparticles to gain an estimate of the weight of the fineparticles in the beam. In the analogous centrifugal procedure since the range of fineparticles present in a beam being used to monitor the line start system is proportional to D_m, the logarithm of the measured transmittance is directly proportional to the mass of the fineparticles, provided that the scattering power is a simple function of K/D, where K is a constant and D is the Stokes diameter of the fineparticle. Experimental data to be presented later in this section would indicate that for many fineparticles down to 0.2 μm, using white light, this relationship is reasonable. In general, it can be stated that most of the experimental work reported in the literature on the performance of centrifugal disk photosedimentometers using white light monitors and a line start strategy has demonstrated that for such equipment, it is often reasonable to take the log of the transmittance of a finite-width monitoring beam as being directly proportional to the weight of the fineparticles in the beam.

Line start–incremental monitoring system centrifugal disk photosedimentometers yield direct information on the largest and smallest fineparticle present in a system. They can be operated in the fingerprint mode; that is, the pen recorder traces for good and bad powders in a quality control situation can be established and compared directly to the traces generated by powder samples taken from production runs. The name "centrifugal disk photosedimentometer" is awkward in frequent use. It is not easy to create an adequate short name for the instrument. Some writers have used the term "photofuge"; this is not a good name since its root words imply that it is something that flees from light. One procedure for shortening the name is to describe the basic disk centrifuge as a *centridisk* and the centrifugal photosedimentometer as a *photocentridisk*. This name has the advantage of drawing attention to the physical operating principles of the instrument.

The first photocentridisk was described by Kaye and co-workers [6–8]. In their equipment a hollow lucite disk was mounted inside a horizontal flywheel mounted on the axis of an ordinary laboratory centrifuge. The equipment was placed in a large wooden enclosure, both as a safety precaution and also to increase the thermal stability of the immediate environment of the centridisk. Suspension could be injected through an axial entry port in the top surface of the disk.

In the original photocentridisk design studies undertaken by Kaye and co-workers, some consideration was given to the possibility of using a servomechanism to achieve stability in the rotation of the disk. However, it was felt that the approach of adding mass to the system in the form of a flywheel not only increased the stability of the system, but avoided problems that could arise from the operation of the servomechanism, thus causing transient speed changes as the spinning disk responded to external controls. In all of the instruments with which Kaye has been directly associated, this approach of using high-mass systems to achieve stability in the equipment has been a basic feature of the instrument. It is possible that some of the discrepancies between the performance of instruments of high mass versus those designed by other workers using feedback control to achieve stability may be due partly to this different approach to the problem of operating at a constant speed. Moreover, by operating with a high-mass centrifuge system, any disturbances caused by the injection of small amounts of suspension are minimized. A massive flywheel also acts as a thermal ballast to improve the thermal stability of the equipment. The movement of fineparticles in the lucite disk was monitored by a beam of white light passing through slots cut in the base of the flywheel.

A more sophisticated photocentridisk based on the physical principles used in the original design evolved by Kaye and co-workers was constructed at the Atomic Weapons Research Establishment, Great Britain. It is known as the *AWRE photocentridisk,* or simply as the *AWRE centrifuge*. This instrument has been used extensively by Burt [9], who has published excellent experimental data on the performance of the instrument. The centrifugal chamber used in the AWRE centrifuge is a composite disk made of glass and metal. The transparent portion of the tank for the light beam consists of arcs cut in the surfaces of the tank and sealed with optical flat glass strips. Entry to the tank is through an axial entry port. The lower axis of the centrifuge has a flywheel to enhance the stability of the rotating system and to minimize the effect of added mass when suspension is injected into the tank. The light beam used to monitor the movement of fineparticles is interrupted by a mechanical chopper that alternately diverts the beam of white light through the sedimentation tank or through an optical wedge filter.

The opacity of the wedge filter is graded so that linear angular rotation of the wedge corresponds to linear increase in the log of the transmittance of the opacity of a suspension being monitored. If the opacity of the suspension in the disk centrifuge differs from that of the optical wedge, an alternating signal is generated at the monitoring photocell, thus initiating rotation of the optical wedge until its opacity matches that of the tank. By using this type of servomechanism, all problems due to the instability of the light source and potential drift in the performance of the photoelectric devices are avoided.

The use of the optical wedge filter has the further advantage that since its angular motion is directly proportional to the log of the transmittance movement, the wedge can be used to drive a pen recorder so that linear motion of the pen is proportional to the weight of fineparticles in the beam.

In Figure 6.3 some of the data obtained by Burt using the line start procedure with the AWRE photocentridisk are summarized. In all experimental work Burt used volume concentrations in the line start suspension below that required to achieve stable line start systems, and this concentration limit was established by a series of experiments at various concentrations. In Figure 6.3*a* data obtained by Burt with the AWRE centrifuge for a titanium dioxide powder is compared with data obtained by V. T. Crowl of the Research Association of British Paint Colour and Varnish Manufacturers by direct measurement of individual fineparticles imaged with the electron microscope.

The data in Figure 6.3*b* show that the use of white light in a centrifuge disk photosedimentometer system for a system of spherical particles with the data transformation used in this system agrees very closely with the size distribution function of the spheres as measured by microscopic examination of the fineparticles. The techniques used by Burt to obtain the data in Figure 6.3 were as follows. The centrifugal study was carried out by sedimenting the spheres through a 46:54 glycerol–water mixture at 750 rpm; the microscopic evaluation was carried out on photomicrographs of the powder taken with a Zeiss photomicroscope, and a thousand fineparticles were measured and the number frequency converted to a weight frequency. The data in Figure 6.3*b* demonstrate the rather interesting fact that the centrifugal system can be used for relatively coarse particles by using a viscous fluid in the centrifuge. Although the fineparticles of this size can be quite adequately measured in gravimetric procedures, the major advantage of using a system such as that in the AWRE centrifugal disk photosedimentometer is that once the fineparticles have passed the monitoring zone, there is no need to stop the centrifuge to clean it. One can immediately inject the next sample of suspension to be characterized. In fact, by using a line start procedure that uses the same fluid in the suspension as in the body of the centrifuge, one can carry out many tests in sequence, and the equipment can easily be automated to study a long sequence of samples overnight without monitoring. (Note, however, that each time a sample of suspension is injected into the centrifuge the height of fall for the next suspension is increased by the width of the line start for the previous sample. This adjustment can be readily incorporated into automated data transformation systems.) In Figure 6.3*c* the Stokes diameter distribution of a close-range spherical polyvinyl toluene latex fineparticle suspension as measured by the AWRE centrifuge is compared to the distribution evaluated by optical microscopic examination.

In the early 1960s the Coulter Electronics Company considered the possibility of developing the AWRE photocentridisk on a commercial basis. They produced several prototypes that were described as Coulter

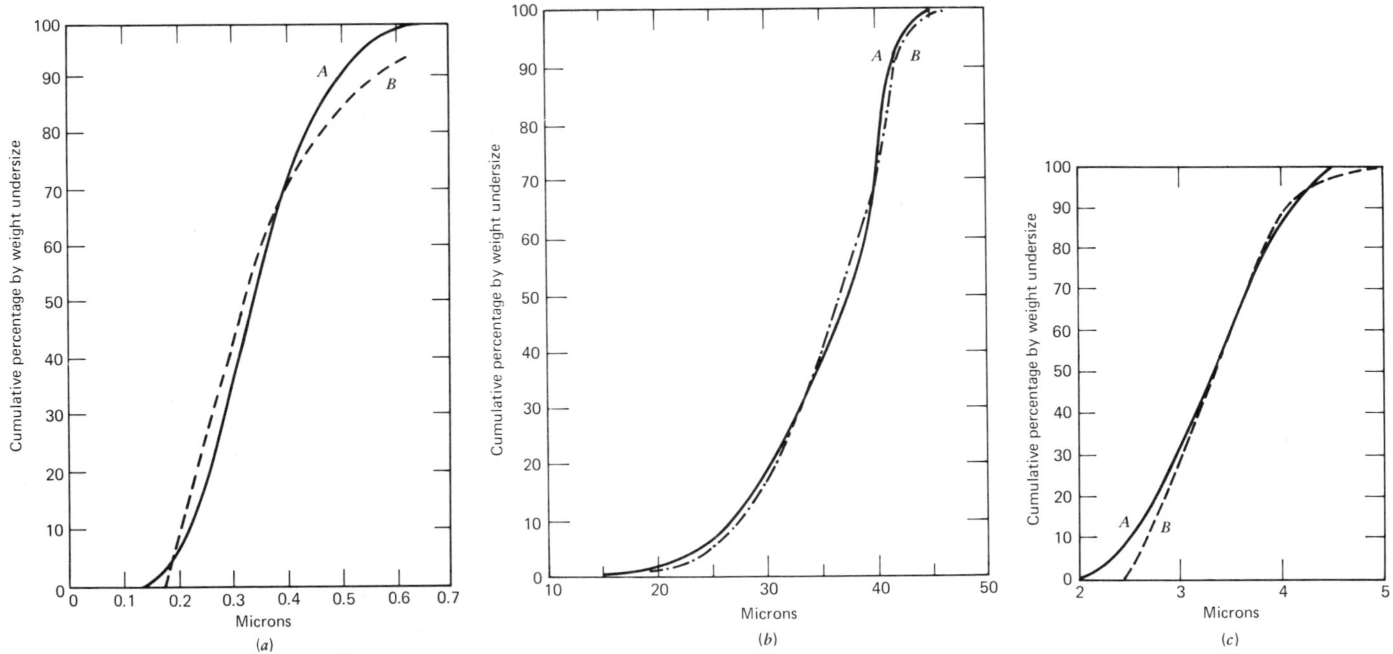

Fig. 6.3 Stokes diameter size distribution data obtained by using a photocentridisk compared with data generated by other methods [9]: (*a*) titanium dioxide powder (*A*, Stokes diameter; *B*, area equivalent diameter from electron micrographs); (*b*) glass beads (*A*, Stokes diameter; *B*, cross-sectional diameter); (*c*) polyvinyltoluene latex (*A*, Stokes diameter; *B*, optical microscope).

centrifuges or as Photofuges®. These instruments retained the use of the servo controlled optical wedge filter, and several investigations describing data obtained with photofuges were published [10–12].

Scarlett and co-workers reported difficulty in obtaining stable line start systems with the Photofuge and introduced a technique that they termed the *three-layer system*. The name is somewhat misleading in that the spin fluid constitutes one of the layers and only two layers are floated onto the vortex of the main fluid. In this technique a low-density fluid is used to prepare the suspension of the fineparticles to be characterized, and a middle layer of fluid of intermediate density is interposed between the spin fluid and the injected suspension. Kaye and co-workers, working in the United States with much more massive centrifuges, did not find it necessary to utilize this system since they were always able to achieve stable line start systems, using fluid identical with the spin fluid to prepare the test suspension at sufficiently low concentrations to avoid fluid suspension breakdown. However, many European workers have reported that the three-layer system is useful with their centrifuges. Burt has compared data obtained with the line start system and the "three-layer system" using the AWRE centrifuge [13]. These data, summarized in Tables 6.1 and 6.2, illustrate the relative merits of the two experimental procedures with that instrument and can probably serve as a guide for probable performance if one opts for the three-line strategy. (It should be realized that sequential analysis of a series of samples is not possible once several fluids are used in the experiments.) Burt reports that for his later work he prefers to use the three-line start because of the smaller injected volume of suspension; however, this conclusion may be specific to his instrument [see later discussion of Illinois Institute of Technology Research Institute (IITRI) photocentridisk].

It is a little difficult to explain the success of the "three-layer" strategy. The writer suspects that the middle fluid plus the injected layer forms a broader initial line start by turbulent mixing that a density gradient restricts to within a small depth of the surface of the vortex of the spin fluid. Within this turbulent mixed zone the operative concentration of solids may be considerably less than that in the injected suspension. The writer's own opinion is that it is theoretically and practically better to seek out an experimental strategy and safe concentration level for the simple line start procedure than to bother with complex mixtures of fluids. A stable line start system ensures that fluid fineparticles interaction effects are eliminated.

It is surprising that some workers working with low-mass centrifuges equipped with feedback-controlled compensated speed regulators have been unable to achieve stable simple line start systems even with very low

TABLE 6.1. Comparative Characterization Data for a Titanium Dioxide Powder Obtained Using the AWRE Centrifuge and Various Suspension Concentrations and Experimental Strategies: Experimentally Determined Distribution Functions[a,b]

Method	Suspension Concentration Percentage by Volume	Sample Volume, ml	Equivalent Stokes Diameter, μm, at Percent by Weight Undersize of				
			95	84	50	16	5
Three-layer	0.05	1	1.05	0.94	0.78	0.60	0.45
	0.05	2	1.02	0.93	0.77	0.56	0.38
	0.05	3	1.08	0.95	0.78	0.57	0.37
	0.025	4	0.99	0.90	0.75	0.58	0.42
	0.025	6	1.03	0.93	0.76	0.55	0.36
	0.025	8	1.06	0.94	0.74	0.54	0.38
	0.01	2	1.02	0.92	0.75	0.53	0.37
	0.02	2	0.97	0.87	0.74	0.56	0.38
	0.04	2	0.98	0.93	0.79	0.61	0.49
	0.08	2	1.03	0.91	0.74	0.50	0.30
Two-layer (three runs)	0.0005	20 (i)	0.97	0.89	0.73	0.50	0.32
		(ii)	0.98	0.91	0.74	0.52	0.34
		(iii)	1.01	0.92	0.74	0.52	0.35

Source. Reference 13.

[a] *Line start system*. The injected suspension is warmed to a few degrees above the temperature of the liquid in the tank so that its overall density is slightly less than that of the liquid in the tank; 20 ml of suspension injected results in a line start layer of 6 mm. Centrifuge speed is 1500. Fluid used is distilled water, containing 0.1% Calgon (a dispersing agent).

[b] *Three-layer system*. Spin fluid distilled water containing 0.1% of Calgon. Middle layer 2 ml of 50:50 mixture of ethanol:water containing 0.1% Calgon. Injected suspension, 80:20 mixture of ethanol:water containing 0.1% Calgon.

TABLE 6.2. Comparative Characterization Data for an Explosive Powder "HMX" Obtained Using the AWRE Centrifuge and Various Suspension Concentrations and Experimental Strategies[a]

Method	Suspension Concentration Percentage by Volume	Sample Volume, ml	Run No.	Equivalent Stokes Diameter, μm, at Percent by Weight Undersize of				
				95	84	50	16	5
Three-layer	1.0	2	1	23.5	18.0	7.7	3.3	2.3
			2	20.5	14.0	6.0	3.0	2.2
			3	23.0	14.0	5.3	3.0	2.2
			4	22.5	15.5	6.1	3.1	2.3
Two-layer	0.05	20	1	27.0	18.5	7.2	3.1	2.3
			2	26.5	18.0	7.4	3.1	2.3
			3	26.0	17.5	7.2	3.1	2.2
			4	27.0	18.5	7.2	3.1	2.2

[a]Materials:

Centrifuge speed	750 rpm
Tank liquid	52:48 glycerol:water (+0.01% Cetavlon)
For three-layer method	
Buffer layer (2 ml)	30:70 glycerol:water (+0.01% Cetavlon)
Suspension layer liquid	20:80 glycerol:water (+0.01% Cetavlon)

Microscopic examination indicates that there are fine particles with diameters of up to about 35 μm. The two-layer traces show maximum fine particle diameters of 32–34 μm, whereas the three-layer traces show maximum diameters of only 26–32 μm.

solids concentration suspensions [14, 15]. One can only suggest either that some transient speed instability in the low-mass centrifuge caused by the injection of the test suspension or that some unspecified aspect of injection procedure is causing problems (e.g., see discussions later in this section of the failure of the spray injection technique tried by Kaye and Jackson [20]).

The Joyce–Loebel disk centrifuge is based on the work of Atherton and Cooper [17]. It is sometimes referred to as the *ICI disk centrifuge* or the *ICI/Joyce–Loebel* disk centrifuge because the original work was carried out at the paint laboratories of Imperial Chemical Industries. This equipment consists of a transparent disk with an axial entry port mounted on a servo controlled motor running at constant speed. The equipment was originally designed to work with physical extraction of the suspension; however, as early as 1969, Peyrade [18] fitted optics to this instrument and used it in conjunction with the line start procedure to size latex fineparticles. When modified in this way, the equipment is essentially a low-mass instrument similar in operational principles to the photocentridisk described by Groves et al. [8]. Later versions of the Joyce–Loebel equipment have optical incremental monitoring as an optional extra. Jones [19] introduced a version of the "three-layer system" that he termed the *buffered layer-start*. In this procedure a transient acceleration is given to the centrifuge after the middle layer has been floated onto the vortex of the spin fluid. It is claimed that this gives a density gradient in the top portion of the vortex of the spin fluid, thus helping to stabilize the injected sample of suspension. This claim would seem to support the "suggestion" made earlier in this section that the "three-layer system" in general is successful because of some turbulent dilution of the injected suspension that is then confined to the top zone of the spin fluid vortex by a density gradient.

In 1964 Kaye and co-workers built and operated a very massive photocentridisk [16, 20]. The rotor of this system is shown in Figure 6.4 with its top protective cover removed. One feature of this instrument is that the entry port can be sealed with a rubber stopper while the centrifuge is running, thus improving the thermal stability of the system by preventing evaporation from free surface of the spin fluid. In Figure 6.4 the rubber stopper can be seen in position on the axis of the rotor. The overall mass of the system was such that after the equipment was cleaned and loaded with clean spin fluid, it required 24 hours of running to reach an adequate level of mechanical and thermal stability. However, this was more than compensated for by the fact that very stable systems were obtained with small-volume injections. As many as 50 samples could by characterized in sequence before stopping and cleaning the centrifuge.

The instrument was built at the Illinois Institute of Technology Re-

Fig. 6.4 The IITRI photocentridisk.

search Institute (an organization usually referred to by its initials, IITRI, and formerly called the Armour Research Foundation). Therefore, the instrument is sometimes known as the *IITRI photocentridisk* or simply the *IITRI centrifuge* [20].

Because of the massiveness of the centrifugal chamber, it was not feasible to make the equipment operate at a range of speeds, and the equipment is driven by a constant-speed synchronous motor. The portholes of the centrifugal chamber act both as a light transmission mechanism and a light chopping system to give an ac signal that can be readily amplified and recorded. It was found that the best method of injecting the test suspension was to use a hypodermic needle and syringe to inject the suspension directly onto the central part of the backplate of the rotating centrifuge chamber. Apparently, by causing the fluid to enter on the axis of the centrifuge where the actual linear motion of the system is relatively low, the fluid was able to flow over the backplate and up the vortex without invading the vortex of the spin fluid.

Initially it was thought that perhaps the creation of an aerosol cloud in the free vortex of the centrifuge would yield a good, stable injected layer.

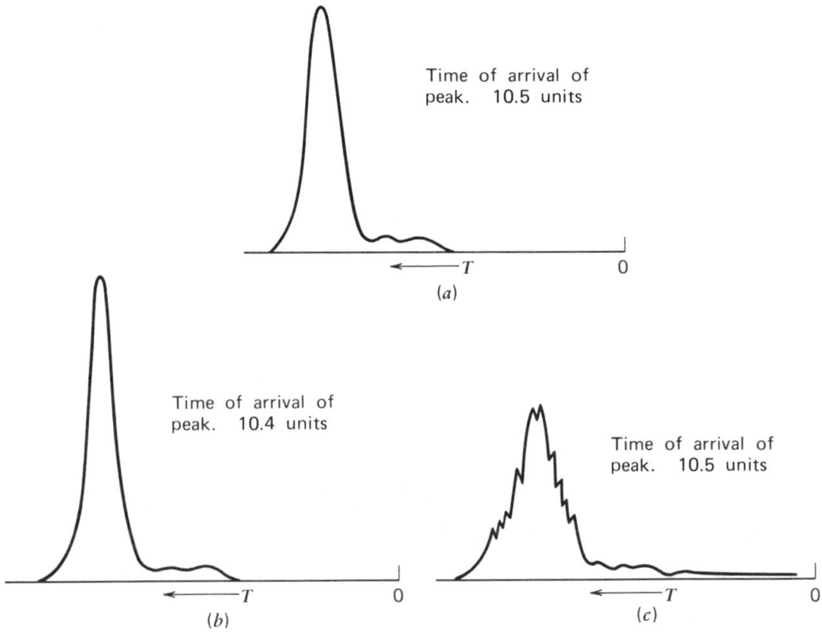

Fig. 6.5 Stokes diameter distributions characterized with the IITRI photocentridisk. Three sequential characterizations of a 2.61-μm spherical latex: (*a*) 2 ml of 0.132% by volume suspension injected with a hypodermic needle (time of arrival of peak, 10.5 units); (*b*) 2 ml of 0.132% by volume suspension injected with a hypodermic needle (time of arrival of peak, 10.4 units); (*c*) 2 ml of 0.132 by volume suspension sprayed as an aerosol into vortex cavity of centridisk (time of arrival of peak, 10.5 units).

However, it was found that such a cloud appeared to give an irregular line start zone as shown in Figure 6.5*c*. It is probable that the bouncing of individual droplets of suspension off portions of the rotating backplate of the chamber near the free vortex gives rise to droplets with relatively high momentum able to invade the vortex surface of the spin fluid for the short distance.

Kaye and co-workers found that the size of the hypodermic needle used in the injection procedure could affect the results. They found that the best results were obtained with the smallest needle compatible with the fineparticles to be injected into the centrifuge. It was felt that the use of fine needles in the injection strategies had the following advantages. First, the high shear rates experienced in the needle as the fineparticles flowed into the centrifuge helped to achieve maximum dispersion of the fineparticles. (Note, however, that with some fineparticles this could be a disadvantage in that it could cause the degradation of agglomerates that normally would maintain their integrity in the fineparticle system to be studied.) Second, the injection of a very thin steady stream onto the backplate of the rotating centrifugal disk appeared to make the suspension flow outward as a thin film over the centrifuge surface to the vortex without gaining enough momentum to be flung off of the surface as drops that would then penetrate the vortex because of their momentum.

As discussed briefly in Chapter 5, the IITRI centrifuge was so massive and stable and its recording system so sensitive that one could use injected layers of several microns thickness containing one or two fineparticles along the radial depth of the injected layer. Because of this very special situation, one could obtain stable line starts with concentrations higher than 0.05% by volume, provided that the injected layers were extremely thin.

The excellent resolution achievable in the IITRI photocentridisk is illustrated by the pen recorder trace reproduced in Figure 6.6. These data are from a sedimentation study of a nominally monosized latex of 2.61 μm diameter, as determined from electron microscopic studies. At the time that the original data were reported, they were somewhat controversial in that they showed the presence of doublets in the supposedly monosized suspension; however, the presence of these doublets in many lattices has now been amply demonstrated by other characterization procedures.

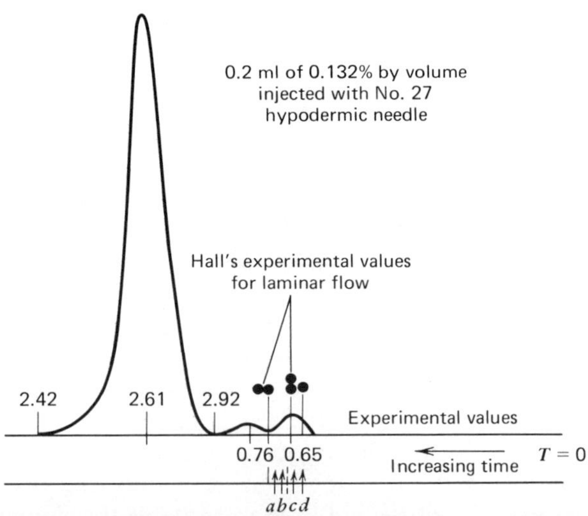

Fig. 6.6 High-resolution traces obtained with the IITRI photo centridisk [20]. Theoretical sedimentation times: *a*, horizontal doublet (touching spheres); *b*, horizontal doublet (rigidly joined spheres); *c*, vertical doublet (touching spheres); *d*, vertical doublet (rigidly joined spheres).

From the breadth of the primary peak obtained in the line start characterization study with the IITRI photocentridisk, it can be seen that the measured range of Stokes diameters in the primary fineparticle population of the latex is 2.42–2.92 μm. The two subsidiary peaks are obviously doublets. The centrifuge has been able to resolve the orientation with which the doublets are falling with respect to the radial direction of the centrifugal force field in the centrifuge. The two subsidiary peaks in Figure 6.6 occur at $0.76T$ and $0.65T$, where T is the sedimentation time of the prime peak into the inspection zone. In Section 5.4 reference was made to Hall's experimental study of the falling speeds of doublets. Hall showed experimentally that two equal-sized spheres forming a doublet aligned in the horizontal plane with respect to the direction of motion had a falling speed just greater than 0.7 of the single sphere, whereas a doublet falling with its orientation along the line of fall had a falling speed of 0.65 of the single sphere. When it is appreciated that random agglomeration of the single spheres of the primary population into doublets could generate doublets ranging in structure from two spheres of 2.4- μm-diameter up to a doublet of two spheres of 2.9-μm-diameter with many mixed intermediaries possible, the spread of velocities for doublets that have horizontal alignment (shown in Figure 6.5) as compared to Hall's experimental value for an ideal doublet is reasonable. The varied size of the components in the doublet would be less important when the doublet is aligned with the direction of flow; hence, it is not surprising that experimental values of falling speeds for the vertical alignment agree well with Hall's data. If two spheres had fused together to form a single larger sphere, such spheres would have traveled into the monitoring beam at the point shown by a slightly enlarged single sphere on the axis shown in Figure 6.6. Theoretical values for rigid and free doublets as calculated by Happel are shown in the diagram for comparison purposes [24]. Typical sets of data showing the effect of a failure to use very low volumes in the injected layer when working with the IITRI photodisk are shown in Figure 6.7 [4].

Groves and Yalabik [22] have used laser beams with photocentridisk equipment. They report good agreement between data obtained using this type of equipment and other characterization procedures down to Stokes diameters as low as 0.02 μm.

The ICI/Joyce–Loebel centridisk is a transparent disk mounted on a servo controlled constant-speed rotation system. The disk is mounted vertically. The front of the disk has an axial entry port. The equipment is partially filled with clean spin fluid and run up to operating speed. A portion of suspension is then injected to create a line start suspension system. The back plate of the Joyce–Loebel centridisk is beveled in an attempt to ensure smooth transition of the injected suspension up to the

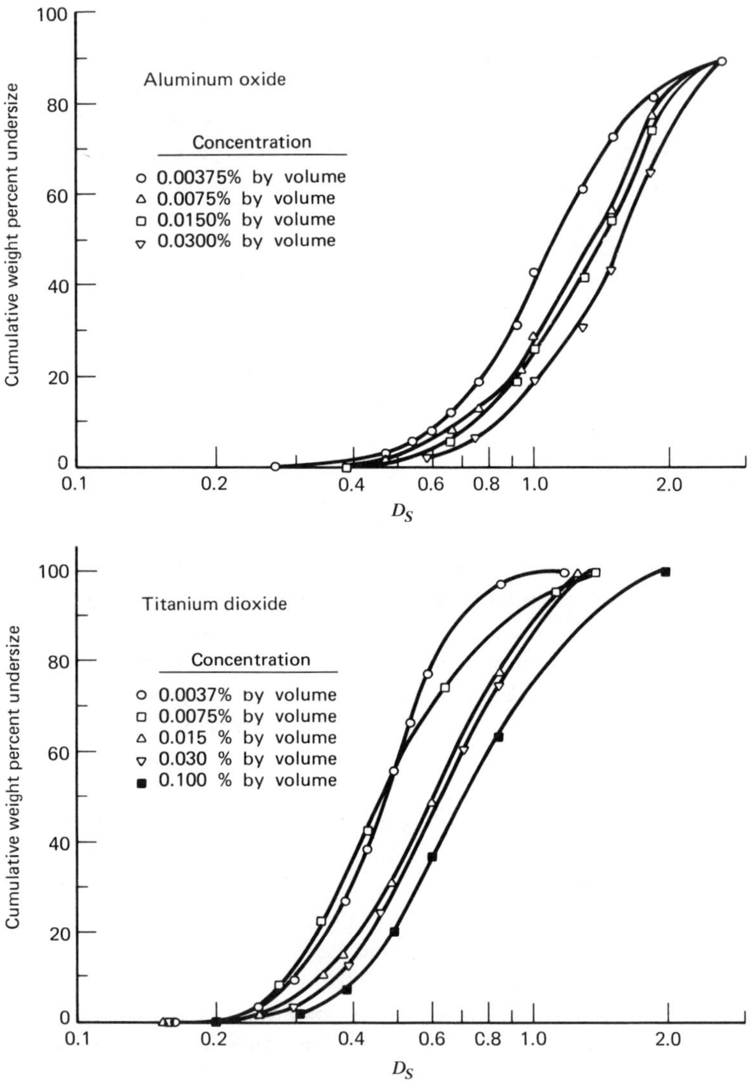

Fig. 6.7 Data on the effect of the concentration of the injected suspension using line start strategy with the IITRI photocentridisk [4] (D_S, Stokes diameter).

vortex of the free fluid. There is a possibility, however, that at the point where the bevel meets the flat surface of the backplate this discontinuity may encourage the shedding of rapidly traveling droplets that could invade the vortex of the spin fluid. There is no *a priori* reason to believe that the beveled backplate is more efficient in guiding the fluid to the surface of the vortex. The movement of thin films of fluid over rapidly rotating surfaces is a difficult subject, and the exact behavior of the film of fluid at the discontinuity between the bevel and the flat surface will depend on the speed of injection and the exact point at which the fluid jet meets the beveled surface.

After the centrifuge has operated for a given time after the injection of suspension, a vacuum-operated extraction probe is inserted through the axial port and fineparticle suspension sucked off from the deposited sediment. It is somewhat difficult to ensure that none of the deposited sediment is picked up during removal of the main suspension because there is slight turbulence over the free surface of the fluid as the extraction probe is operated. The performance of this equipment could perhaps be improved by placing a thin strip of rigid sponge plastic around the base of the disk centrifuge so that deposited fineparticles could be trapped in the sponge and relatively safe from turbulent resuspension as the extraction probe is moved close to the surface of the sponge. The use of such a sponge should at least minimize turbulent resuspension of deposited fineparticles even if it did not completely eliminate the problem (see discussion of Donaghue–Bostock stepped-bowl centrifuge in Section 6.5). The concentration of fineparticles in the extracted fluid is determined and hence the amount of fineparticles settled on the base of the centridisk determined. The experiment is repeated at a series of centrifuge times and/or speeds, and the data used with the equations developed in Section 6.3 calculate the Stokes diameter distribution of the injected fineparticles. The overall procedure is quite time consuming [23].

6.5 CENTRIFUGAL SEDIMENTATION USING HOMOGENEOUS SUSPENSIONS

If one uses large-radius centrifuges with short sedimentation distances at the extremities of the arms of the centrifuge, then from equation 6.1, we obtain

$$V_c = \frac{(\rho_P - \rho_L)}{18\eta} D_S^2 \omega^2 H$$

Thus

$$D_S^2 = \sqrt{\frac{18\eta \cdot V_c}{(\rho_P - \rho_L)\,\omega^2 H}} \tag{6.5}$$

If sedimentation distances are studied such that the change in distance from the center of the centrifuge ΔH is negligible, the fineparticles can be considered to sediment outward under a force of $\omega^2 H'$. For such systems, we can write

$$D_S = \frac{K \cdot V_c}{\omega^2 H'} \tag{6.6}$$

where D_S = Stokes diameter of the fineparticle as measured in the centrifuge

K = a constant equal to $18\eta/(\rho_P - \rho_L)$

$H' = \dfrac{H_1 + H_2}{2}$, where $H_2 - H_1 = \Delta H$, distance moved by sedimenting fineparticles.

Even when equation 6.6 is not exactly valid, it is often used to interpret centrifugal sedimentation data because of the extremely difficult theory required to make full account of the dilution effect and the variable acceleration effect experienced by a large array of fineparticles moving through a varying centrifugal force field. When operating with a long-arm centrifuge using short sedimentation distances at the extremities of the arms, one can effectively use all the data transformation procedures employed in the equivalent gravity sedimentation procedure by employing equation 6.6 to adjust the force field constant from g to the appropriate value in the data transformation equations.

Naughton and Spiel [27] used a long-arm bucket centrifuge to characterize the Stokes diameter distribution of ceramic fineparticles using hydrometers to follow density-gradient changes in the sedimenting suspension. Jacobson and Sullivan [30] used the long-arm–short sedimentation distance and measured the accumulation of fineparticles at the base of the sedimentation tube by stopping the centrifuge and decanting the supernatant suspension. They calculated the Stokes diameter distribution of the fineparticles using the modified force field of $\omega^2 H'$ in the equations developed for sedimentation balances in Chapter 5.

Gupta [28] developed an analogue to the Andreasen sedimentation procedure for use with the long-arm–short fall system in which he mounted sedimentation tubes at the extremities of a bicycle wheel. He used an ingenious system of spring-loaded syringes and hypodermic needles to extract samples from the body of the suspension after it had been centrifuged for a given period. The spring-loaded syringe was held in position by a metal wire that was melted with an electric current while the centrifuge was still running. The melting of the wire freed the spring loading to drive the extraction of a sample into the body of the syringe. The centrifuge was stopped and the solids content of the sample of suspension determined [28].

Hildreth and Patterson [29] used a light beam with a disk centrifuge. They treated their data as if they had used a gravitational photosedimentometer with the fineparticles falling under the influence of the increased force of the centrifugal field.

Muschelknautz and co-workers [24] developed a long-arm centrifuge based on the concepts of a differentially operated dual sedimentation balance. The basic system used in the Muschelknautz system can be appreciated from the system shown in Figure 6.8. The two arms of a sedimentation balance are placed in two sedimentation columns, one of which is filled with suspension to a slightly greater height ΔH than the other. Simple balance pans are shown in Figure 6.8, but in the actual equipment used by Muschelknautz all the recommended modifications to the simple sedimentation balance proposed by Leschonski (described in detail in Chapter 5) are incorporated for each balance pan. The net weight registered by the twin sedimentation system is equivalent to the fineparticles that would have been deposited on a single balance pan if a line start of thickness ΔH at height H on clear fluid had been used to characterize the fineparticles. The system has the added advantage that nearly all problems due to powder lost to the walls, any residual problems due to net seepage of fineparticles around the balance pan edges, and so on are automatically compensated for. Thus all the advantages of the line start system are achieved without the difficulties of physically creating the line start system. A fineparticle suspension system such as that used in the differential sedimentation balance can be described as a pseudo–line start strategy.

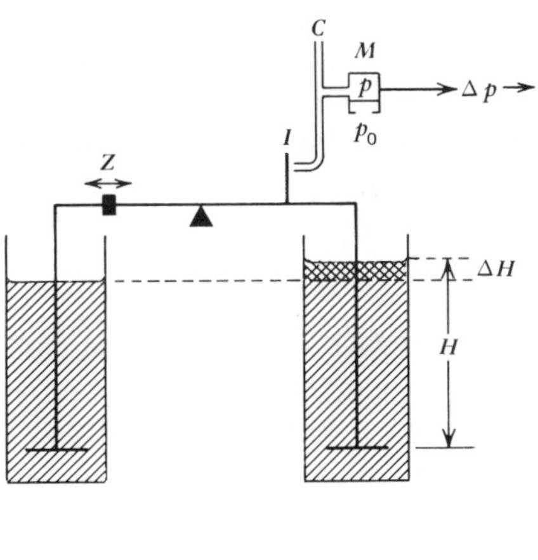

Fig. 6.8 Differential sedimentation balance described by Muschelknautz [24, 25]. Symbols: Z, zero adjustment rider M, pressure manometer I, impact plate C, compressed air; ΔH, pseudo–line start system.

As shown in Figure 6.8, there is a small rider on the beam balance that is adjusted at the beginning of an experiment to make all forces on the system equal immediately before the fineparticles start to settle. Muschelknautz chose to use a pneumatic method to measure the net weight of fineparticles deposited on the right-hand balance pan. The conceptual design of his pneumatic system is illustrated in Figure 6.8. Compressed air is directed onto a small plate mounted on the right-hand side of the beam balance. When there is an increase in the weight on the right-hand-side balance pan, this tends to move the plate closer to the orifice of the compressed air jet. This increases the resistance experienced by the air emerging from the tube so that with the pneumatic system employed there is actually a slight increase in the pressure inside the compressed-air tube. Details of the actual pneumatic system used can be found in the work by Bürkholz [25]. The device labeled M in Figure 6.8 is a differential manometer able to record the difference between the internal pressure of the tube and the atmospheric pressure. This differential pressure is calibrated and used to record the net weight of fineparticles accumulating on the right-hand balance pan. The advantage of using a pneumatic system to record the weight of the fineparticles is that it can be used in many factory situations where compressed air is available and where it is more difficult to operate electronic feedback mechanisms because of the hostile environment.

Two centrifugal adaptations of Muschelknautz's differential gravity sedimentation balance are shown schematically in Figure 6.9. For the sake of simplicity, the feed mechanisms that allow suspension to be put into the centrifugal buckets while the centrifuge is operating at the speed used in the characterization study are not shown in Figure 6.9.

In the system shown in Figure 6.9*a* two balance pans are opposed to each other on the opposite end of a straight shaft that is carried on air bearings. The equipment is sometimes operated as a pseudo–line start system, or alternatively clean fluid is placed in one bucket and suspension is placed in the other buckets. Muschelknautz refers to this latter operational system as the *unilateral mode operation*. One advantage of this manner of operation is the simplification afforded by the use of a short-fall trajectory on a long-arm centrifuge, thus allowing the data to be regarded as moving in a uniform force field of magnitude $\omega^2 H'$.

Whether operated unilaterally or with a pseudo–line start system, the excess of fineparticles deposited on the balance pan on the right-hand side of Figure 6.9 causes the movement of the shaft carrying the balance pan. This movement is opposed by the jet of air shown in the diagram. Again, a differential manometer records the increase in pressure caused by the movement of the fineparticles. The air jet portion of the pneumatic surface

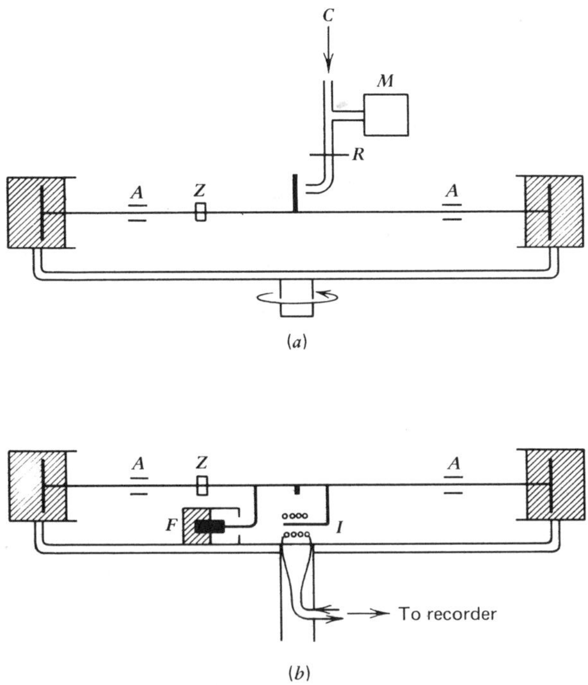

Fig. 6.9 Two different centrifugal versions of Muschelknautz's differential sedimentation balance [24]: (*a*) pneumatic evaluation of deposited fineparticles; (*b*) inductive recording of deposited fineparticles. Symbols: *A*, air bearings; *Z*, zero adjustment rider; *M*, differential manometer; *R*, linkage mechanism to permit airtube to rotate with balance and buckets; *F*, float; *I*, induction coil and movable magnetic core; *C*, compressed air.

has to rotate with the centrifuge head, and this is shown schematically in Figure 6.9*a* by line *R*, which is intended to indicate an interface between the static part and the rotating parts of the pneumatic circuit.

Bürkholz, who has given detailed descriptions of operating procedures for using the system shown in Figure 6.9*a*, found it necessary in some instances to place 5 cm^3 of oil on the top of each bucket to prevent surface evaporation from the system being characterized. He points out that the enhanced force field from the centrifugal action tends to accentuate any problems due to density differences caused by evaporative cooling. In operation, the two buckets are partially filled with clear fluid, and at the appropriate time a small amount of suspension is fed into the buckets while they are rotating at the characterization speed. He reports that the incoming suspension immediately mixes turbulently with the fluid in the buckets to create an initially homogeneous suspension with minimum

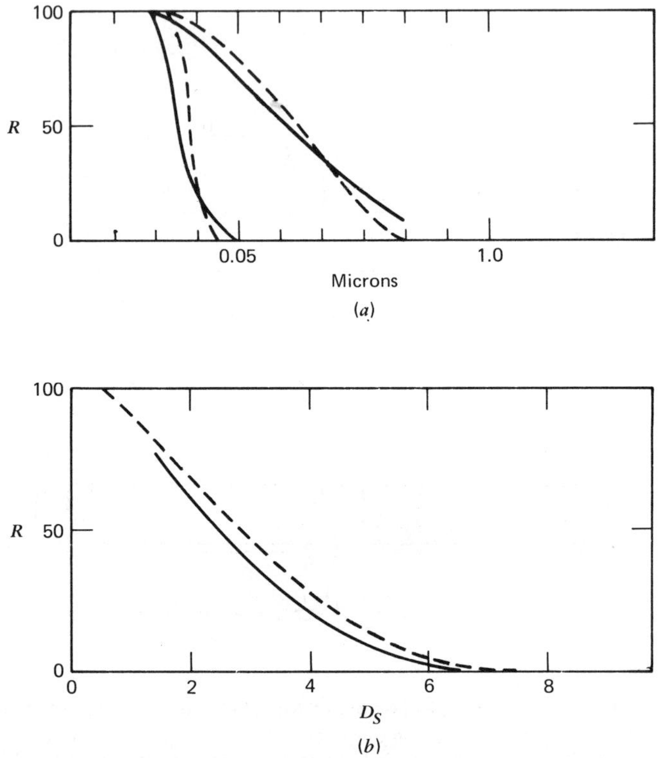

Fig. 6.10 Comparative data on the performance of the Muschelknautz equipment [25]: (*a*) size distribution of two sets of silver bromide crystals (----Stokes diameter by centrifugal sedimentation; balance; —— projected area distribution from electron micrographs); (*b*) Stokes diameter distributions of "organic dust". —— Sartorius sedimentation balance; ---- Muschelknautz centrifugal equipment (pneumatic recording—unilateral operation). Symbols: *R*, cumulative percentage less than stated size, D_S Stokes diameter.

uncertainty in the time of centrifugation [25]. Data reported by Bürkholz using the system shown in Figure 6.9*a* are summarized in Figure 6.10.

In a later version of the Muschelknautz equipment, shown in Figure 6.9*b* [24], the pneumatic system has been dispensed with, presumably because of the difficulty of operating with a rotating pneumatic supply. The movements of the shaft bearing the opposed balance pans are now recorded by movement of the core of a small induction coil mounted directly on the centrifuge. The inductance of the coil is measured automatically while the centrifuge is running by means of electrodes that pick up the appropriate signals directly off the shaft of the centrifuge. The movement of the core of the recording induction coil is balanced by an equivalent system connected to a float valve as shown in Figure 6.9*b* for the dynamic balance of the centrifuge to be maintained. This device also serves as a viscous damping system to dampen out any transient vibrations in the system that may be created during the operation of the equipment.

The pseudo–line start system invented by Muschelknautz could be employed with advantage in other centrifugal systems for evaluating the Stokes diameters of fineparticles.

Two possible systems for operating photocentridisks in a pseudo–line start mode are suggested and are shown in Figure 6.11. In the first system an initial study is carried out with the disk filled to a distance *H* from the center of rotation. The equipment could be filled with suspension as the disk was operating at characterization speed with the assumption that the general turbulence from the filling act would maintain general agitation until the tank contained the necessary amount of suspension and that centrifugation time for the fineparticles could be reasonably calculated as being from the termination of the act of filling or shortly thereafter. A stiff rod could be inserted into the disk to maintain general turbulence until the act of filling was over, with this rod swiftly withdrawn when the centrifuge was loaded with the necessary suspension. The signals from the monitoring light beam would be stored in a computer and the centrifuge stopped after the smallest fineparticle had passed the beam. The experiment would then be repeated with a small increase in the volume of suspension placed in the disk so that the difference between the new set of signals and that from the previous experiment would be due to the fineparticles from the pseudo–thin layer created by the extra volume of suspension added in the second case. The computer manipulation of the data could simulate the real-time operation of comparative measurements from the two suspensions, yielding data from an equivalent pseudo–line start system.

The lower portion of Figure 6.11 shows a dual-disk centrifugal system that could perform the same characterization study in real time. Since the data from the two disks would be utilized, a solid-state integrated circuit for measuring light levels could be used to yield the logarithm of the net transmittance that would be directly proportional to the logarithm of the transmission of the pseudo–line start system. Such an equipment would be able to directly generate with appropriate electronic manipulation of the data a direct distribution evaluation of the Stokes diameter of the system. The dual-disk system in Figure 6.11*b* would also have the advantage that everything could be directly balanced and the optical signals would be traded off to allow for any smear or dirt that might appear on any of the disks during the filling operation.

A major theoretical objection to a system such as that shown in Figure

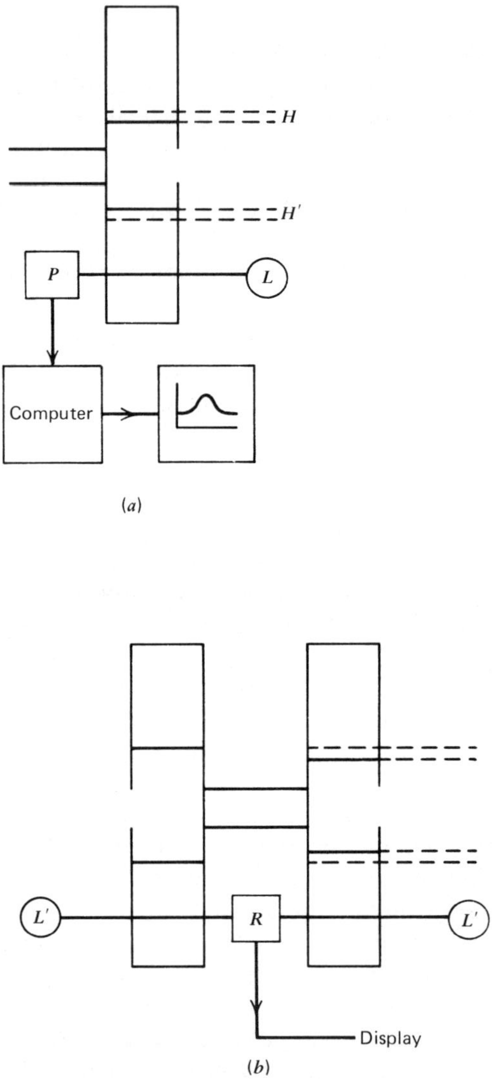

Fig. 6.11 Pseudo–line start systems can be either simulated with a computer or achieved in real time by using a duodisk system: (*a*) dual run + computer storage–manipulation of the data (*H*, suspension depth for first run; *H*′, suspension depth for second run; *P*, photocell; *L*, light); (*b*) duodisk system gives pseudo–line start in real time [*L*′, fiber-optic light source (one bundle divided into two to provide two sources); *R*, receiver incorporating logic circuits].

6.11*a* is the problem of coping with the acceleration up to constant speed of the system and problems associated with the automatic filling and emptying of the disks. Figure 6.12 shows an alternative system that could prove to be very useful in both a research situation and a quality control environment [26]. The base of the centrifuge could be a solid metal disk with appropriate slots cut in its outer rim to permit the passage of the monitoring beam of light. Alternatively, the disk could be an open wheel such as a bicycle wheel. The suspension to be characterized is placed in four tubes, with two of them slightly longer than the others, as indicated in Figure 6.11. It is envisaged that these tubes could be completely sealed, thus eliminating all problems associated with the presence of an open surface in the centrifuging system. The tubes would be mounted in matching pairs so that the dynamic balance of the centrifuge would be maintained at all times. It is envisaged that the tubes would be clipped into position on disk cradles able to spin independently until the operating speed of the centrifuge is achieved. When the operating speed is reached, a centrifugally operated locking pin would be triggered to lock the cradle into position. In this way, the tubes full of suspension would spin in a

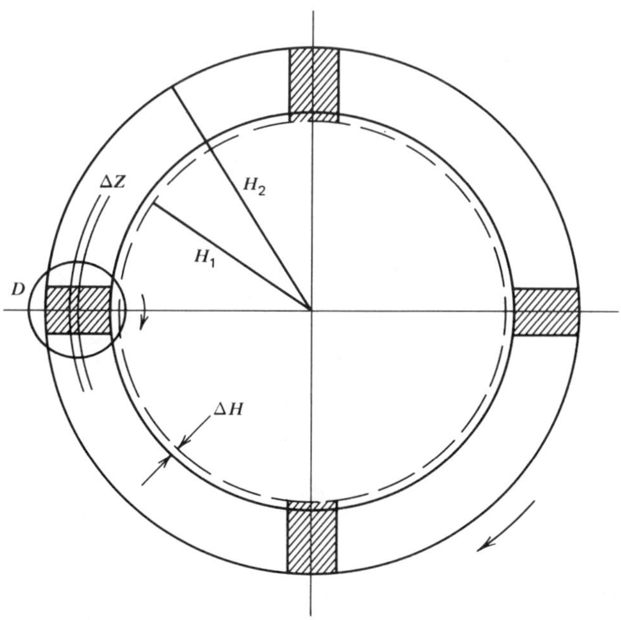

Fig. 6.12 Possible scheme for a pseudo–line start strategy with a long-arm centrifuge. Symbols: ΔZ, optical monitoring zone; ΔH, width of pseudo–line start zone; H_2, distance from center to base; H_1, distance from center to free surface; D, disk cradle able to spin until operating speed is achieved.

planetary-type motion until the centrifuge reached constant operating speed. It might be necessary to add small flags to the cradle to create enough windage to spin the disk cradles until the operating speed is achieved. By using these subsidiary disk cradles, it would be possible to prevent sedimentation in the tubes until the operating speed of the system was achieved. Once the locking pins had operated, the system would operate as a pseudo–line start photodisk system having a suspension layer of thickness ΔH monitored by a light beam of thickness ΔZ. Thus one would gain all the advantages of the line start photocentridisk procedure while using an initially homogeneous suspension. Furthermore, appropriate designs of the slots through which the monitoring beam moved could ensure that significant signals occurred at half of the frequency of the rotation of the centrifuge, whereas any error signals due to dirt on the individual tubes or irregularities in the glass of the tubes would yield signals at the frequency of rotation of the wheel. They could, therefore, be filtered out electronically from the signals carrying the significant information. This would greatly improve the flexibility and lower the cost of the system in that it would remove the need for exactly matched systems from an optical transmission point of view. The system could also be used with relatively small volumes of suspension while utilizing the greater stability and the lower cost of a large-radius, low-mass centrifuge. It is envisaged that the optimum performance of the instrument would occur if the plane of operation was such that the disk appeared vertical to the external observer. In a research situation flat-sided optical glass tanks such as those used in a spectrometer could be clipped directly onto the cradles. In a quality control situation it may be possible to operate with ordinary glass tubes. The equipment is potentially low cost and could be multiplexed in an industrial environment. It could be operated by compressed air.

It can be shown that a general relationship governing the weight deposited at the base of a centrifuge distance H_2 from the center of rotation when the centrifuge is filled with suspension to a height of H_1 is given by

$$P = W_{DM} + \int_0^{D_m} \frac{H_2^2}{H_2^2 - H_1^2} \{1 - \exp(2kD^2t)\} F(D)\,dD \qquad (6.7)$$

where P = weight deposited at depth H_2 at time t

H_1 = distance of top of the suspension from center of rotation

D_m = Stokes diameter of fineparticles that have traveled from H_1 to H_2 in time t

$F(D)$ = distribution function of Stokes diameters of fineparticle system

$$k = \frac{(\rho_P - \rho_L)\,\omega^2}{18\eta}$$

W_{DM} = weight of fineparticles with diameters greater than D_m

It can be seen from this equation that a major problem in the data transformation procedure is that the distribution function to be evaluated appears as a parameter governing the deposition rates. It is useful to rewrite this equation, writing β for H_2 and α for H_1 because in some experimental strategies they are treated as variables. Thus

$$P = W_{DM} + \int_0^{D_m} \frac{\beta^2}{\beta^2 - \alpha^2}\{1 - \exp(2kD^2t)\}F(D)dD \tag{6.8}$$

Romwalter and Vendl [31] attempted to derive a general solution of equation 6.8 by differentiating the equation with respect to time and substituting the results back into the original equation. They then used this equation to determine the distribution function of Stokes diameters by carrying out a series of deposition studies with fixed α and β to obtain a set of variables of P with t [31]. However, Brown [32] has shown that Romwalter and Vendl made a mistake in their original derivation.

Martin and Robinson [33, 34] derived an approximate solution for the general case and used the following equations to relate P to W_{DM}:

$$W_{DM} = \int_0^{D_m} F(D)dD$$

$$= 1 - \left[\frac{\gamma(6 - \gamma)}{8}P - \frac{\gamma D_M}{8}\cdot\frac{dP}{dD} + \frac{\gamma(\gamma - 2)(\gamma - 4)I(D_m)}{8}\right]$$

where

$$I(D_m) = \frac{1}{D^\gamma}\int_0^{D_m} PD^{\gamma-1}d(D)$$

and

$$\gamma = \frac{4(\beta^2 - \alpha^2)}{\alpha^2 \ln(\beta/\alpha)}$$

Brown carried out a series of deposition studies by using a given sedimentation time and varying the amount of suspension placed in the centrifuge. This is equivalent to treating the term α in equation 6.8 as a new variable; thus

$$W_{DM} = P - \frac{\beta^2 - \alpha^2}{2\alpha}\cdot\frac{\delta P}{\delta\alpha} \tag{6.9}$$

This latter equation can be used in conjunction with the operation of the Joyce–Loebel centrifuge. The centrifuge is filled initially with homogeneous suspension, and it is assumed that the time required to reach operating speed is a negligible proportion of the total centrifugation time employed in the series of measurements. This strategy has proved effective when studying very fine pigments such as carbon blacks and ultrafine titanium dioxide or the very fine ceramic powders used in some modern processes. It has enabled centrifugal sedimentation to be pushed down to limits of well below 0.1 μm utilizing the Joyce–Loebel centridisk.

Another approach to utilizing equation 6.8 is to treat the quantity β as a new variable. When this approach is adopted, the data transformation equation becomes

$$W_{DM} = P + \beta \frac{(\beta^2 - \alpha^2)}{2\alpha^2} \cdot \frac{\delta P}{\delta \beta} \tag{6.10}$$

This equation was used by Donaghue and Bostock with an instrument called the *stepped-bowl centrifuge* [35]. The stepped bowl used by these workers in effect constituted a series of disk centrifuges stacked one above the other, each step of the bowl representing a disk with a different value of β. The system was operated by filling it with suspension, and it accelerated rapidly to top speed. After centrifuging for a given time, a flexible hose was pushed down onto the central hollow spindle as the centrifuge continued to operate at a fixed speed. Note that the flexible hose is made of some kind of low-friction plastic so that a good seal is achieved without increasing the load on the centrifuge significantly. The central spindle was connected to a series of drainage channels close to the edges of the individual steps. In the original equipment used by Donaghue and Bostock, the deposited particles were collected on metal strips placed around the steps of the bowl. After the supernant suspension had been sucked off through a central spindle, the speed of the centrifuge was increased for a short time to consolidate the deposits onto the strips, which were then removed from the equipment and dried so that the weight of the fineparticles could be determined. The instrument was not widely used because of problems associated with turbulence at the metal strips as the supernant suspension was removed from the centrifuge. Today low-cost thin rigid plastic foam is available, and this would prove to be a more suitable material for lining the steps of the centrifuge. The fineparticles being deposited would be trapped in the foam mesh and not so vulnerable to the turbulent action accompanying the withdrawal of the supernant suspension. The instrument has the advantages of low cost and multiple evaluation of points on the distribution function with one operation of the centrifuge.

Several disk centrifuge systems have been developed to operate with incremental monitoring of an initially homogeneous suspension. For this situation, the mathematics of the general solution is very difficult, and the three commercial instruments based on this experimental strategy are provided with either graphical techniques based on partial solutions of general equations for data transformation or complex proprietary software for computer transformation of experimental data. The essential experimental strategies of these instruments are briefly reviewed, and the interested reader should consult the specialized literature for the details of the complex data transformations.

Berg [36] was one of the first to attempt a general solution for incremental monitoring of the dynamics of a homogeneous suspension subjected to varying centrifugal forces. He used divers to follow sedimentation gradients in the centrifuge. Kamak attempted to use an Andreasen-type strategy in a centrifugal environment. He constructed sector-shaped sedimentation tubes to avoid problems associated with the impingement of fineparticles on the walls of cylındrical vessels as they followed radial trajectories [1]. Figure 6.13 shows the essential system of commercial equipment that Slater and Cohen evolved from the original system of Kamak [37]. This instrument is variously known as the *Slater–Cohen disk centrifuge,* or the *Simcar centridisk.* Although its data transformations and experimental strategy are related to the work of Kamak, physically the equipment appears to have evolved from the stepped-bowl centrifuge of Donaghue and Bostock. Thus, as in that equipment, samples of suspension are withdrawn from the centrifuge chamber by drainage channels in the base of the disk through the hollow spindle, which can be connected to an external flexible hose while the equipment is operating. When the characterization study is under way, the centrifuge is filled with suspen-

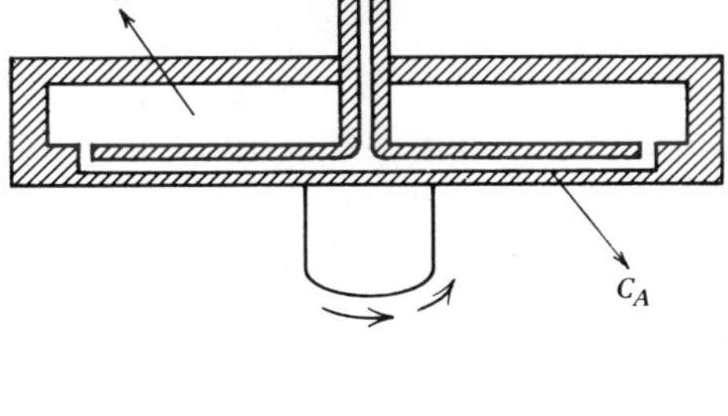

Fig. 6.13 Basic structure of the SIMCAR centridisc equipment [37, 38]. Symbols: *F,* flexible hose can be pushed onto spindle and used to extract sample while centrifuge is running; C_A, channel for removing a sample of the suspension; C_H, centrifuge chamber filled with suspension.

TABLE 6.3. **Comparative Data Reported by Kaye and Burt on Stokes Diameter Distributions Measured in the AWRE Photocentridisk and the Simcar Centridisk**

Zinc Oxide				Barium Sulfate				Titanium Dioxide (Rutile)				Silica			
Stokes Diameter, μm	Cumulative Proportion Undersize Percent, w/w			Stokes Diameter, μm	Cumulative Proportion Undersize Percent, w/w			Stokes Diameter, μm	Cumulative Proportion Undersize Percent, w/w			Stokes Diameter, μm	Cumulative Proportion Undersize Percent, w/w		
	A	*B*	*B'*[a]		*A*	*B*	*B'*		*A*	*B*	*B'*		*A*	*B*	*B'*
4.0	90	94	95	2.0	96	98	98	2.0	95	98	98	8.0	92	96	95
3.0	87	91	92	1.5	92	95	95	1.5	92	96	95	5.6	80	86	85
2.0	80	85	86	1.0	82	80	79	1.0	81	86	85	4.0	68	71	70
1.5	74	74	75	0.9	77	74	73	0.9	77	79	78	2.8	53	53	52
1.0	61	48	46	0.8	72	67	66	0.8	70	71	60	2.0	36	33	33
0.9	56	40	37	0.7	64	57	56	0.7	61	62	60	1.4	23	16	16
0.8	50	32	28	0.6	53	46	44	0.6	48	48	47	1.0	13	9	10
0.7	43	25	21	0.5	41	34	32	0.5	37	30	31	0.7	6	4	5
0.6	34	18	15	0.4	27	22	20	0.4	17	18	19	0.5	3	2	2
0.5	25	13	10	0.3	13	11	12	0.3	5	10	10	—		—	
0.4	16	8	6	0.2	3	5	4	0.2		4	5	—		—	
0.3	7	5	3	—		—		—		—		—		—	
0.2	2	2	1	—		—		—		—		—		—	

[a] *A* = characterization as determined on the Slater–Cohen disk centrifuge (Simcar centridisk); *B* and *B'* = characterization on the centrifugal disk photosedimentometer. The data for the Slater–Cohen disk centrifuge are published by permission of the management of Simon Carves Ltd. Experimental conditions for photocentridisk characterization are as follows. All four powders were characterized in the AWRE disk centrifuge by using water containing 0.01% of Cetavlon (commercial dispersing agent) as the dispersed phase and spinning in a centrifuge at 750 rpm. The value of the solids concentration used was determined by carrying out successive characterization at increasing solids dilution until the measured distribution was independent of the concentration used. The concentrations used in these experiments were:

	Percentage by Volume
Barium sulfate	0.002
Silica	0.008
Titanium dioxide	0.0015
Zinc oxide	0.003

sion and after a given operating time a small amount of suspension is withdrawn through the base channels. Centrifugation is continued, with successive samples being withdrawn from the centrifuge at a series of times. These data are then transformed to a Stokes diameter distribution [38, 39]. Table 6.3 shows some comparative data for Stokes diameter characterizations carried out with the AWRE photocentridisk and the Simcar centridisk. Also, Figure 6.14 summarizes some comparative data on the Stokes diameter distribution of a silica powder as measured by three different techniques. These data were reported by Burt [9]. For many purposes, the congruence of the three measured distributions would be quite adequate. The data from the MSA centrifuge system gave the finest measured distribution. This would seem to indicate that probably the resuspension effects created by the deceleration of the tube centrifuge dominate any potential coarsening of the measured distribution from fluid–fineparticle interaction from the line start system.

Allen [40] has developed centridisk equipment similar to that of the Simon Carves instrument but adapted to operate with much smaller

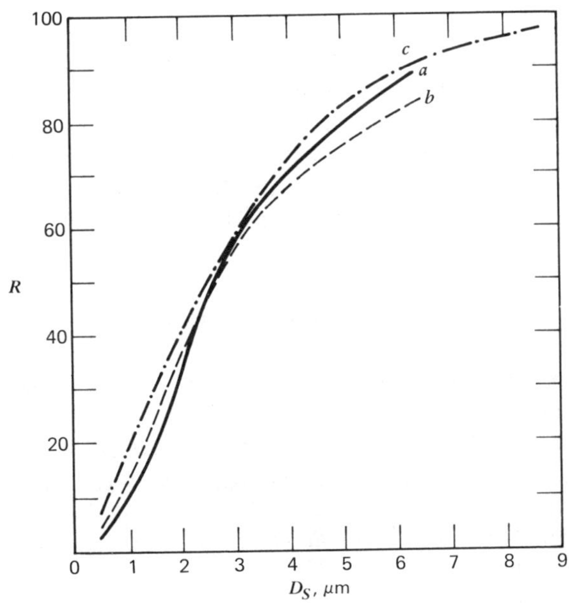

Fig. 6.14 Comparative data on the Stokes diameter distribution of a silica powder [9]: (*a*) Stokes diameter distribution measured with AWRE photocentridisk; (*b*) Stokes diameter distribution measured with Simcar centridisk; (*c*) Stokes diameter distribution measured by Whitby–MSA centrifuge system. Symbols: *R*, cumulative percentage greater than stated size; D_S, Stokes diameter.

amounts of suspension. Allen has reported on the extensive data manipulations required to convert the information from the pipette withdrawal–centridisk equipment [41, 42]. Svarovsky and Svarovska [43] have recently described an analogue computer that can be built for the data transformation of information obtained using homogeneous sedimentation in a disk centrifuge.

In the LADAL x-ray centrifuge an isotope source is used to generate x-rays to monitor the radial migration of sedimenting fineparticles in a disk centrifuge. Allen [40] reports that the characterization procedure is limited to powders with components of atomic number greater than 13, that is, to high atomic number elements capable of absorbing the x-ray radiation used in the monitoring strategy. Allen reports that normally it is necessary to use powder concentrations in the range 0.2–1% by volume to obtain measurable attenuations of the x-ray beam. The equipment can either be used with a static monitoring x-ray probe, or, alternatively, the x-ray probe can be moved along the radius of the centrifuge to decrease characterization study times the data.

Recently Naylor has suggested the use of a line of fiber optic probes mounted along the radius of centridisk equipment to accelerate the collection of characterization data. Kaye has pointed out that such equipment could be used with various suspension strategies, including line start, pseudo–line start, and general homogeneous sedimentation. Naylor has pointed out that in this latter situation sequential scans along the radius of the centrifuge at times separated by a minute could be used to remove the ambiguity in the general solution of the homogeneous sedimentation equations [see discussion in reference 26].

NOTES AND REFERENCES

1. For an introduction to the mathematics of centrifugal sedimentation, see H. J. Kamak, *Anal. Chem.* 23 (1951), 844.
2. C. E. Marshall, *Proc. Roy. Soc.*, Section A, 126 (1930), 427.
3. K. T. Whitby, "A Rapid General Purpose Centrifuge Sedimentation Method for Measurement of Size Distribution of Small Particles," *ASHAE J.* (*"Heating, Piping & Air Conditioning"* section) (January and June 1955). Commercial information from MSA.
4. R. Davies, private communication.
5. C. R. D. Treasure, Whiting and Industrial Powders Research Council, Welwyn, England, Technical Paper No. 50, 1964.
6. B. H. Kaye, *Improvements in/or Relating to Centrifuges*, Br. pat. 717358 (applied for 1958; granted May 2, 1962).

7. B. H. Kaye, *Physical Problems of Particle Size Analysis*, Ph.D. thesis, University of London, 1962.

8. M. J. Groves, B. H. Kaye, and B. Scarlett, "Size Analysis of Subsieve Powders using a Centrifugal Photo Sedimentometer," *Br. Chem. Eng.*, 9 (1964), 742–744.

9. M. W. G. Burt, "The Accuracy and Precision of the Centrifugal Disc Photosedimentometer Method of Particle Size Analysis," *Powder Technol.*, 1 (1967), 103–115.

10. Coulter Electronics, 590 20th Street, Hialeah, Florida 33010.

11. B. R. Statham, *Proc. Soc. Anal. Chem.*, 9 (1972), 40–43.

12. B. Scarlett, M. Rippon, and P. J. Lloyd, *Particle Size Analysis*, proceedings of a conference held at Loughborough University, England; published by the Society for Analytical Chemistry, London, p. 242.

13. M. G. W. Burt, private communication, Atomic Weapons Research Establishment, Aldermaston, England.

14. J. Beresford, "Size Analysis of Organic Pigments Using the I.C.C.–Joyce Loebel Disc Centrifuge," *J. Oil Colour Chem. Assoc.*, 50 (1967), 594–614. For a critique of some of the data reported in this paper, see reference 15.

15. B. H. Kaye, "Statistical Analysis of Recent Data on Concentration in Suspensions," *Powder Technol.*, 3 (1969–70), 78–82.

16. B. H. Kaye and M. R. Jackson, "Size Analysis of Latex Emulsions Using a Centrifugal Disc Photosedimentometer," *Powder Technol.*, 1 (1967), 81–88.

17. E. Atherton and A. C. Cooper, Br. pat. 983,760 (1962). Commercial information from Joyce Loebel Ltd., Princes Way, Team Valley, Gateshead 11, Durham, England.

18. Peyrade, J., private communication, Pechiney-Saint Gobain, Centre de Recherches d'Aubervilliers, 12–14 Rue des Gardinoux, 93 Aubervilliers, France.

19. M. H. Jones, *Proc. Soc. Anal. Chem.*, 3 (1966), 116.

20. B. H. Kaye, M. R. Jackson, and R. Kahrun, "The Stability of Line Start Suspension Systems in the Centrifugal Disc Photosedimentometer," *Powder Technol.*, 2 (1968–69), 290–300.

21. The IITRI photocentridisk system was developed commercially by the Martin Sweets Company. Inquiries regarding the commercial availability of the Martin Sweets analyzer should be directed to the Diano Corporation, Mid-West Technical Center, 1866 Production Drive, Louisville, Kentucky.

22. M. J. Groves and H. S. Yalabik, "The Size Analysis of Dispersed Submicrometre Pigments and Lattices," *Powder Technol.*, 17 (2) (1977), 213–218. The laser centrifugal photosedimentometer used by Groves and co-workers was designed by Technord Developments Ltd., 17 Pine Grove, Twyford, Berkshire, and the equipment is distributed by Northey International Systems Ltd., 5 Charles Lane, St. John's Wood, High Street, London NW8 7SV.

23. See comments made by W. Carr in his plenary lecture at The Particle Size Analysis Conference held at Salford, September 1977. Proceedings published by the British Society for Analytical Chemistry, London.
24. E. Muschelknautz, "Uber die Entwicklun einer Fliehkraftssedimentationswaage," *Dechema-Monographien*, Band 79, 1975, NR1589–1615 Teil B 267–277.
25. A. Bürkholz, "Particle Size Determinations in the Range from 0.1 to 10 microns with a Centrifugal Sedimentation Balance," *Staub-Reinhalt Luft*, 30 (1) (January 1970) (available in English translation from the U.S. Department of Commerce).
26. B. H. Kaye, "Monitoring Particulate Solid Properties in Automatic Control Loops, Past, Present and Future," in *Proceedings of the Conference on Instream Measurements of Particulate Solid Properties*, Bergen, Norway, August 1978. Information from Chris Michelsen Institute, Bergen.
27. F. H. Naughton and S. J. Spiel, *J. Am. Ceramic Soc.*, 21 (1938), 89.
28. A. K. Gupta, *J. Appl. Chem.*, 9 (1959), 487.
29. J. D. Hildreth and D. Patterson, 1964, *J. Soc. Dyers Colour.*, 80 (1964), 474.
30. A. E. Jacobson and W. G. Sullivan, *Anal. Chem.*, 19 (1947), 855.
31. A. Romwalter and M. Vendl, *Kolloidzeitschrift*, 72 (1935), 1.
32. C. Brown, *J. Phys. Chem.*, 48 (1944), 246.
33. S. W. Martin and H. E. Robinson, *J. Phys. Colloid Chem.*, 42 (1948), 845.
34. S. W. Martin and H. E. Robinson, *J. Phys. Colloid Chem.*, 53 (1949), 860.
35. J. K. Donaghue and W. Bostock, *Transact. Inst. Chem. Eng.*, 33 (1955), 72.
36. S. Berg, *Ingen Vidensk Skr.*, Part B (2) (1940).
37. C. Slater and L. Cohen, *J. Sci. Instrum.*, 39 (1962), 614.
38. Commercial information on the Simcar centridisk is available from Simon Carves Ltd., Stockport, Lancashire, England.
39. For a worked example of a data transformation used in conjunction with the Simon Carves centridisk, see reference 40.
40. T. Allen, *Particle Size Measurements*, 2nd ed., Chapman and Hall, London, 1974.
41. Commercial information is available from Microscal Ltd., 20 Mattock Lane, Ealing, London.
42. Commercial information is available from Ladal Scientific Equipment Ltd., Warlings, Warley Edge, Warley, Halifax, Yorkshire HX27RL, England.
43. L. Svarovsky and J. Svarovska, "Particle Size Distribution Measurements by Homogeneous Sedimentation in a Disc Centrifuge," *Dechema Monograph*, 79 (1589–1615), Part b (1975), 293–308.

CHAPTER

7

FINEPARTICLE CHARACTERIZATION BY ELUTRIATION

7.1 BASIC PRINCIPLES OF ELUTRIATION

Elutriation is the process of separating fineparticles into fractions of different mean diameter by suspending the fineparticles in a moving fluid. Forces operating on fineparticles in an elutriator can be combinations of gravity, viscous drag, and electrostatic and centrifugal forces. In an elutriation characterization study the population of fineparticles is split into two groups at a given Stokes diameter and the relative magnitude of the two fractions used to establish a point on the Stokes diameter distribution function of the fineparticle system. The complete distribution function is evaluated by splitting the fineparticles into many different-sized fractions. It is generally assumed in the literature of elutriation processes that all flow conditions in the elutriator are laminar. Elutriation studies do not readily yield information on the largest and the smallest fineparticles present in a fineparticle system, but they can yield direct gravimetric estimates of weights greater or smaller than a specified Stokes diameter.

In a vertical elutriator the fluid containing the suspended fineparticles flows upward. For a specific fluid velocity, fineparticles smaller than a given size move upward with the fluid whereas the larger fineparticles settle to the bottom under the influence of gravitational force. The Stokes equation (see Chapter 3) is used to calculate the Stokes diameter of the fineparticle just balanced by the upward-moving fluid. This size is defined as the cut size of the elutriator when operating at the specified fluid flow velocity. Thus from the Stokes equation, we have

$$d_c = \sqrt{\frac{18\eta \cdot v}{(\rho_P - \rho_L)g}}$$

where η = coefficient of viscosity
v = average velocity of fluid flowing upward through the elutriator
ρ_P = density of fineparticle

ρ_L = density of liquid
g = acceleration due to gravity
d_c = cut size of elutriator

Theoretical difficulties associated with the performance of a vertical elutriation are:

1. The fluid flow velocity across the cross section of the elutriating chamber is not constant. Thus in a cylindrical chamber the distribution is parabolic (see discussion of fluid flow through a pipe in Chapter 3). Random fluctuations in fluid flow cause migration of larger fineparticles just supported by the higher flow in the center regions to the outside of the chamber where the lower fluid flow is no longer able to sustain their buoyant weight. These fineparticles then fall and create a return flow of fluid that further disturbs the overall flow pattern of fluid. Attempts have been made to even out the flow rate across the main tube of an elutriator by placing a coarse grid across the tube [1].
2. If an elutriator is operated for long periods of time, the cut size associated with maximum flow along the center of the elutriator column dominates the fractionation achieved with the elutriator rather than the cut size calculated from the average flow.
3. To obtain steady flow conditions, long feed pipes to the elutriator are necessary. If there are any sharp bends in the feed pipework leading to or from the main tube, flow conditions are hard to predict and fractionation performance has to be determined empirically.

A practical problem associated with the operation of the elutriator is that the solids concentration of the suspension affects the performance of the elutriator; usually the higher the concentration of solids, the less predictable the fractionation efficiency. If the fluid flow rates required to fractionate the powders are very low, requiring several days to achieve fractionation, it is difficult to achieve thermal isolation and stability. It is also difficult to generate, maintain, and monitor the low flow rates required to operate the system.

In horizontal elutriators a stream of suspension is passed over a reservoir of clear fluid. The larger fineparticles leave the suspension stream and are collected at the bottom of the settling chamber. The fractionation efficiency of this device is not good since some fineparticles of all sizes will leave the suspension stream as it moves over the clear fluid in the collection chamber. The solids concentration of the feed stream must be kept low; otherwise, bulk penetration by the feed layer into the fluid in the container will occur (see discussion of the stability of the line start suspension system in Chapter 5). It is possible to use a more dense fluid in the reservoir to improve the stability of the system, but this system has not been used in size characterization studies. Fineparticles that drop through the full height of the fluid stream in the time taken for the fluid stream to move across the elutriating chamber are completely removed from the feed stream.

Cyclones, the operational principles of which were discussed in Section 3.2, are centrifugal elutriators. The efficiency of separation and the cut size of cyclones depend on the exact flow pattern established in the cyclone. Cyclones are widely used for removal of fineparticles from gaseous streams. They have not been used extensively for the preparation of well-defined, short-size-range powder fractions basically since flow conditions inside the cyclone are often unknown and are unpredictably affected by small changes in body design.

The empirical cut size of an elutriator is defined as that size at which 50% of the fineparticles (by weight) are in the undersize fraction. A measure of the efficiency of fractionation is the "coarse grade-efficiency curve" or Tromp's curve, which is calculated from the distribution functions of the undersize and oversize fractions. It is constructed by calculating the fractional weight of the feed fineparticles of a given size, which is to be found in the oversize fraction (see Figure 7.1). The coarse grade-efficiency curve is a dimensionless function that runs from the lower limit of the oversize fraction to the upper limit of the undersize fraction [2].

The variable shown as x_{25} in Figure 7.1*c* is the Stokes diameter of the fineparticle for which the coarse grade-efficiency curve has reached 0.25. Again, x_{75} is defined as the Stokes diameter for which the grade-efficiency curve has reached the value 0.75. The "index of fractionation efficiency" is defined as the ratio $x_{25}:x_{75}$. It would have a value 1 for an ideal elutriator. Because elutriators split the fineparticle system to be characterized into fractions and since the act of fractionating a fineparticle system is often referred to as *classification* (i.e., placement of the various-sized fineparticles into several classes), elutriators are sometimes referred to as "classifiers." We use the term "classifier" when it is the usual commercial name of an instrument but normally use the term "elutriator" for such instruments.

7.2 VERTICAL AIR ELUTRIATORS

A vertical elutriator described by Gonell [3] is shown in Figure 7.2. The powder to be fractionated is placed at the base of the elutriating column.

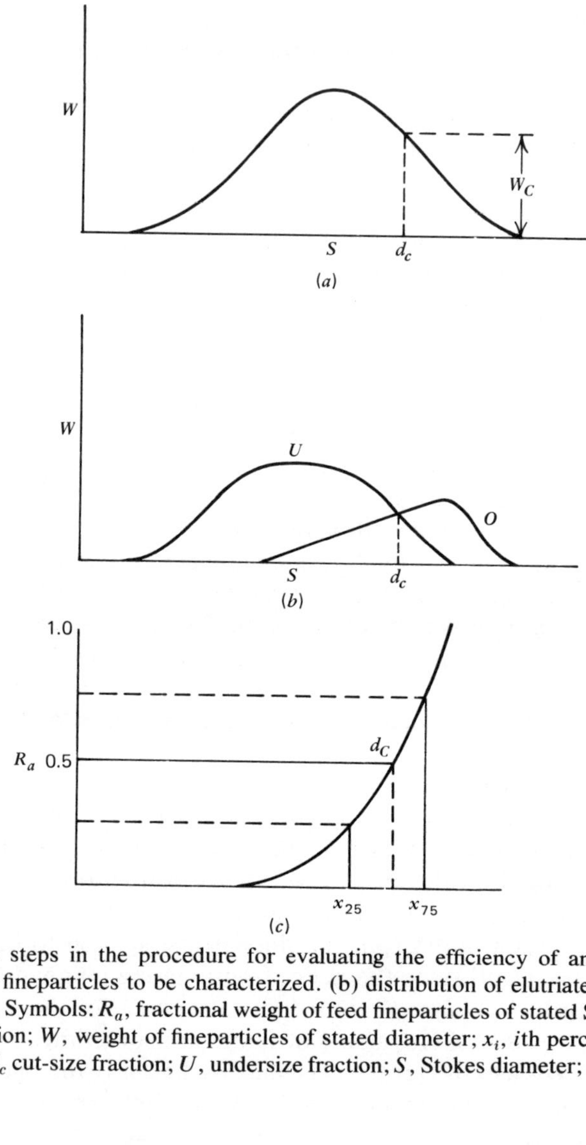

Fig. 7.1 Basic steps in the procedure for evaluating the efficiency of an elutriator: (*a*) distribution of fineparticles to be characterized. (b) distribution of elutriated fractions. (*c*) Tromps curve. Symbols: R_a, fractional weight of feed fineparticles of stated Stokes diameter in coarse fraction; W, weight of fineparticles of stated diameter; x_i, ith percentile; O, oversize fraction; d_c cut-size fraction; U, undersize fraction; S, Stokes diameter; W_c, feed weight of size d_c.

Air from the jet impinges down into, and through, the powder fluidizing it upward through the column. The long column is used to suppress flow disturbances created by the conical entrance of the elutriating chamber. A general problem of air-operated elutriators is that movement of dry powders around with dry air generates electrostatic charges, this causes powder to cling to the side of the elutriating chamber. Rapper hammers

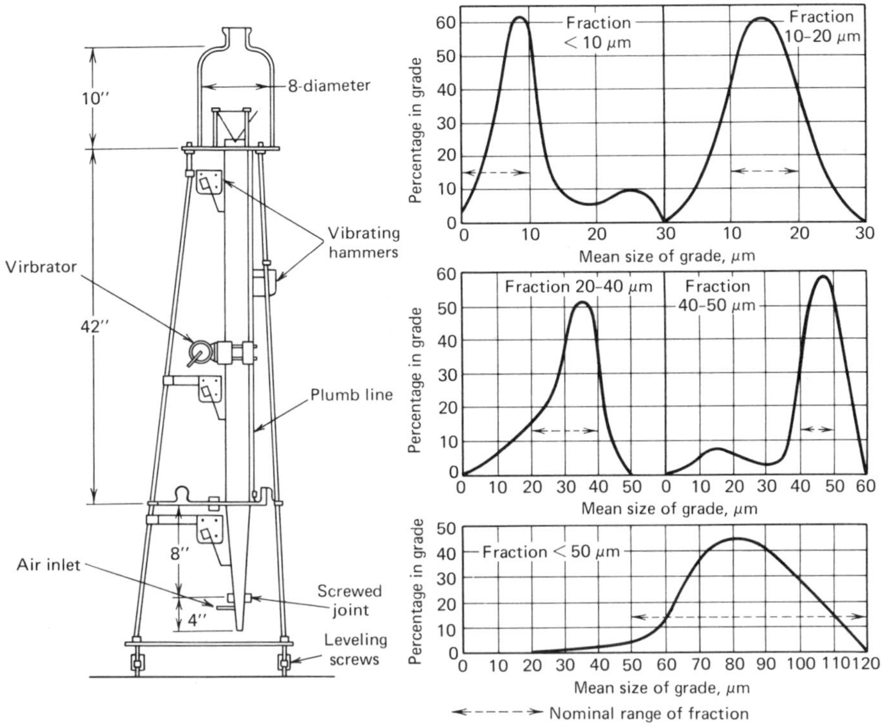

Fig. 7.2 Performance data for the Gonell elutriator [4]. Grade characterization obtained with the Phototrack sedimentometer (see Section 5.4). All micron sizes are Stokes diameters.

are used in the Gonell elutriator to remove powder from the side of the chamber.

Data on the performance of the Gonell elutriator, reported by Stairmand [4], are summarized in Figure 7.2. Using four different velocities of air, he fractionated a powder into groups shown. The theoretical size limits of the fractions calculated from fluid flow theory are shown. The size distributions of the fractions were obtained by using the Phototrak sedimentometer (see Section 5.4). Since the elutriation process is based on the Stokes equation, this accurate sedimentation method should provide comparable data. It can be seen that apart from an oversize fraction of greater than 50 μm, the fractionation efficiency is not good. However, compensatory errors help to yield a fair estimate of the Stokes diameter distribution of the powder if the data are combined to generate a distribution function.

The Alpine Company of Germany developed the Gonell elutriator.

Their version of the equipment has been described by Lauer [6]. In the Alpine Company literature it is stated that they do not provide flow accessories for fractionating at theoretical cut sizes smaller than 12 μm Stokes diameter since they find experimentally that below this size agglomeration of the powder prevents efficient fractionation. The literature also states that when dealing with powders as small as 12 μm fractionating times for high-efficiency separation can be as high as 5 hours. Table 7.1 compares

TABLE 7.1. Comparative Data on Elutriator Performance Reported in British Standards Specification No. 3406, Part 3, 1963, "Air Elutriation Methods": Size Characterization Data Obtained Using a Silica Powder

	Percentage Weight in Range			
Stokes Diameter, μm	Pipette Characterization, Andreasen Equipment, (Two Independent Laboratories)		Gonell	Microscope
105–75	—	—	—	3
75–53	7	8	8	5
53–37	18	14	17	13
37–26	14	10	15	17
26–19	12	15	13	18
19–13	13	13	11	14
13–9.4	8	8	6	10
9.4–6.6	4	8	6	6
6.6–4.7	7	6	3	5
4.7–3.3	4	3	4	3
3.3–2.3	3	3	2	1.5
2.3–1.7	2	1	13	2.5
1.7	8	11	13	2.5

data obtained with the Gonell elutriator with pipette sedimentation data generated by use of the British Standards characterization procedures.

Using a special version of the Gonell elutriator, Weilbacher [5] investigated the flow conditions within the elutriation chamber. He established that strong turbulence and unstable flow conditions characterize the air inlet region of the chamber of the Gonell elutriator and states that "the flow conditions of the lower end of the tube govern the separation achieved with the instrument but here the flow regime is not only indeterminate, but varies with time."

Weilbacher confirmed that the parabolic laminar velocity distribution forms near the upper end of the Gonell elutriator. He noted that the arrangement at the top of the tube for the removal of fineparticles fractionated in the chamber moved them through the boundary region of low velocities at the side of the tube, thus causing resedimentation into the boundary region and down the elutriation column and prolonging the separation process. The system shown in Figure 7.3, which uses only a short, straight fractionating column in a conical elutriating chamber, was designed to overcome these unsatisfactory features of the Gonell elutriator. The ratio of the height of the fractionating region of the chamber to the diameter of the elutriating chamber is 0.1. The base of the elutriating chamber is equipped with a high-impedance air filter. This dampens out any turbulence caused by fluctuations in the air supply to the base of the elutriating chamber. The top of the equipment can be readily separated from the base so that the material to be fractionated can be inserted on top of the filter.

Air fluidizes the material on the filter, and the fines are carried up the fractionation column. After spending a short time in the straight portion, they reach the angle of the collection cone region of the elutriating chamber. Here the flow rate starts to increase so that any fineparticles that are able to move to the top of the fractionating column are then swept out of the equipment and into the collection region at the top. The fines are collected on a filter as shown. Performance data for this new elutriator, reported by Leschonski and Rumpf, are summarized in Figure 7.3 [7, 8].

It is common practice in the air elutriation of fine powders to add what is known as a *deagglomerating agent* or a *flow agent* to improve the flow characteristics of the powder. A frequently used flow agent is colloidal silica. The exact mechanisms by which the improvement of flow is achieved by the addition of these very fine powders is not fully understood [9].

In their discussion of the performance of the elutriator shown in Figure 7.3, Leschonski and Rumpf state that the equipment should not be used to

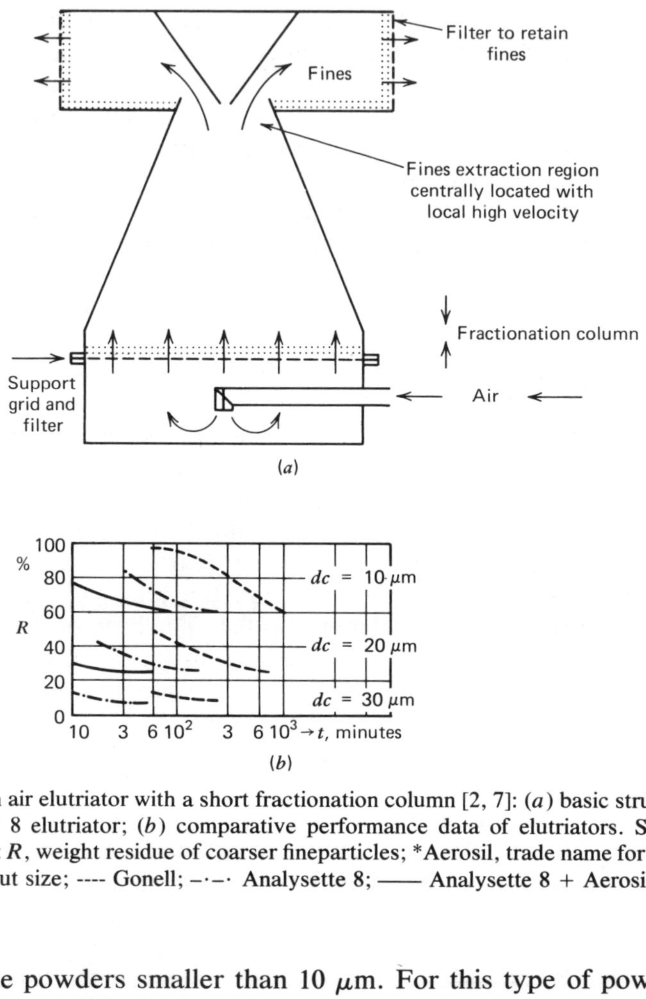

Fig. 7.3 An air elutriator with a short fractionation column [2, 7]: (*a*) basic structure of the Analysette® 8 elutriator; (*b*) comparative performance data of elutriators. Symbols and tradenames: *R*, weight residue of coarser fineparticles; *Aerosil, trade name for a silica flow agent; *dc*, cut size; ---- Gonell; –·–· Analysette 8; —— Analysette 8 + Aerosil*.

fractionate powders smaller than 10 μm. For this type of powder, they recommend the use of the centrifugal counterflow elutriator described in Section 7.5.

The Roller elutriator shown in Figure 7.4 was first described in the late 1930s [10]. It is widely used and is specified in both ASTM [11] and British Standards [12] procedures for characterizing fineparticles. A special feature of the Roller elutriator is the "goose neck," a flexible feed tube to the elutriating chamber that is vibrated rapidly during the operation of the elutriator. Fines passing through the elutriator are captured on a special filter. Data on the fractionation efficiency of the Roller elutriator are summarized in Figure 7.4. The performance of the Roller elutriator is

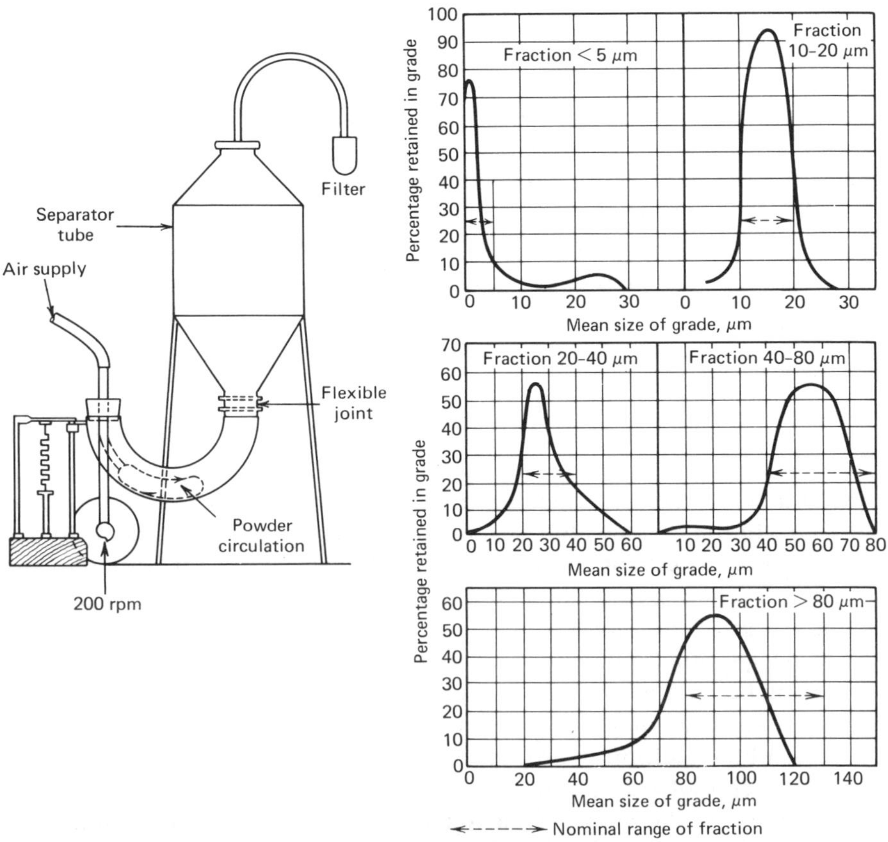

Fig. 7.4 Data on the performance of the Roller elutriator [4]. All sizes quoted are Stokes diameter. Grade fraction characterization carried out with Phototrack sedimentometer (Section 5.4).

regarded as adequate for many industrial purposes, and it is widely used in the ceramics and metallurgical industries [13].

The Gonell and Roller elutriators are single-chamber instruments. With both instruments, different cut sizes can be achieved by changing the air velocity. To extend the range of operational cut sizes, the manufacturer of the Roller elutriator supplies several chambers of varying diameter. The Roller equipment is noisy because of the mechanical system used to jolt the goose neck, and the equipment is usually operated in a sound-muffling box.

The Aminco classifier is an adaptation of the Roller elutriator in which eight different-sized elutriating chambers are linked together in series [14].

The Haultain infrasizer is a cascade air elutriator [15]. This equipment was developed by Professor Haultain at the University of Toronto in the late 1930s. The special feature of this apparatus is the dispersion system at the bottom of each of the conical elutriation chambers. It utilizes a golf ball. Air entering the elutriating chamber elevates the ball slightly, creating a narrow annulus around the ball. Shearing action caused by the rotation of the rough surface of the golf ball aids the deagglomeration of fine powders [16].

7.3 LIQUID GRAVITY ELUTRIATORS

The Schone elutriator shown in Figure 7.11 was one of the first liquid-operated elutriators described in the scientific literature. Data on the performance of the Schone elutriator are summarized in Figure 7.5 [4]. It can be seen that the equipment is relatively efficient in the separation of the oversize fraction greater than 80 μm but that the efficiency of fractionation at 10 μm is not good.

The design of a liquid elutriator that can be used to make narrow-size-range fractions, the Blythe elutriator, is shown in Figure 7.6 [18]. A unique feature of this elutriator is that the technologist can follow the progress of the elutriation process by taking samples from the sump at the base of each elutriating column. Paddle stirrers are used to agitate the suspensions at the base of each elutriating column. Each elutriating chamber plus exit tube is a link in a syphon chain. The total flow through the syphon chain is established by feeding the sump of the first chamber with a constant-pressure apparatus and by adjusting the screw clip on the final exit tube of the system. When the system is operating, there is a slight oscillation of fluid in the individual elutriating chambers superimposed on the through flow of fluid because of the use of the syphon mechanism to draw suspension through the system. Thus the syphon action tends to draw fluid out of the container immediately below an elutriating chamber. This flow of fluid into the next sump builds up a reverse pressure that attempts to alter the flow direction of the syphon. Fluid will flow in one direction until the reverse pressure generated is sufficient to cease the flow of fluid and reverse its direction. When flow is reestablished in the reverse direction, fluid will flow until the redeveloped reverse pressure stops and reverses the flow.

The elutriator system is initially filled with clear liquid and the powder

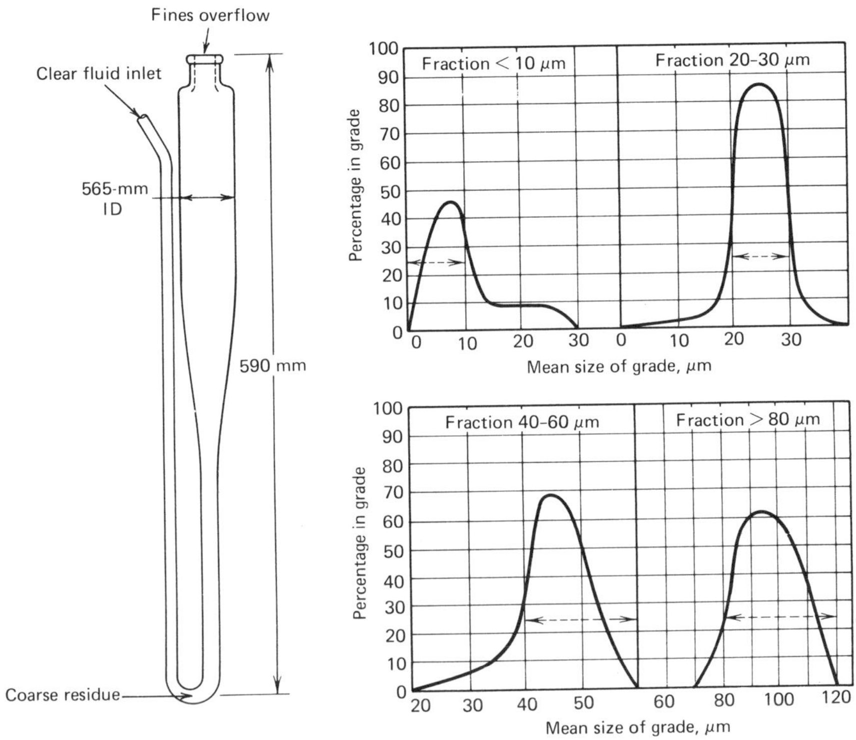

Fig. 7.5 Data on the performance of the Schöne elutriator. *Powder*—sand. All sizes quoted are Stokes diameters in microns; fraction characterization by Phototrack (ID = inner diameter).

to be fractionated placed in the first sump. The moving boundary of the suspension climbing the elutriator is clearly visible in the initial stages of elutriation. When this boundary reaches the top of the column, turbulent flow conditions established at the outlet from the elutriator column are often clearly visible. If the top of the elutriating chamber is not carefully designed, a density-driven convection current down the side of the elutriating chamber forms at the exit and the movement of fineparticles down the side of the elutriating chamber are clearly visible.

The Blythe elutriator can be used with relatively low flow rates operating over a period of 3 days, to fractionate fineparticles as small as 2–3 μm Stokes diameter. The size fractionation limit is governed by both the density of the fineparticles to be fractionated and the difficulties of providing a thermally stable environment for the equipment.

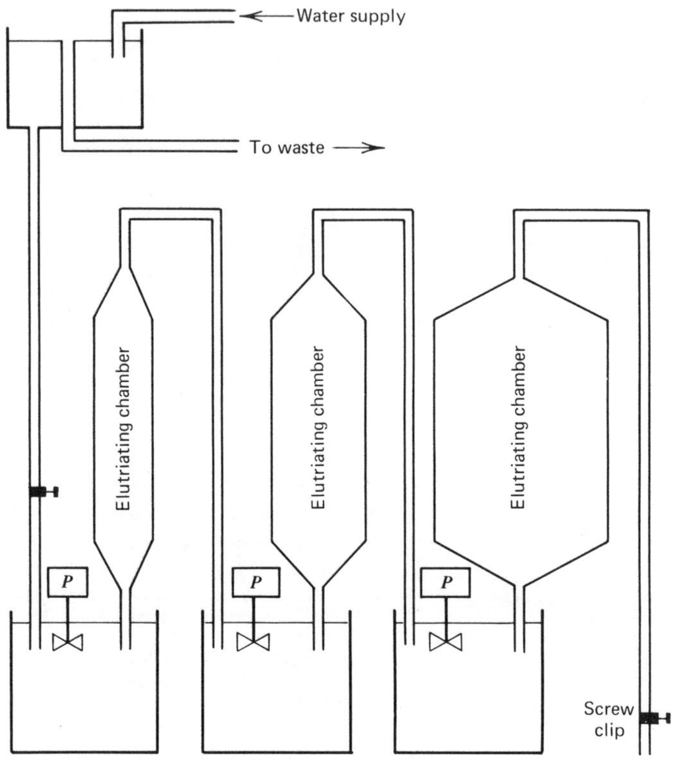

Fig. 7.6 The Blythe elutriator [18] (*P*, paddle stirrer).

7.4 HORIZONTAL ELUTRIATORS

The Hexlet sampler used to sample aerosols is a horizontal elutriator that splits the aerosol fineparticles into two fractions. Its design, shown in Figure 7.7, is based on the work of B. M. Wright at the Pneumoconiosis Research Unit of the Medical Research Council, Cardiff, Wales [19]. The intake of the equipment is of a stack of parallel plates. Fineparticles starting at the entrance to the rectangular duct formed by adjacent plates sediment toward the bottom plate as the aerosol stream moves through the equipment. Fineparticles above a certain critical size are all deposited onto the plates as the air stream reaches the end of the rectangular duct. The sedimentation velocity of the fineparticles in the vertical direction is

---

Fig. 7.7 The Hexhlet aerosol sampler [19]: (*a*) physical design; (*b*) performance data. Symbols: *S*, Stokes diameter in microns; *PE*, percent penetration of stated size through elutriator plates. (See p. 239 for illustration.)

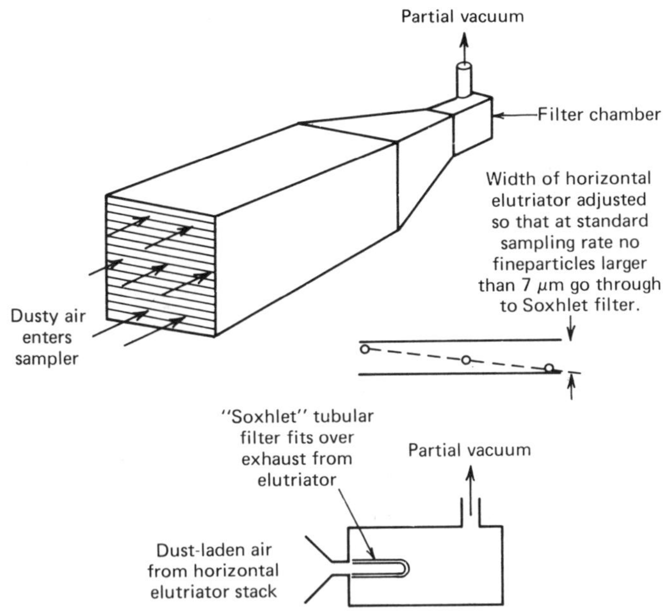
Partial vacuum
Filter chamber
Width of horizontal elutriator adjusted so that at standard sampling rate no fineparticles larger than 7 μm go through to Soxhlet filter.
Dusty air enters sampler
"Soxhlet" tubular filter fits over exhaust from elutriator
Partial vacuum
Dust-laden air from horizontal elutriator stack

(*a*)

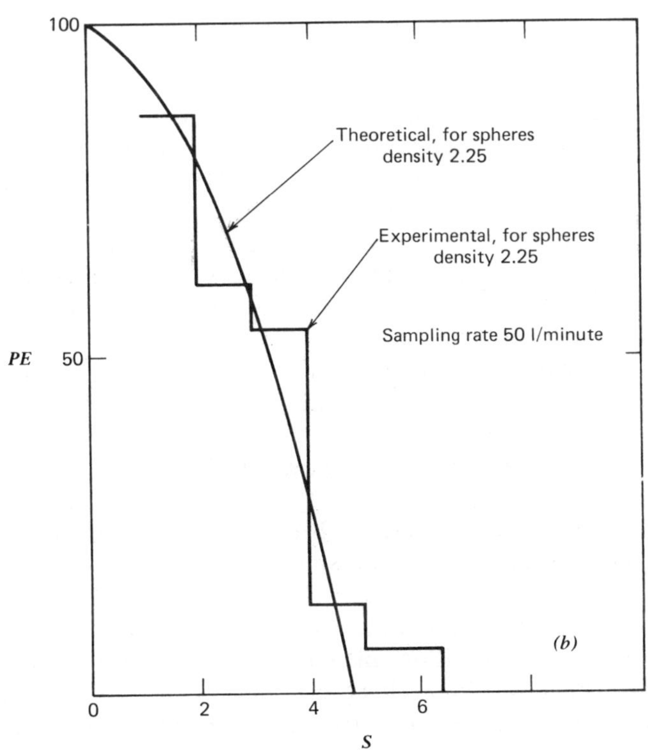
100
Theoretical, for spheres density 2.25
Experimental, for spheres density 2.25
Sampling rate 50 l/minute
PE
50
(b)
0
2
4
6
S

calculated by using Stokes equation. The penetration of the horizontal elutriator chambers for spheres of density 2.25 when the equipment is operating at 50 l/minute is illustrated in Figure 7.7. Spheres of density 2.25 were used in studying the penetration of the elutriator since the original work was concerned with coal dust, and 2.25 is representative of the density of the type of dust encountered in coal mines.

It can be seen that under these operating conditions, fineparticles of Stokes diameter of greater than 7 μm are all precipitated onto the plates of the elutriator. The bottom plates of the elutriator ducts are oiled to prevent reentrainment of fineparticles into the air stream leaving the elutriator. The reason for choosing 7 μm as the cutoff in the elutriator design is that in Great Britain, fineparticles larger than this are not considered dangerous to the lungs. There is some controversy over the magnitude of this size limit for dust capable of causing damage to the lungs, and European and North American standards opt for a slightly lower critical size.

Fineparticles leaving the elutriation stage of the Hexlet air sampler are captured by a high-efficiency tubular filter known as a *Soxhlet thimble filter*. In mine safety practice the weight of powder retained by the Soxhlet filter in a given period is described as the "burden" of dangerous dust in the working environment. The flow rate through the horizontal elutriator can be altered to achieve different cut sizes.

In the Timbrell aerosol spectrometer [20], shown in Figure 7.8, the aerosol to be characterized is aspirated into the center of a flow of clean air at point *B*. As this stream passes through the elutriation chamber *W*, the fineparticles sediment to the bottom. The largest are deposited at the entrance to the chamber. The flow rate through the system drops as the aerosol moves along the wedge-shaped elutriation chamber to give the smaller fineparticles longer residence time to facilitate their deposition on the base of the chamber. The commercial equipment is provided with charts showing the location along the duct at which fineparticles of a given Stokes diameter will be deposited for a given flow rate through the system. The number of fineparticles deposited at a given location is evaluated by microscope inspection.

The Whitby aerosol analyzer [21], shown in Figure 7.9, is an elutriator in which the significant force applied to the fineparticle in the moving viscous fluid is electrostatic in origin. It is used to fractionate very small aerosol fineparticles of Stokes diameter of 0.015–2.0 μm. The effects of gravitational forces on such fineparticles can be neglected. The aerosol is sucked into the ionization chamber at the point shown in Figure 7.9. The ionization conditions within this chamber are controlled so that the fineparticles of the aerosol are charged in such a way that the electrical mobility of the fineparticle is proportional to the size of the fineparticle.

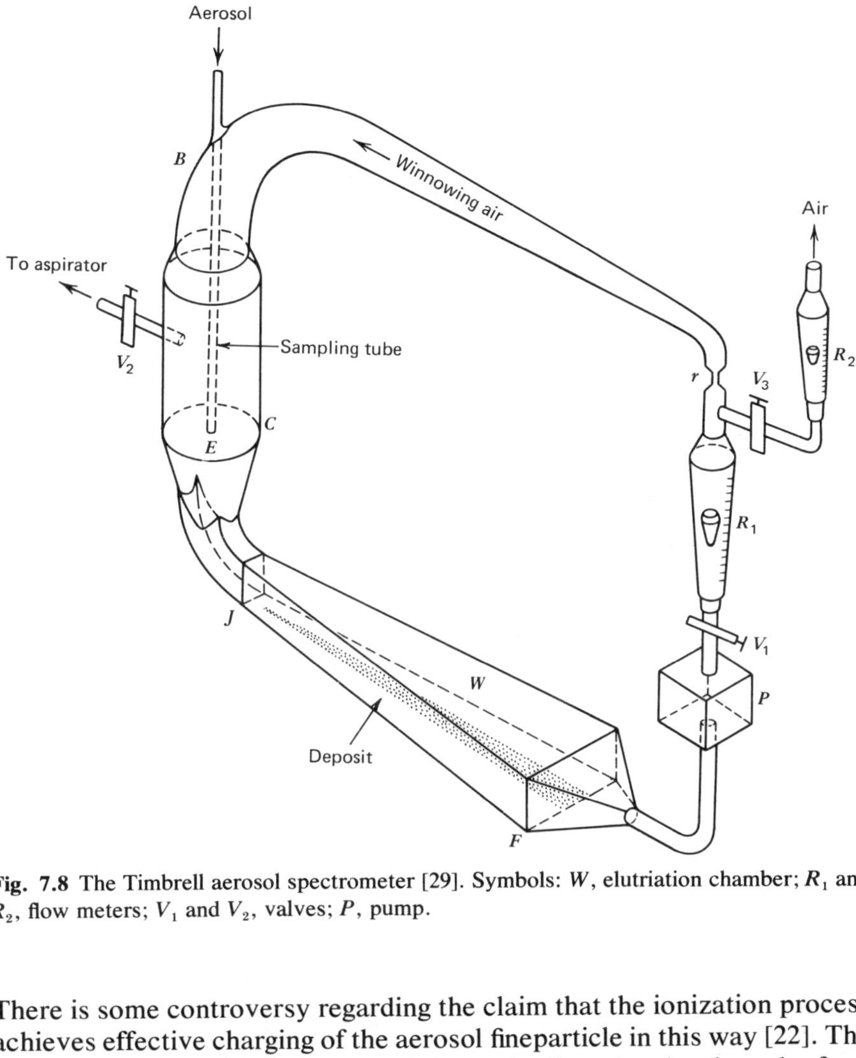

Fig. 7.8 The Timbrell aerosol spectrometer [29]. Symbols: W, elutriation chamber; R_1 and R_2, flow meters; V_1 and V_2, valves; P, pump.

There is some controversy regarding the claim that the ionization process achieves effective charging of the aerosol fineparticle in this way [22]. The flow of charged fineparticles out of the ionization chamber into the fractionation chamber is monitored as shown in Figure 7.9. Down the center of the cylindrical chamber is a rod that can be charged to a high potential. An auxiliary system shown in the top right-hand side of Figure 7.9 is used to create a clean air sheath around the rod. The outer limits of this sheath are indicated in Figure 7.9 by dotted lines.

The charged aerosol fineparticles moving through the fractionation chamber are attracted to the central rod. Depending on the electric potential of the rod, all fineparticles less than a given size are collected by the

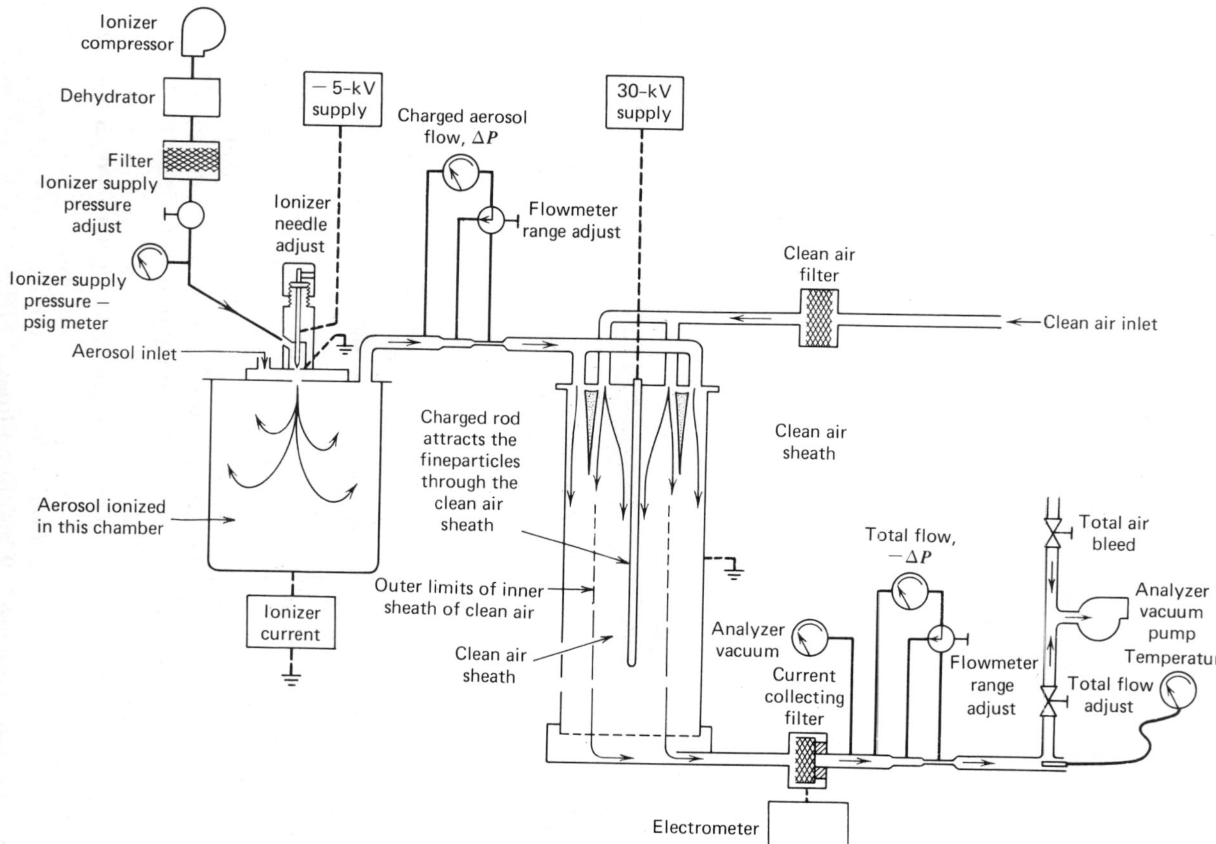

Fig. 7.9 The Whitby aerosol analyzer [21].

rod whereas the larger ones escape to the charge collection filter. The magnitude of the current generated at the collection filter is a measure of a number of fineparticles reaching the filter. An aerosol is sized by carrying out a series of experiments measuring the charge on the current-collecting filter at different collection potential differences applied to a central rod. The equipment is calibrated using aerosols of known characteristics.

The cross-flow classifier shown in Figure 7.10, developed by various investigators [23–25], is an industrial-scale horizontal elutriator. Fineparticles to be fractionated are entrained in a high-speed air jet, which is fired across a winnowing stream of air. Typical velocities used are 60 m/second for the high-speed air jet and 20 m/second for the winnowing air stream. The dispersed array of fineparticles created by the interaction of the high-speed jet and the winnowing air is split into eight fractions by using sharp-edged partitions as shown in Figure 7.10*a*. The largest fineparticles have the most kinetic energy and are caught in compartment 8, with the smallest fineparticles being blown into container 1. Performance data are given in Figure 7.10*b*.

Leschonski and co-workers used a photometer to evaluate the fineparticle concentration across the dispersion fan created by the interaction of the high-speed air jet and the winnowing air. In Figure 7.10*c* data generated by a typical track of the photometer across the dispersion fan is illustrated. The Lambert–Beer law is used to relate the transmission drop at any position within the dispersion fan to the concentration of fineparticles at that location. Leschonski and co-workers calibrated the photometer readings by using a powder of known size distribution as determined by the Sedigraph 5000 x-ray sedimentometer (see Chapter 5). Leschonski reports that with equipment calibrated in this way, good agreement is obtained between size characterization by photometric inspection of the dispersion fan in the cross-flow classifier and characterization by Sedigraph 5000. (*Note:* In essence, this is a direct empirical correlation between two experimental procedures, and the agreement between the two sets of data gives no information on the presence of any systematic errors in either procedure, only on the invariance of the cross correlation when the two procedures operate under certain conditions.) Leschonski reports that the photometer scan of the dispersion fan required only 3 seconds and that operating at throughputs of 2–5 kg/hour, size characteristic distributions were reported every minute, with each distribution the average of 10 sets of individual measurements [25].

The simple horizontal elutriator shown in Figure 7.11 could prove useful in the mineral processing industries. A very thin veil of fineparticles is allowed to fall across a horizontal stream of air. The air is fed into the rectangular duct through a glass-wool filter so that the air velocity is

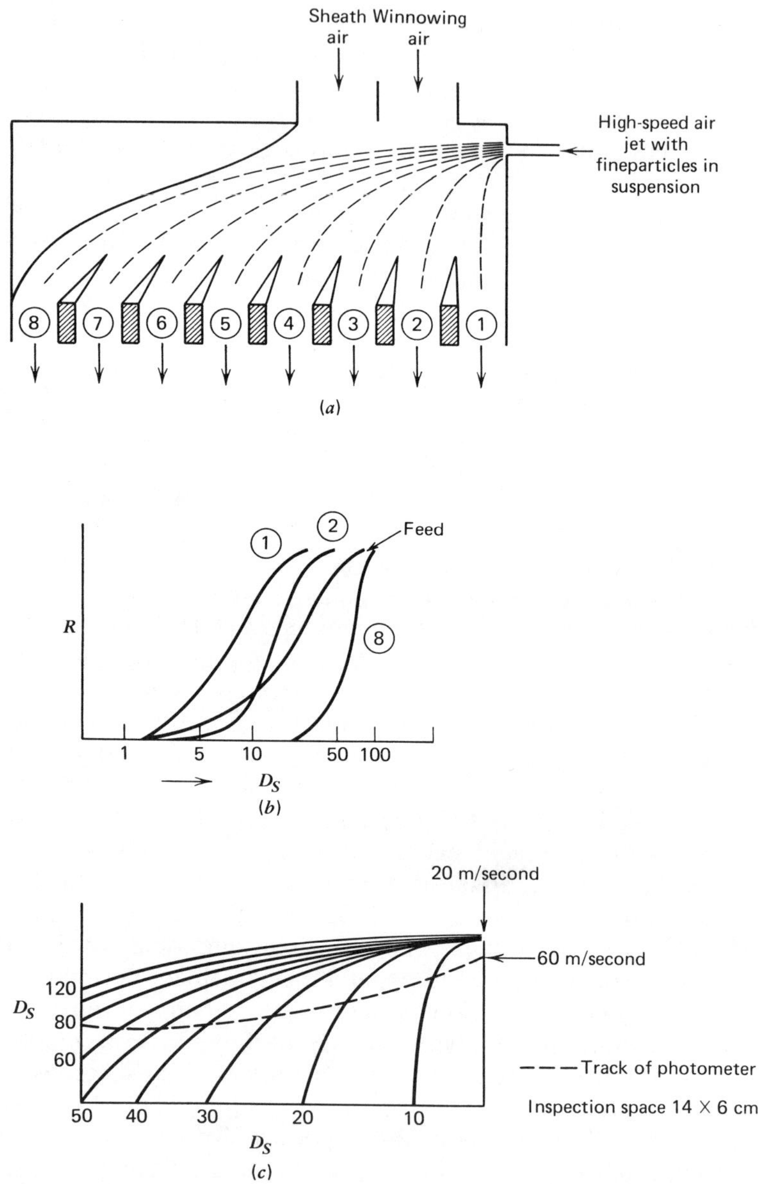

Fig. 7.10 The cross-flow classifier: (*a*) basic system (fineparticles split into eight fractions); (*b*) typical performance data; (*c*) fineparticle size location within the dispersion "fan"; *R*, percentage less than stated size; D_S, Stokes diameter.

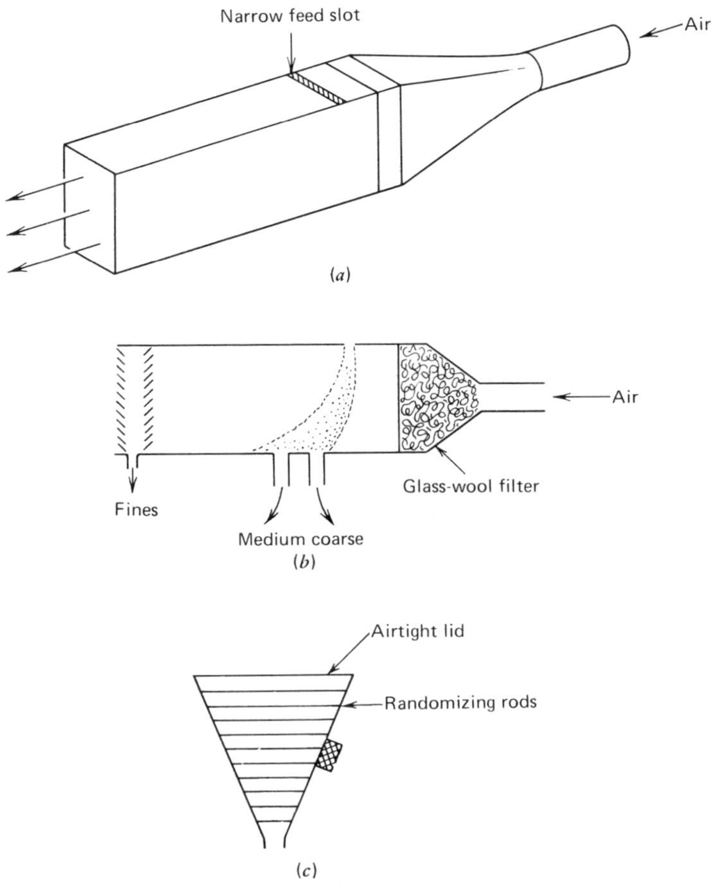

Fig. 7.11 The Winsift is a simple horizontal elutriator: (*a*) side view; (*b*) section view; (*c*) special feed hopper.

uniform across the duct. The fineparticles in the falling veil are winnowed by this air. The fines are blown along the duct and collected at the end by a filter. The exit slots for coarse and medium fineparticles are provided with reservoirs closed to the air during the operation of the equipment to enhance the stability of the winnowing action. The hopper used to feed the fineparticles into the winnowing duct has a close-fitting lid and is equipped with a vibrator. The hopper contains many randomizing rods so that the fineparticles have no momentum when vibrated into the air stream. Two or three small holes are drilled in the lid to prevent a buildup of a back pressure within the hopper.

7.5 AIR-OPERATED CENTRIFUGAL ELUTRIATORS

Centrifugal elutriation of fineparticles has been used in the fractionation of aerosols. The first instrument of this type, the Conifuge shown in Figure 7.12, consists of a rotating solid cone to which is attached a rotating conical shell. These cones are housed in a closed container. The rotation of the cones, typically at 3000 rpm, creates a pumping action, drawing the aerosol down into the space between the cones. A normal sampling rate would be about 25 cm³/minute. A small valve on the outside container is adjusted so that part of the air moving over the cones is recirculated. This recirculating air is free of fineparticles larger than those

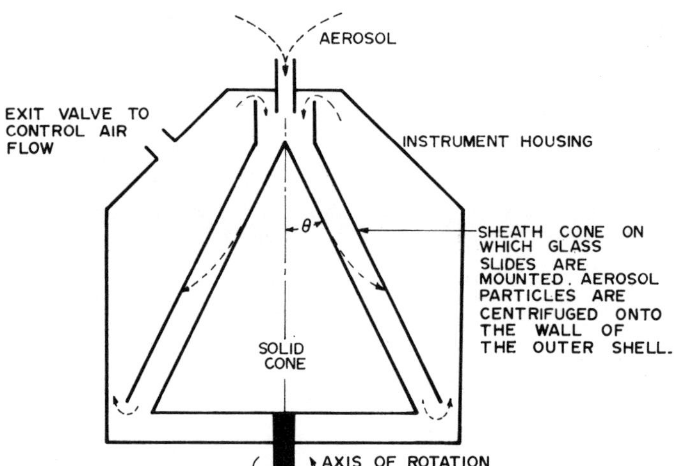

EQUATIONS:

$$y = \frac{\pi^{1/2} \tan\theta \, \exp\left[A^2(x+a)^2\right] \left\{\left[\operatorname{erf} A(x+a)\right] - \operatorname{erf}(Aa)\right\}}{2A}$$

WHERE $A = \left[(\pi a \omega^2 U_s \tan\theta)/Qg\right]^{1/2}$

y AXIS ALONG CONE AXIS.

θ ANGLE OF CONE AS SHOWN

$a = (D/2)\cot\theta$

D = DISTANCE BETWEEN CONES (GAP WIDTH)

$$\operatorname{erf}\beta = \frac{1}{\sqrt{\pi}} \int_0^{\beta} e^{-\beta^2} d\beta$$

Fig. 7.12 The Conifuge: (*a*) basic system; (*b*) graphical summary of deposition locations for various aerodynamic diameters.

that can be deposited on the cone and forms a sheath for the incoming aerosol streams as indicated. The centrifugal action of the spinning cone causes the fineparticles in the aerosol to move outward through the air sheath as they move downward. Thus the fineparticles are deposited on the outer cone according to their Stokes diameter, with the largest at the top end of the cone. Two glass slides are fitted along the length of the inside of the outer cone to collect the fineparticles for subsequent examination and evaluation. Because the overall flow rate varies as the aerosol moves down over the inner cone and the centrifugal force experienced by the fineparticles in the cone gap is constantly changing, an exact solution relating deposition location for a given Stokes diameter to the operating characteristics of the Conifuge is difficult. Sawyer and Walton [26] have provided a graphical summary of their solutions to the general equations of motion in the Conifuge for specified conditions. Typical deposition data are summarized in Figure 7.12 [6].

The basic system of the Goetz aerosol spectrometer used for sampling and characterizing aerosols is shown in Figure 7.13. Two deep helical grooves are cut in the surface of the conical rotor. Just above the rotor is a detachable cup, to the inner surface of which is clamped a removable metal foil. The fineparticles moving down through the helical grooves are

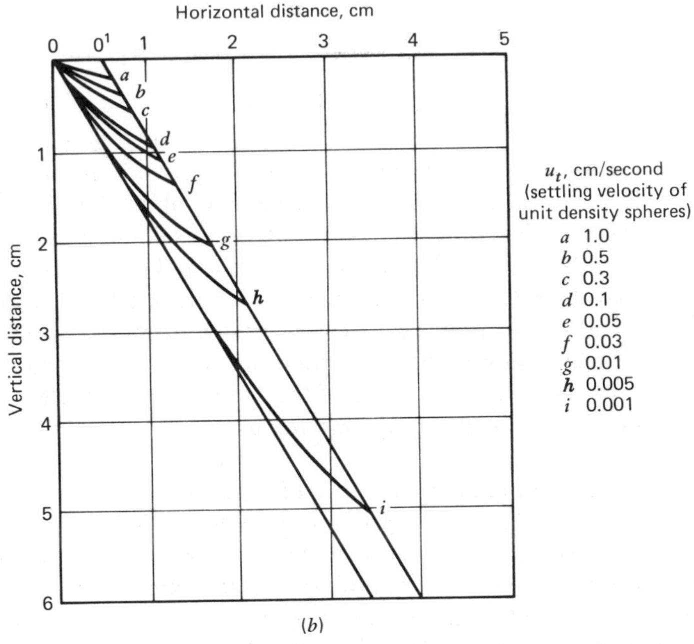

(*b*)

Fig. 7.12 (*Continued*)

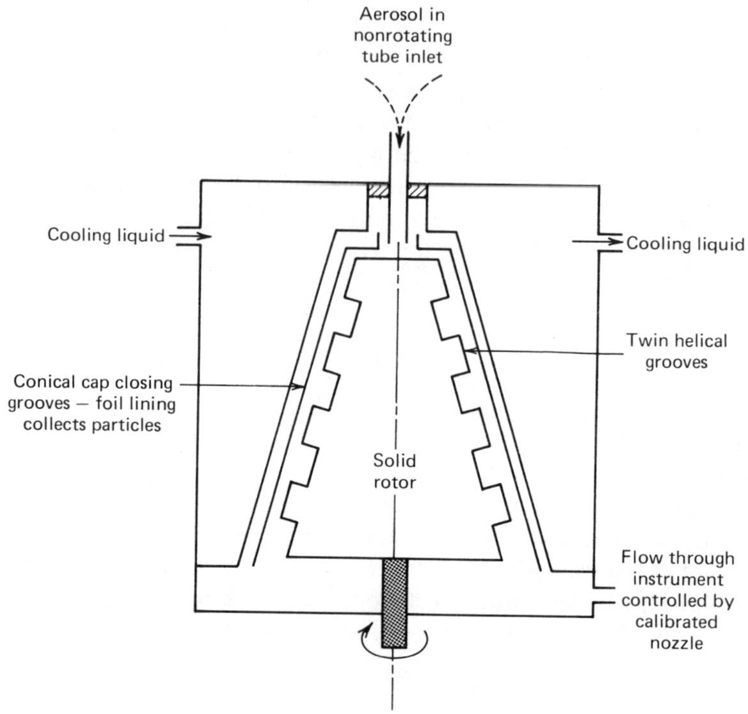

Fig. 7.13 The Goetz aerosol spectrometer [27].

centrifuged toward the wall. When the foil is removed and laid flat, the fineparticle deposit on the foil is a pair of helical bands. The fineparticles can be examined microscopically. For fineparticles of dimensions of less than 1 μm, dark-field microscopy has to be used. The success of dark-field microscopy depends on the degree to which the light scattered by the background is suppressed. Since dark-field microscopy is a sophisticated technique, the background skills of which are not always available in a given laboratory, the manufacturers of the Goetz spectrometer supply a specialized microscope viewing system for use with the foils taken from the spectrometer.

Goetz carried out extensive tests to find a suitable foil for use with the instrument. One of the best foils proved to be highly polished chromium. Although the finest polish grooves in the foil caused a disturbing light scattering level, if the polishing technique is such that all the grooves run parallel over the entire foil, their scattering contribution can be minimized by inserting a special optical filter into the illuminating beam of the microscope. The optical filter has to be put in a special position so that the back-scattered light from the foil grooves is eliminated.

The outer casing of the Goetz spectrometer can be cooled or heated so that the internal temperature of the centrifuge can be controlled within very close limits so that the temperature conditions within the centrifuge can be made to match those of the sampled aerosol cloud. Special versions of the equipment that have high-capacity thermal control systems for use in arctic or tropical conditions have been constructed [27].

The centrifugal elutriator for sampling and characterizing aerosols designed by Hochreiner and Brown is illustrated in Figure 7.14. In this instrument the whole "centrifuge–air sheath generating mechanism" rotates, thus obviating the need for rotating air seals. This is a major advan-

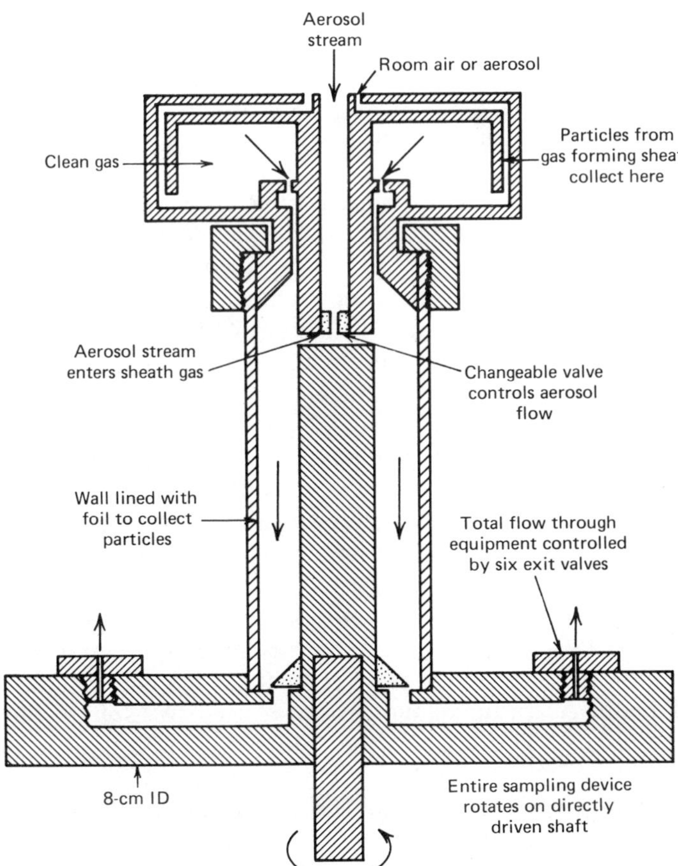

Fig. 7.14 Cylindrical centrifuge sampler described by Hochreiner and Brown [29].

tage of instrument design from an engineering point of view. The aerosol stream to be characterized is drawn through the apparatus by the self-pumping action of the rotating centrifuge head. Adjacent to the aerosol intake is a set of holes that take air from the atmosphere into the upper air-cleaning chamber. This chamber has a larger dimension than the cylinder in which the aerosol fineparticles are to be collected for examination. Therefore, all fineparticles bigger than those of interest are sedimented onto the walls of the upper chamber. The resultant clean air is then drawn down through the main fractionating chamber to provide a clean sheath for the incoming aerosol stream. The aerosol stream enters the fractionating chamber through a changeable valve that can be used to control the relative flow of aerosol to clean air. The wall of the main cylinder chamber is lined with foil to collect the deposited fineparticles for subsequent examination. Residual air makes its exit from the instrument through the valves shown. These valves are further out from the axis of rotation than any other portion of the cylinder to maintain self-pumping action of the air through the centrifuge as it rotates. These exit valves can be changed to control the volume flow rate through the instrument [28, 29].

One disadvantage of the cylindrical-type instrument is the comparatively long cylinders required to deposit the smallest fineparticles of interest. Hochreiner and Brown [28] have described another version of their centrifuge in which the classification chamber is conical as in the Goetz spectrometer and the Conifuge. In this equipment, as the fineparticles move lower down the classification chamber, the volume flow rate falls off so that the fineparticles spend a longer time in the centrifugal field and experience greater centrifugal forces [29].

As discussed in Chapter 3, impaction and impinger aerosol sampling devices are centrifugal elutriators in which the deflection of a jet of aerosol by a collector glass slide creates a quarter turn of a centrifuge. The basic operational theory of the jet impactor was given in Chapter 3. The Konimeter, manufactured by Sartorius, is a widely used jet impactor in which a fixed volume of air is drawn rapidly through a nozzle placed 0.5 mm from a collection disk [30]. The sharp change of direction imposed on the air stream by the collection plate centrifuges the fineparticles onto the collection plate. A spring-loaded piston is used to aspirate the sample through the nozzle. This enables the equipment to be used in industrial areas without the risk of generating electrical sparks which could ignite potentially explosive atmospheres. The equipment has been widely used in the sampling of dust in coal mines. In English-speaking countries the Konimeter is often known as a *Conimeter*.

In cascade impactors the aerosol flow is arranged so that it passes through a series of jets of decreasing orifice size. Thus in the Sciotec cascade impactor, a nest of interlocking cylinders, each of which contains a jet and a slide collector, are assembled with the widest-mouthed jet at the top. Smaller and smaller fineparticles are deposited at each successive stage of the instrument. As normally operated, it has six stages of impaction. Aerosol is sampled at the rate of 12.56 ft^3 of air per minute. Calibration nomographs are provided with the system so that aerodynamic diameters deposited at any given stage with any specified flow rate through the instrument can be rapidly calculated [31].

Another widely used cascade impactor, the Anderson sampler [32], is constructed from a set of nesting cylindrical sections. The impaction jets are holes drilled in the base plates of the cylinders. There are 400 holes in the base of each element of the sampler. A shallow glass dish, known as a petri dish, is placed under the jets. These dishes are either smeared with Vaseline to help retain the deposited aerosol or can be filled with agar jelly so that, if biological aerosols are being sampled, the bacteria colony can be grown directly on the collecting medium. As the aerosol streams move down through the column the air passes through smaller and smaller jets. The flow rate through the system can be set at any required level.

The structure of the May cascade impactor is illustrated in Figure 7.15. The microscope slide used for collecting deposited aerosol at each stage can be removed from the body of the impactor by removing the rubber cap. During the deposition operation the microscope slide is held in position by spring clips (not shown in Figure 7.15). The aerosol to be characterized is drawn through four successive stages of the impactor, and the jet orifice decreases at each stage of the impactor. The equipment is sometimes used with a fifth stage containing a membrane filter. The percentage penetration of various sized fineparticles of the various stages, for normal operating conditions, are summarized in Figure 7.29. It can be seen from Figure 7.29 that fineparticles not normally considered a respiration hazard are deposited by the stage 1 jet [33].

In the Lundgren cascade impactor system the fineparticles are collected on revolving drums placed opposite the impactor jets [34]. Each drum has a total surface of 10 $in.^2$ In one model, each drum rotates once every 24 hours, and this is a useful feature in air-pollution studies. In another version of the equipment eight speeds of rotation of the collector drums, varying from one revolution per minute to one revolution per day, can be selected. The drums can be covered with either foil or paper for collecting the impacted fineparticles. One reason for using foil is that this can be used with dark-field illumination conditions to view fineparticles less than 1 μm in size (see discussion of the Goetz aerosol spectrometer earlier in this section).

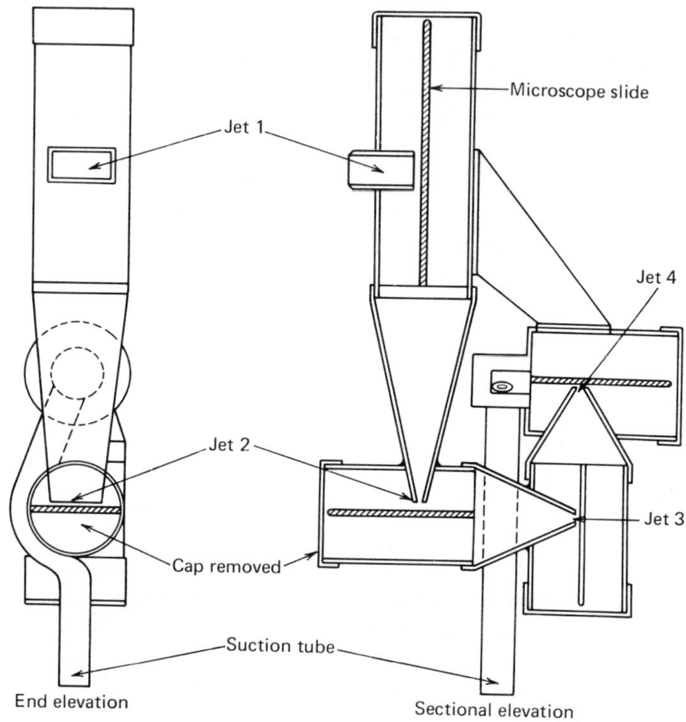

(*a*)

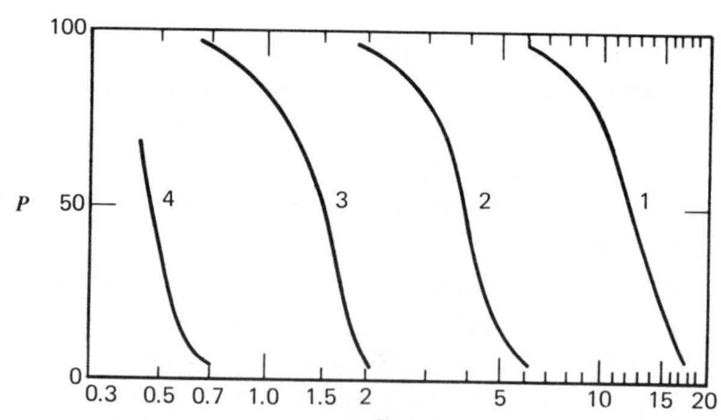

(*b*)

Fig. 7.15 May cascade impactor [33]: (*a*) construction; (*b*) penetration performance. Symbols: *P*, percentage penetration of indicated stage at a flow rate of 17.5 l/min; 1, 2, 3, and 4 denote stage of the impactor; S_A, aerodynamic diameter (i.e., Stokes diameter of unit density spheres) in microns.

In the Unico cascade impactor the impacted fineparticles are collected on glass slides placed opposite to rectangular slits. At the final stage residual aerosol is filtered on a membrane filter. Each collection plate is attached to a geared wheel that can advance the plates simultaneously by $\frac{1}{16}$ in. to provide a fresh collection surface whenever necessary. A nomogram is provided with this instrument for calculating the penetration efficiency per stage under various operating conditions for various types of aerosol [35].

As pointed out in Chapter 3, the use of solid collection surfaces in impactors creates problems of bounce and reentrainment. To minimize this type of experimental uncertainty in the operation of impactors, the virtual impactor surface, in which the aerosol fineparticles sink into a virtual surface created from a pocket of still air, was developed. As described in Chapter 3, experimental investigation established that better results were obtained with virtual surfaces if there was a slight airflow through the collection reservoir, typically $\frac{1}{50}$th of the airflow past the top of the virtual surface. The Centripeter® impactor [36] uses virtual collection surfaces. The construction of this instrument is illustrated in Figure 7.16 [37]. In this device the aerosol stream is forced to turn sharply at the orifice entrance. The centrifugal action created by this sharp diversion is greater than that in a normal jet impactor of comparable orifice diameter. The word "centripetal" is based on two Latin words that mean "center seeking," and hence the name of this device in which the fineparticles are forced to seek the center of the airflow because of the forces created in the sharply turning air in the feed nozzle of the equipment. At each stage of the centripetal impactor the oversize fineparticles impacted into the reservoir are slowly drawn to the collection filter by a low flow rate of air down through the reservoir forming the virtual surface. The undersize fineparticles move with the airstream diverted around the collection orifice down to the next stage.

Celesco Industries manufactures a cascade impactor in which the deposition surface is a piezoelectric crystal oscillator [38]. The frequency of this oscillator varies with the mass of fineparticles deposited on the crystal. With this equipment one can continuously record and, if necessary, transmit the relevant information to a remote receiver, the arrival of fineparticles on the deposition stages of the impactor. A piezoelectric crystal alters its shape when it is subjected to an electrical field. Thus, if an alternating electrical field is applied to such a crystal, the crystal expands and contracts in response to this electric field; that is, it acts as an oscillator. The frequency at which a piezoelectric crystal oscillates is related to the mass of crystal. If the effective mass of the crystal alters—as a result of the deposition of fineparticles onto its surface—while

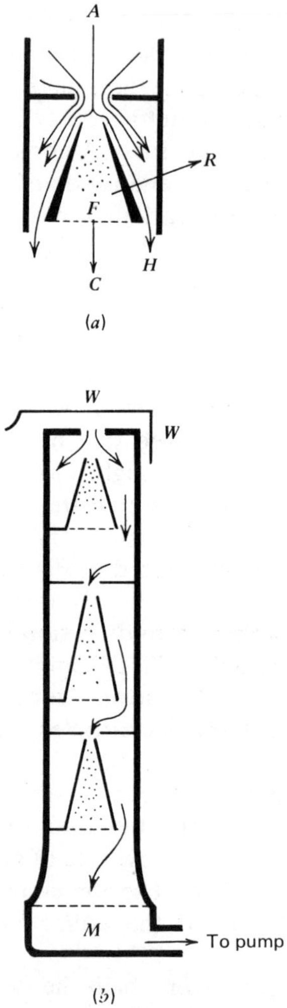

Fig. 7.16 The Centripeter [37]: (*a*) details of virtual surface reservoir. (*A*, aerosol; *R*, reservoir; *F*, filter; *C*, low flow rate; *H*, high flow rate); (*b*) cascade assembly (*W*, wind and rain shield; *M*, membrane filter).

it is oscillating, the frequency at which it oscillates shifts. This frequency shift can be related to the mass of fineparticles deposited on the crystal.

In the "3200 particle mass monitor system" the collection surface is a quartz crystal coated with metal on two sides connected into an oscillator circuit. The aerosol being studied is deposited onto the crystal by a combination of electrostatic forces and impaction forces. The change in oscillation frequency is proportional to the mass of deposited aerosol fineparticles. In this instrument a reference crystal is used to allow for any drift in the temperature of the instrument. The frequency of the deposition surface crystal is constantly compared to that of the reference crystal [39].

Sometimes impactors have been used to size liquid droplets. This is not advisable because of the high shear rates and the varying pressure conditions on a typical trajectory through a cascade impactor. The relationship between measured size distribution for a spray using a cascade impactor and its original distribution in the "free-cloud" state may be arbitrary and dependent in a fickle way on the operating conditions. Naylor has fitted glass sides to the body of a Unico cascade impactor and studied the fineparticle population approaching and leaving an impaction stage by laser light-scattering techniques [44]. He has been able to detect the existence of larger than cut-size droplets, probably formed by turbulent agglomeration, in the aerosol stream leaving an impactor stage. Although it is not advisable to use impaction devices with liquid droplets, sometimes there is no other reasonably priced equipment available, and if all operating conditions are maintained constant, relative information for assessing the performance of spray-generating equipment can be obtained by use of such instruments.

The spiral duct centrifuge aerosol spectrometer is used to sample and size aerosol fineparticles. Its construction is illustrated in Figure 7.17. The main structural feature of the instrument is a spiral duct machined in a heavy-metal rotor. This rotor is shown without its drive mechanism or lid in Figure 7.17*a*. This spiral duct passes through the center of the rotor and extends some distance beyond the center in the opposite direction. The aerosol dynamics in this instrument are outlined in Figure 7.18. The self-pumping action created by the rotation of the centrifuge draws both clean air in at the end of the extension duct and a thin stream of aerosol through a hole on the axis of the centrifuge. The clean air drawn in at the end of the extension duct passes through louvers to ensure stable laminar flow along the duct. As the clean air passes by the small hole at the center of the rotor, through which aerosol to be characterized is being aspirated, it helps to confine the aerosol to a thin stream along the inner wall of the spiral duct. As the aerosol stream moves around the duct, the aerosol fineparticles are subjected to centrifugal forces and are forced across the duct onto the metal foil lining the outer wall of the duct. This metal lining is an integral part of the lid. It is shown in some detail in Figure 7.17*b*. The centrifuge is calibrated by using an aerosol made from a liquid suspension of monosized latex fineparticles [40].

Stober and associates have built a version of the spiral-duct centrifuge aerosol spectrometer in which piezoelectric crystals are located along the duct wall. These are connected to oscillator–transmitter circuits built into

(a)

(b)

Fig. 7.17 The spiral duct centrifuge aerosol spectrometer: (*a*) rotor of centrifugal aerosol spectrometer showing helical deposition channel; (*b*) technician holds deposition foil, complete with quartz piezoelectric transmitters (note static cooling coils around the rotor of the centrifuge); (*c*) complete equipment ready for lid with deposition foil.

(c)
Fig. 7.17 (*Continued*)

the lid of the centrifuge so that aerosol deposition rates for select-sized fineparticles can be reported while the centrifuge is operating (see Figure 7.19).

The "centrifugal particle classifier" used to fractionate aerosol fineparticles is shown in outline in Figure 7.20. The air inside the centrifuge rotates under forced vortex conditions; that is, the air rotates with the rotating cylinder en masse, without slip. The centrifuge chamber is con-

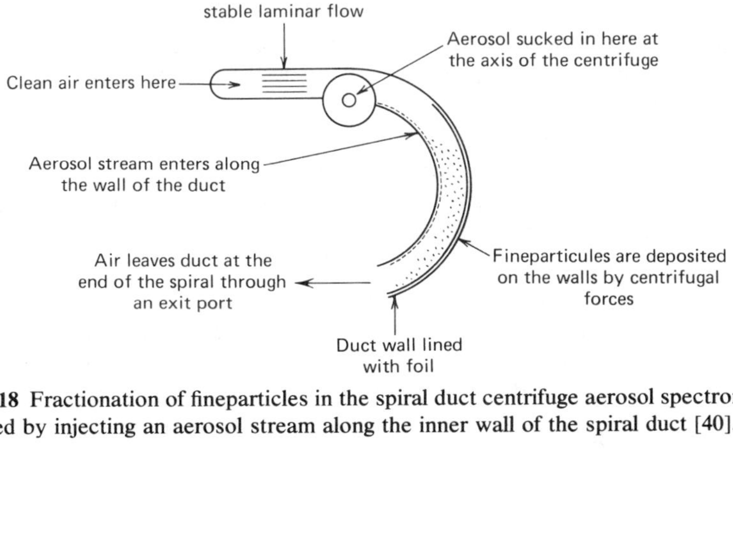

Fig. 7.18 Fractionation of fineparticles in the spiral duct centrifuge aerosol spectrometer is achieved by injecting an aerosol stream along the inner wall of the spiral duct [40].

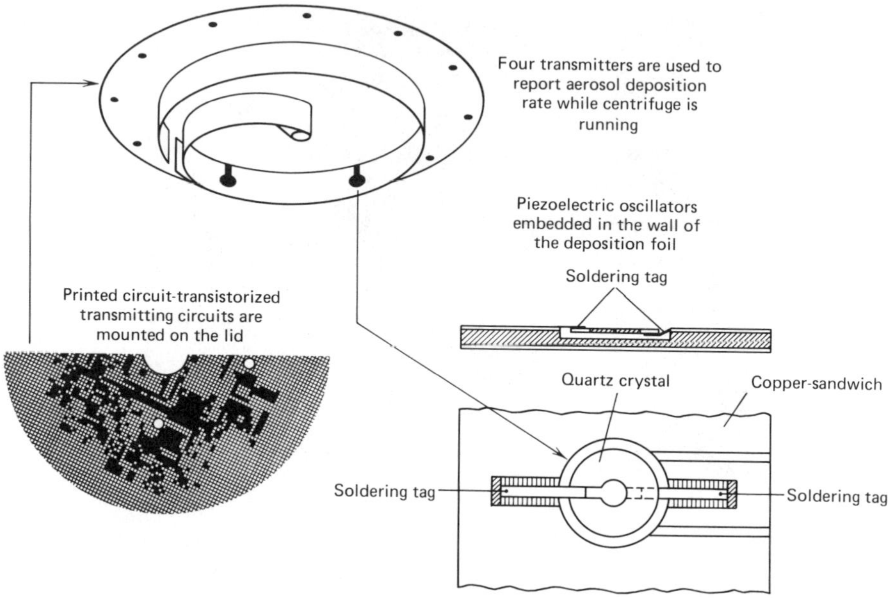

Fig. 7.19 Miniaturized piezoelectric oscillator–transmitters used in the spiral duct aerosol spectrometer.

structed out of a single piece of high-strength aluminum alloy. Two small holes are drilled in the rotor wall every 10° around the periphery of the rotor. The flow of air into this chamber and out through the holes can be varied from forced to almost free vortex conditions, and from laminar to quite turbulent flow, simply by sealing or opening some or all of the holes in the rotor wall. The top plate of the instrument is made from Plexiglas. The equipment is provided with several covers so that the height of the chamber, and hence the flow rates of aerosol into the chamber, can be controlled. The aerosol to be characterized is injected through a non-rotating cylinder coaxial with the axis of rotation of the classifying chamber. Ejection of a small amount of gas through the small holes as the rotor spins is sufficient to draw the aerosol into the chamber. To maintain the dynamic balance of the rotor, the centrifuge chamber is divided into two symmetrical regions by a plastic sheet extending from the top cover to the bottom of the chamber as shown in Figure 7.20. It should be noted that the conditions between the hollow rotating shaft and the coaxial, nonrotating stainless-steel tube ensures high-shear conditions in the final zone before the aerosol is drawn into the centrifuge chamber. As the fineparticles enter into the centrifuge chamber, they move outward under

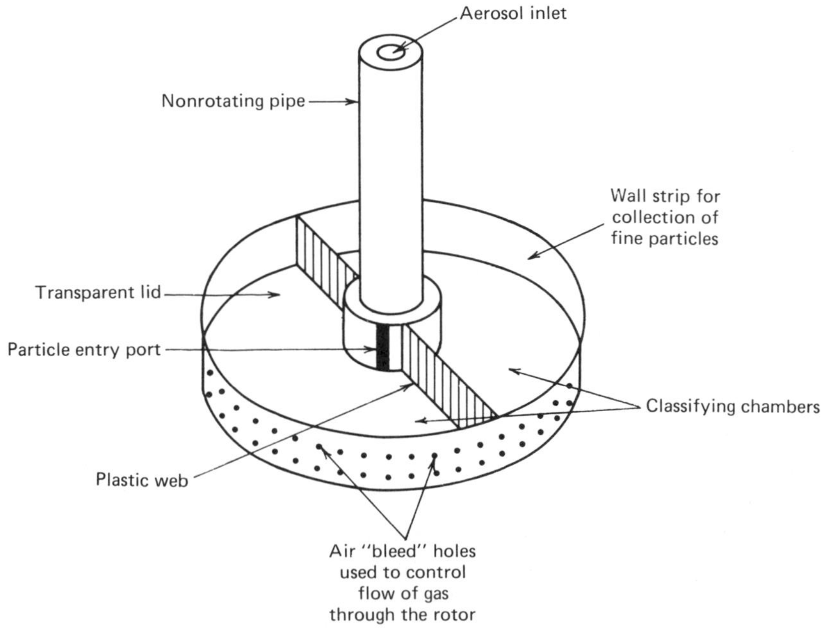

Fig. 7.20 Basic system used in the "centrifugal particle classifier."

centrifugal forces and at the same time are carried around by the air in the rotor. The net effect is that the fineparticles are fractionated and spread around the periphery of the rotor chamber and on the plastic wall dividing the chamber. The inner wall of the fractionating chamber and the dividing wall are covered with plastic strips that can be removed easily for subsequent examination or utilization of the deposited fineparticles [41].

The prediction of the deposition position for a fineparticle of known Stokes diameter onto a given position along the periphery depends on the flow conditions within the centrifuge that, in turn, depend on the operating conditions. The "bleed" rate of aerosol into the centrifuge chamber chosen in any given situation usually is a compromise between a sufficiently large flow rate to yield significant quantities of deposited fineparticles and a flow rate low enough to maintain essentially "forced vortex conditions" within the centrifuge. To obtain a useful separation between fineparticles of different Stokes diameters, Burson et al. [41] found it necessary to operate the centrifuge at speeds such that flow conditions around the moving fineparticles in the centrifuge were not truly laminar, that is, where the Reynolds number for the flow conditions were greater than 0.1 (see discussion of Reynolds number in Section 3.1). In analogue computations of the trajectories of fineparticles of a given Stokes diameter, they found it necessary to use more complex relationships for the drag coefficient than that used when Stokes law holds. For example, for fineparticle Reynolds number up to 800 they used the semiempirical relationship

$$C = \frac{24}{Re}\left\{1 + 0.150\, Re^{0.687}\right\}$$

Another type of centrifugal elutriator employing air or gas as the fluid medium is the "counterflow classifier." The basic system used in this type of elutriator is illustrated in Figure 7.21. The air suspension of fineparticles to be fractionated is injected tangentially as shown. The air stream spirals into the exit hole at the center. As the fineparticles move with the air, they are subjected to centrifugal forces that tend to move them outward to the periphery of the chamber and drag forces that tend to move them inward and out through the axial exit port. The fluid drag forces are proportional to the surface of the fineparticles, whereas centrifugal forces are proportional to their mass. The competition between these two sets of forces result in a cut size such that fineparticles larger than this cut size are collected at the periphery of the fractionation chamber whereas smaller ones leave with the air stream through the axial exit port. The cut size achieved in any particular counterflow centrifugal

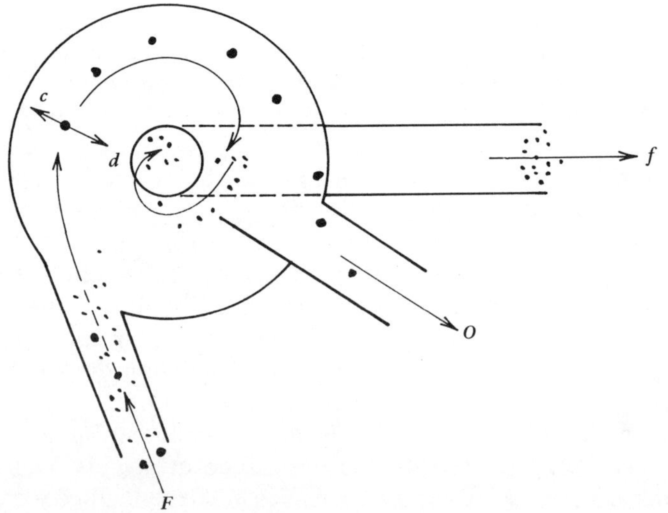

Fig. 7.21 Operational principles of a counterflow centrifugal elutriator. Symbols: f, fines; O, oversize; C, centrifugal force; d, drag force; F, stream of fineparticles.

elutriator is usually established experimentally. Centrifugal counterflow elutriators are often referred to in the commercial literature as *spiral flow classifiers*.

The BAHCO classifier is a counterflow centrifugal elutriator. It is named after the Swedish company that developed it [42, 43]. The basic operating system of the BAHCO is illustrated in Figure 7.22. Except for the powder hopper, the whole of the system shown in Figure 7.44 rotates. Blades angled radially on the underside of the top plate (not shown in Figure 7.22) act as a fan to draw air through the throttle at the base of the rotor. Entering air passes through a stack of radial disks. The narrow gap between these disks ensures that the air is spiraling as it moves into the fractionation chamber immediately above the disks. In this chamber the air-entrained powder feed emerges from a small orifice situated approximately half way between the periphery and the axis of the rotor, as shown in Figure 7.22. As the fineparticles acquire the motion of the spiraling air, they experience the opposing centrifugal and drag forces. The flow conditions at the point where the fineparticles enter the fractionation chamber determine the cut size of the elutriator because airflow drops and centrifugal force increases outward from the orifice, thus ensuring that fineparticles already moving outward will continue to do so. Similarly, fineparticles that start to move inward under the influence of the drag

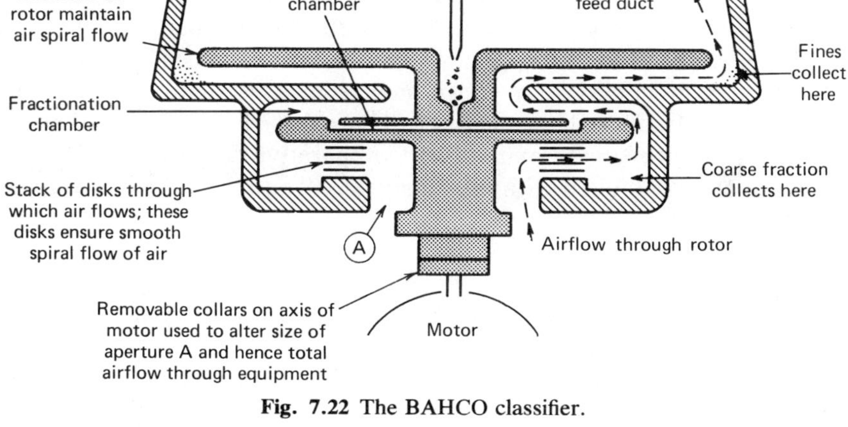

Fig. 7.22 The BAHCO classifier.

forces are in an accelerating airflow and experience decreasing centrifugal forces and increasing drag forces so that they are swept out of the fractionating chamber. At the outer edge of the top plate the airflow rate drops drastically. At the same time, the centrifugal force is at a maximum. Therefore, fines collect at the point shown in Figure 7.22. A series of fractions, either for specific studies or for purposes of characterizing a powder, is achieved by operating the instrument using various flow rates. To change the flow rate the flow of air through the equipment is altered by means of collars that control the aperture at the entrance to the rotor attached to the base of the rotor.

Weilbacher and Rumpf [5] investigated factors that control the performance of counterflow centrifugal elutriators and pointed out that in an instrument such as the BAHCO, a fineparticle is only in the operating fractionation zone determining which way it will move for only a short time. The effects of any instability in the fluid flow that caused a fineparticle to start to move in the wrong direction is, therefore, difficult to reverse. As a result of their studies, Weilbacher and Rumpf [5] developed a new type of centrifugal counterflow elutriator. The commercial version of their equipment is known as *Analysette*® 9 [44]. The operation of this equipment is outlined in Figure 7.23. Powder to be fractionated is entrained into air fed tangentially into the fractionation chamber. The exit port on the far side of the periphery of the chamber is closed during the initial operation of the instrument. The spiraling air stream carrying fineparticles smaller than the cut size leaves the fractionation chamber at the center. The fineparticles larger than cut size move to the outer wall of the fractionation chamber and spin along the wall. When they have completed one circuit of the fractionation chamber, they are met by the incoming jet of air. This reentrains them, and the fierce turbulence at this point tends to disperse any residual aggregates in the oversize fineparticles. It also flushes out any smaller than cut-size fineparticles that may have been attached to or trapped by the oversize fineparticles. This feedback mechanism can make oversize fineparticles circulate 2000 times per minute in the fractionating chamber. The oversize fraction is removed intermittently through the exit valve. Comparative data for the efficiency of fractionation achieved with the Analysette 9 and the BAHCO are summarized in Figure 7.23.

7.6 LIQUID-DRIVEN CENTRIFUGAL ELUTRIATORS

The basic operation of a cyclone, which is a centrifugal elutriator that uses liquid, was discussed in Chapter 3 (see Figure 3.7). A factor limiting

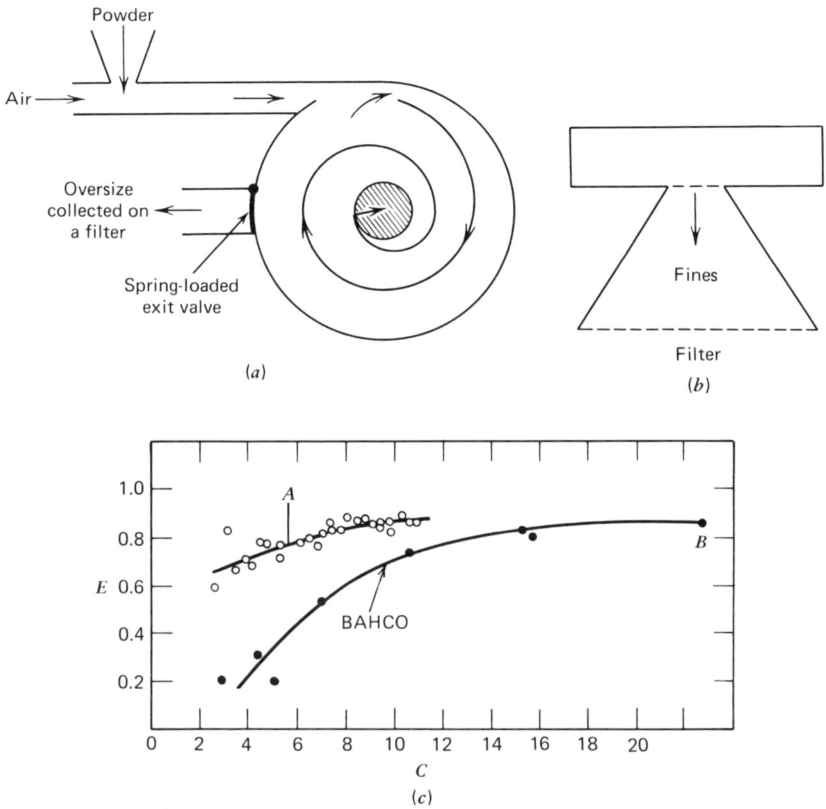

Fig. 7.23 The Analysette 9 counterflow centrifugal elutriator: (*a*) plan; (*b*) elevation; (*c*) performance data. Symbols: $E = x_{25}/x_{75}$, where x denotes percentile on Tromps curve; C, cut size (Stokes diameter in microns); A, data for Analysette 9; B, data for Bahco classifier.

the efficiency of a fractionation achieved with liquid cyclones is the fact that the fineparticles spend only a short time in the high-shear region between the outer layer of fluid and the inner vortex. Therefore, aggregates of fineparticles may not be dispersed and can drop down into the oversize fraction.

To overcome this problem, Kelsal and McAdam [45] inverted the cyclone as shown in Figure 7.24. In the inverted cyclone the oversize fraction thrown to the periphery of the cyclone migrates under the influence of gravity back into the high-shear region of the cyclone [45]. Therefore, any aggregates are given repeated opportunities to break down and participate in the fractionating action of the cyclone. The Warman cyclosizer shown in Figure 7.24 is based on the innovations of Kelsal and McAdam. It

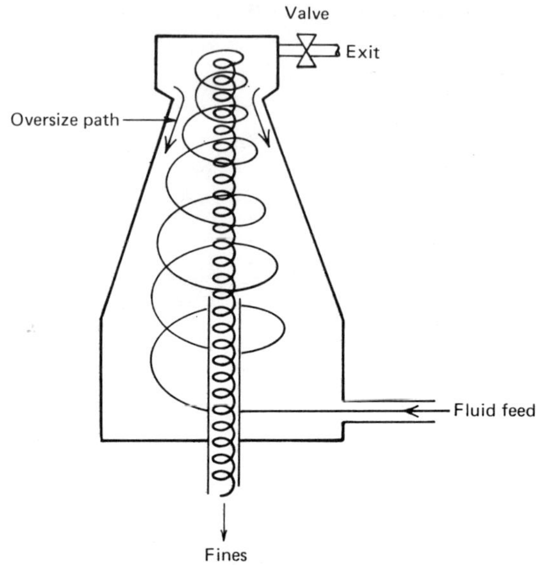

(*a*)

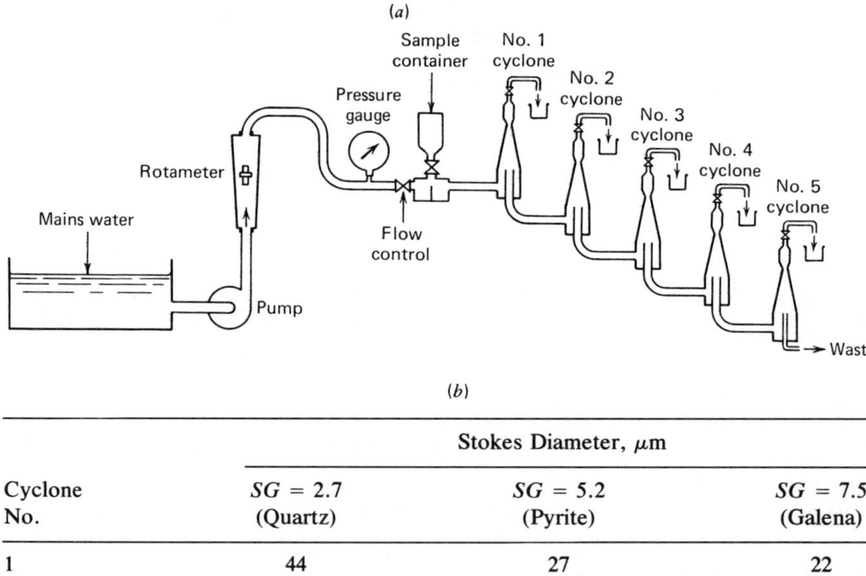

(*b*)

	Stokes Diameter, μm		
Cyclone No.	SG = 2.7 (Quartz)	SG = 5.2 (Pyrite)	SG = 7.5 (Galena)
1	44	27	22
2	35	22	18
3	23	14	11
4	15	9	8
5	9	6	5

Fig. 7.24 The Cyclosizer® system [45]: (*a*) inverted cyclone recirculates oversize fraction back into the active zone of the cyclone; (*b*) the Warman cyclosizer; (*c*) operating characteristics of the Warman cyclosize (upper limit of fineparticles retained by a given cyclone under normal operation conditions). Symbols: SG, specific gravity.

utilizes a cascade of inverted cyclones to fractionate a suspension of powder. To operate the equipment, a slurry of the fineparticles to be fractionated is fed into the system as shown in Figure 7.24. An operating period of 10 minutes is usually sufficient to achieve efficient fractionation. The instrument was developed for use in the mining industry which is concerned with relatively coarse fineparticle populations. Performance data for various mineral powders under normal operating conditions are summarized in Figure 7.24. The manufacturers of the instrument claim that the effective range of the instrument is 5-80 μm for material of specific gravity similar to quartz. For particles of higher specific gravity, such as Galena, the manufacturers suggest that efficient and satisfactory fractionation may be achieved down to 4 μm Stokes diameter [45].

A counterflow centrifuge elutriator has been described by Colon et al. [46], whose system is illustrated in Figure 7.25. A fractionation chamber is cut into the body of a flywheel. Elutriating liquid is fed to the base of this fractionation chamber through a channel cut in the body of the rotor. The suspension to be fractionated is fed into a channel that emerges in the middle of the fractionation chamber. The elutriating fluid moves up the fractionating chamber and mingles with the suspension midway along the chamber. The oversize fineparticles move against the elutriating fluid down to the base of the chamber, where they accumulate until they can be removed at the end of the run through the drain provided. Fineparticles smaller than the cut size move upward with the elutriating fluid and exit through a fines drain shown in Figure 7.25. Later models of this equipment are able to operate continuously with the fines suspension collected by a drain placed underneath the rotor of the instrument, and the oversize slurry is sprayed into the outer chamber and drained away. Figure 7.26 shows an improved fractionating chamber incorporated in later versions of the Colon counterflow centrifugal elutriator. The walls of the improved chamber are shaped so that the flow rate of elutriating fluid moving up the fractionating chamber varies in such a way that the centrifugal and drag forces are in equilibrium over the whole length of the fractionating chamber. Performance data for the instrument are summarized in Figure 7.26. The variable E in this data summary is the ratio of the percentiles from Tromp's curve (see Figure 7.1).

7.7 ELUTRIATORS FOR QUALITY CONTROL SITUATIONS

The efficient design of feedback control loops in the fineparticle-processing industries requires devices capable of rapidly assessing the characteristics of fineparticles carried in a stream of fluid. For example, in

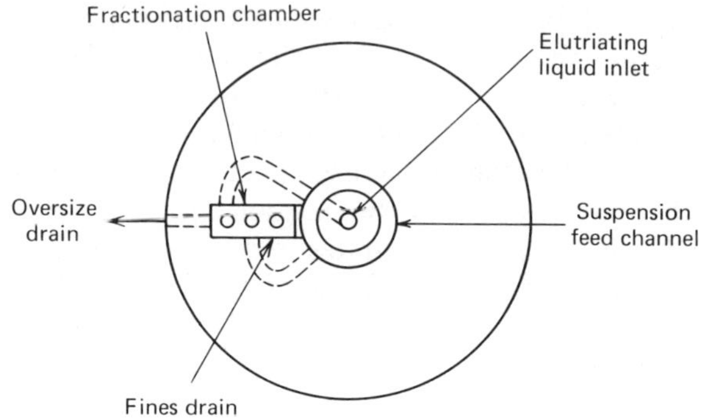

(*a*)

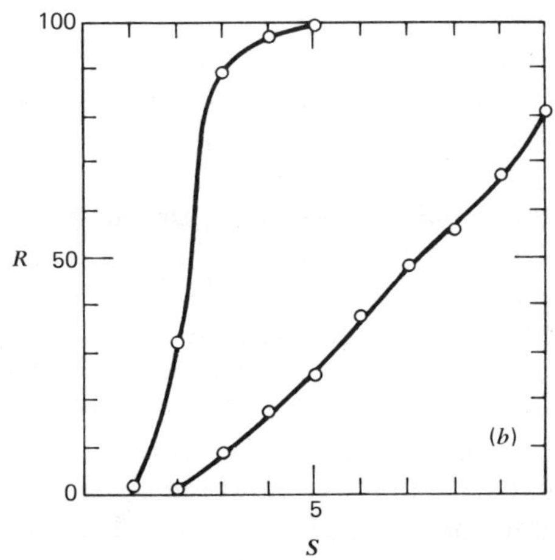

Fig. 7.25 A counterflow liquid centrifugal elutriator [46]: (*a*) basic system; (*b*) oversize and fines distribution of fractions of a glass sphere powder fractionated by using system shown here. Symbol: R, percentage weight less than stated size. S = Stokes diameter in microns.

the wet grinding of many mineral ores and coal, the effluent from a continuously operating ball mill has to be continuously monitored. Engineers concerned with such problems often refer to the output from the ball mill as either a *slurry* or a *mill pulp*. The overall system of ball mill plus monitor and any recirculation of the mill pulp is referred to as a *grinding circuit*.

It should be appreciated that the solids concentrations of most mill pulps emerging from a grinding circuit is usually too high for direct use even in a one-point-check quality control system, and elaborate precautions have to be taken to sample and dilute the emergent mill pulp

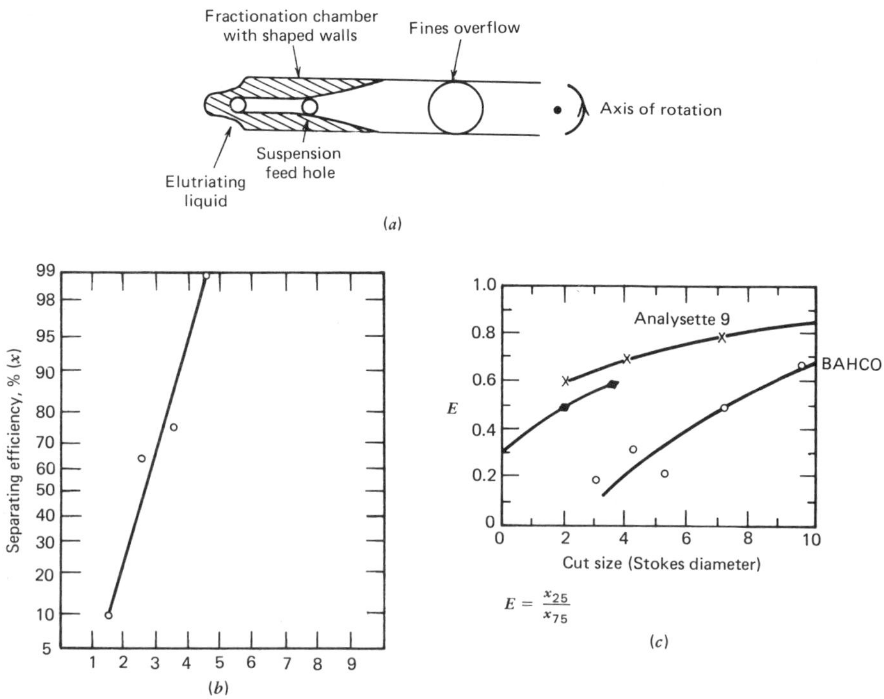

Fig. 7.26 Performance data for a counterflow centrifugal elutriator with an improved fractionation chamber: (*a*) improved "shaped fractionation chamber"; (*b*) fractionation data for experiments with improved cell (glass spheres), cut size = 2.6 μm, $E = x_{25}/x_{75} = 0.61$; (*c*) comparative data with other counterflow centrifugal elutriators.

adequately. Engineers use the term "dilution shock" to describe any physiochemical changes that can occur in a mill pulp when subjected to dilution. For example, the addition of diluent to a mill pulp could change the nature of the physicochemical environment in the pulp with the subsequent flocculation of the fineparticles in suspension. A useful review of sampling and diluting procedures for use with mill pulps has been given by Hinde and Lloyd [47], who also review simple "one-point" instruments for following changes in the size distribution function of industrial mill pulps.

NOTES AND REFERENCES

1. J. T. Scholz, D. R. Uhlmann, and B. Chalmers, "Elutriation Particle Separator," *Rev. Sci. Instrum.*, 36 (12) (December 1965), 1813–1816.
2. For a discussion of the use of Tromp's curve, see K. Leschonski and

H. Rumpf, "Principle and Construction of Two New Air Classifiers for Particle Size Analysis," *Powder Technol.*, 2 (March 1969), 175–186.

3. H. W. Gonell, "Die Bestimmung der Kornzusammensetzung, Staubforminger Stoffe ins Besondere von Zement," *Ton Industrie Zeitung*, No. 13 (1929); *Ton Industrie Zeitung*, No. 25 (1935).
4. C. J. Stairmand, "Some Practical Aspects of Particle Size Analysis in Industry," in *Symposium on Particle Size Analysis* (1947), *Suppl. Transact. Inst. Chem. Eng.*, 25 (1947).
5. M. Weilbacher and H. Rumpf, *Aufbereitungstechnik*, 9 (7) (1968), 323230.
6. O. Lauer, *Grain Size Measurements on Commercial Powders*, Alpine AG, Augsburg, West Germany (available in English and German).
7. Alfred Feitsch and Company, D-6581, Idar-Oberstein, 1, West Germany.
8. Commercial information on the improved elutriator is available from Geoscience Instrument Corporation, 435 East Third Street, Mount Vernon, New York 10553.
9. B. H. Kaye, "Silica Flowagents and the Movement of Powder," in *Proceedings of Technical Programme International Powder and Bulk Solids Handling and Processing*, May 1977, pp. 251–256; published by Industrial and Scientific Conferences Management, Inc., 222 West Adams Street, Chicago, Illinois, 1977.
10. P. S. Roller, *Ind. Eng. Chem.* (April 1931).
11. "Subsieve Analysis of Granular Metal Powders by Air Classification," *ASTM*, Part B, 293–360.
12. "British Standard Methods for the Determination of the Particle Size of Powders," Part 3, "Elutriation Methods," British Standards No. 3406, Part 3, 1963.
13. The Roller Elutriator is available from the American Instrument Company, 8030 Georgia Avenue, Silver Spring, Maryland.
14. The Aminco classifier is available commercially from the American Instrument Company; see reference 13.
15. H. E. T. Haultain, *Transactions of the Canadian Institute of Mining and Metallurgy*, 40 (1937), 229; E. W. Price, *Ind. Eng. Chem.*, Process Design and Development, 1962, p. 79.
16. Information on the Haultain infrasizer is available from the Department of Mining, University of Toronto, Ontario, Canada.
17. E. Schöne, "Uber Schlammanalyse und Einen Neuen Schlammapparat," quoted in reference 10.
18. H. N. Blythe, E. G. Pryor, and A. Eldridge, "Recent Developments in Mineral Dressing," *Inst. Mining Met.* (*Lond.*) (Symposium 23 to September 5, 1952), p. 11.
19. Commercial information on the Hexhlet Aerosol Sampler is obtainable from the Casella Corporation, Regent House, Britannia Walk, London, NI.
20. V. Timbrell, "The Terminal Velocity and Size of Airborne Dust Particles," Suppl. No. 3, *Br. J. Appl. Phys.* (1954). Commercial information is available

from Flemming Instruments Ltd., Caxton Way, Stevenage, Hertfordshire, England.

21. K. T. Whitby and W. E. Clarke, "Electric Aerosol Particle Counters and Size Distribution Measuring System for the Range 0.015 to 1 Micron Size," *Tellus*, 18 (1966), 573–585. Commercial information on the Whitby aerosol analyzer is available from Thermo Systems, Inc., 2500 Cleveland Avenue North, St. Paul, Minnesota 55113.
22. W. H. Marlow, "A Limitation on Electrical Measures of Aerosols," NBS Publication 464, 1977, p. 213.
23. H. Rumpf and K. Leschonski, *Chemie-Ingenieur-Technik,* 21 (1967), 1231.
24. K. L. Metzger and K. Leschonski, "Untersuchung Eines Quernstromsichters zur Online-Teilchengrössenanalyse," *Dechema Monographien*, 79 (1589–1615), Part B (1975), 77–94.
25. K. Leschonski, K. L. Metzger, and U. Schindler, "A New Online Particle Size Analyzer," paper presented at the Salford Conference on Fineparticle Characterization, September 1977; Proceedings, M. J. Groves, Ed., Heyden & Son, London, 1978.
26. K. F. Sawyer and W. H. Walton, *J. Sci. Instrum.*, 27 (1950), 272–276.
27. A. Goetz, "The Aerosol Spectrometer, A New Instrument for the Analysis of Airborne Particles in the Sub Micron Range," *Public Works Magazine*, (February 1959), 91–93. Commercial information on the Goetz aerosol spectrometer is available from the Zimney Corporation, 1900 South Myrtle Avenue, Monrovia, California 91016.
28. D. Hochreiner and P. M. Brown, *Env. Sci. Technol.*, 3 (1969), 813.
29. D. Hochreiner, *J. Colloid Interface Sci.*, 36 (1971), 191.
30. Information on the Konimeter is available from Sartorius, 34 Weenderlandstrasse, Göttingen, West Germany 96102.
31. Sciotic Corporation, 4037 Robert Road, Columbus, Ohio 43228.
32. A. A. Anderson, "A Sample for Respiratory Health Hazard Assessment," *Am. Ind. Hyg. Assoc. J.*, 27 (1966), 160–165. Commercial information on the Anderson sampler is available from "2000, Inc.," 5899 South State Street, Salt Lake City, Utah 84107.
33. K. R. May, *J. Sci. Instrum.*, 22 (1945), 187. Commercial information on the May cascade impactor is available from Casella Corporation; see reference 21; for BGI Incorporated; see reference 42.
34. D. A. Lundgren, *J. Air Pollution Control Assoc.*, 17 (1967), 225. Commercial information is available from Sierra Instruments, Box 909, Village Square, Carmel Valley, California 93924.
35. S. M. Lippmann, *Ind. Hyg. Assoc. J.*, 22 (1961), 348. Commercial information is available from Unico Environmental Instruments, Inc., P. O. Box 590, Fall River, Massachusetts 02722.
36. R. F. Hounam and R. J. Sherwood, "The Cascade Centripeter, a Device for Determining the Concentration and Size Distributions of Aerosols," *Am. Ind. Hyg. Assoc. J.*, 26 (March/April 1965), 122–131.

37. Commercial literature on this equipment is available from BGI Incorporated, 58 Guinan Street, Waltham, Massachusetts 02154 or Bird and Tole Ltd., Bledlow Ridge, High Wycombe, Buckinghamshire, England.

38. R. D. Cadle, *The Measurement of Airborne Particles*, Wiley, New York, 1975, p. 312.

39. J. G. Olin, F. J. Sem, and D. L. Christenson, *Am. Ind. Hyg. Assoc. J.*, 32 (1971), 209. Commercial information is available from Thermo-Systems, Inc.; see reference 24.

40. W. Stöber and H. Flachsbart, *Environmental Science and Technology*, 3 (1969), 1280. Recent information is available from the Institute for Aerobiology, Grafschaft, Sauerland, West Germany.

41. J. H. Burson, A. Y. H. Keng, and C. Orr, Particle Dynamics in Centrifugal Fields, *Powder Technol.*, 1 (1968), 305–316. The elutriator developed by Burson and colleagues was made available commercially by the Micromeritics Corporation under the name "micromeritics aerosol sampler model 1001." Information is available from Micromeritics Instrument Corporation, 800 Goshen Springs Road, Norcroft, Georgia 30071.

42. The BAHCO classifier was developed at the laboratories of Hjorth and Company, Bahco, Sweden. See H. Ashley, "Particle Size Analysis by Elutriation and Centrifugation, as Exemplified by the BAHCO Microparticle Classifier," in *Particle Size Analysis*, J. D. Stockham and E. G. Fochtman, Eds., Ann Arbor Science Publishers Incorporated, P. O. Box 1425, Ann Arbor, Michigan 48106. The BAHCO classifier is known in the United States as the *Microparticle classifier* and is available there from the Harry W. Dietert Company, 9330 Roselawn Avenue, Detroit, Michigan 48204.

43. Committee Number TA5 of the Air Pollution Control Association (dust, fumes, mists, and fog collectors) has recommended that the BAHCO classifier be adopted as a standard instrument for determining the size characteristics of fineparticles when specifying and testing industrial dust-collecting devices. A standardized operation procedure for use with the BAHCO for this purpose has been prepared by the American Society of Mechanical Engineers. The procedure is entitled "Determining the Properties of Fine Particulate Matter," Power Test Code PTC 28, American Society of Mechanical Engineers, New York, 1965.

44. See catalog of Alfred Fritsch and Company (reference 7).

45. D. F. Kelsall and J. C. H. McAdam, "Design and Operating Characteristics of a Hydraulic Cyclone Elutriator," *Transact. Inst. Chem. Eng.*, 41 (1963), 84. Commercial information is available from Warman International Proprietary, Ltd., 18–26 Dixon Avenue, Artarmon, Sydney, Australia.

46. F. J. Colon, J. W. Van Heuven, and H. M. Van der Laan, "Centrifugal Elutriation of Particles in Liquid Suspension," in *Proceedings of the Conference on Fineparticle Characterization*, Bradford, England, 1970, British Society for Analytical Chemistry, Saville Row, London, 1971.

47. A. L. Hinde and B. J. D. Lloyd, "Realtime Particle Size Analysis in Wet Closed Circuit Milling," *Powder Technol.*, 12 (1975), 37–50.

CHAPTER

8

STREAM METHODS FOR CHARACTERIZING FINEPARTICLES

8.1 BASIC CONCEPTS

In stream methods the fineparticles to be characterized are passed through an interrogation zone. Changes in the physical properties of the interrogation zone caused by a fineparticle are related to the size of the fineparticle. In this book stream methods for characterizing fineparticles are classified according to the type of physical signal used to monitor the interrogation zone behavior. Thus, if the changes in electrical resistance of the interrogation zone are monitored, the method is described as a resistazone stream counter. The resistazone principle has been widely exploited in such instruments as the Coulter counter® and other instruments described in Section 8.2. Stream counters in which an optical signal is used to measure the changes in the interrogation zone are described as photozone counters; those operating with an acoustical signal are referred to as *sonizone counters*.

Some stream methods require special preparation of the fineparticles before they are presented to the stream counter. For example, if a resistazone stream counter is used, the fineparticles have to be dispersed in a conducting fluid, and inappropriate techniques at this stage can alter the size distribution of the fineparticles. This aspect of preparatory technique is usually specific to a given fineparticle system, and the writer does not feel that it is useful to give a discussion of the many different types of fineparticle that may be encountered and the problems of dispersing them in particular fluids. Many manufacturers of instruments for characterizing fineparticles by stream methods are happy to give advice and assistance in selecting appropriate fluids and dispersion technology prior to the use of their proprietary instrument to characterize a fineparticle system.

In many stream methods for characterizing fineparticles it is not economical to measure the physical properties of the interrogation zone over all the zone for all the time used in the study. One economic compromise that is widely used in the design of stream counters is to design the system so that the statistical probability of multiple occupancy of the interrogation zone is negligible. When this assumption is made, all changes in the physical properties of the interrogation zone are assumed to be caused directly by the presence of a single fineparticle. In practice, some stream methods do not always achieve single occupancy of the zone. Multiple occupancy of the interrogation zone can give rise to three different errors in the measured size distribution of the fineparticle system: (1) primary count loss, (2) secondary count gain, and (3) shadow loss. ''Primary count loss'' means that two or more fineparticles in the zone together results in the underestimation of the population frequency of the smaller fineparticles. Sometimes this primary count loss is accompanied by the fact that the multiple occupancy results in a false signal of a pseudolarge fineparticle, thus resulting in the measured frequency of these larger fineparticles being higher than in the real population. This apparent enrichment of the larger fineparticle population is known as *secondary count gain*. Sometimes the presence of one fineparticle desensitizes the interrogation zone for a short time with a subsequent failure to register the presence of a closely following fineparticle. This type of error is described as *shadow loss*. For example, in the acoustical stream counter described in Section 8.3, there appears to be strong evidence that during the period when the primary sonic signal from the passage of a fineparticle is decaying, the system is unable to register the presence of any other fineparticles entering the interrogation zone. Most manufacturers of stream counters for characterizing fineparticles recommend carrying out a series of measurements at a series of increasingly diluted suspensions until further dilution leaves the measured size distribution of the fineparticle system unchanged, at which stage it is assumed that multiple-occupancy effects, including close-following shadow effects, are no longer significant. Some workers have used statistical theory to predict the potential primary count loss and secondary count gain when using dilutions above those at which single occupancy can be assumed. However, since the predictive statistics require a primary knowledge of the distribution function of the system being measured, this can only be a procedure of successive approximations utilizing initial measurements to refine the estimates of the underlying distribution function by statistical manipulation of data. Wherever possible, the technologist should opt for extreme dilution since this greatly simplifies the transformation of the appropriate experimental data to the desired size characteristic distribution.

Fractional occupation of the interrogation zone can also distort the measured fineparticle distribution function. For example, in some light-scattering stream counters, the interrogation zone is not restricted physically but is a region defined by the optical illumination system. In such systems it is possible to have partial occupancy of the zone by a fineparticle traveling near the outer limits of the optically defined region. Designers of aerosol stream counters using optical interrogation zones have attempted to deal with this problem by using a sheath of clean air to confine the fineparticles well within the optically sensitive region of the interrogation zone.

When interpreting the data from a stream method, one must be concerned for the fact that the physical properties of the interrogation zone may not be isotropic. For example, in the literature on the Coulter counter, a well-known resistazone counter, it was assumed for a long time that the signal generated in the instrument was independent of the location of the fineparticle within the zone. More recent work, however, has established that if the fineparticle is close to the wall of the cylinder of fluid in the interrogation zone, the signal will differ from that generated by the same fineparticle moving down the axis of the interrogation zone. For high-resolution and accuracy work, the Coulter Electronics Corporation has introduced what it calls an "electronic editor" to reject signals from fineparticles not traveling down the center of the interrogation zone.

8.2 RESISTAZONE STREAM COUNTERS

The physical change in the electrical resistance of a well-defined cylinder of conducting fluid caused by the presence of a fineparticle is the basis of the resistazone stream counter. Commercially available resistazone stream counters are the Coulter counter, the electrozone® counter, and the Nanopar® analyzer [1–3]. The basic system used in a resistazone stream counter is shown in Figure 8.1. The fineparticles to be characterized are suspended in an electrolyte and made to pass through a cylindrical orifice placed between two electrodes. The presence of a fineparticle

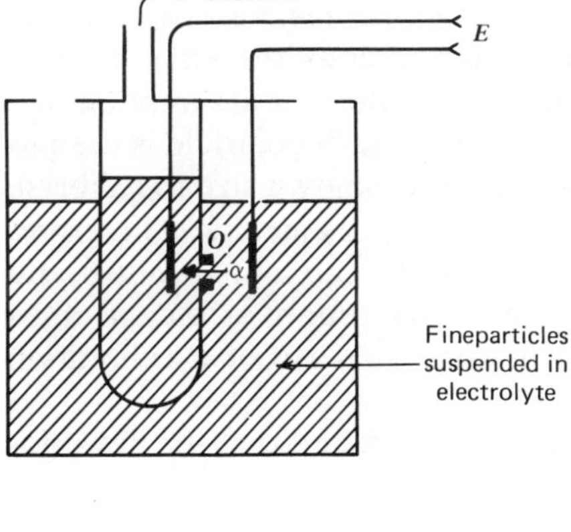

Fig. 8.1 Basic systems of a resistazone stream counter. Symbols: E, electrodes; α, stream of fineparticles; O, orifice.

in the orifice between the electrodes causes a voltage pulse, and this pulse is amplified and recorded. Commercial instruments based on this basic principle became available in the early 1950s. Since then, the sophistication of the electronics used to evaluate and record the electronic pulses generated by a resistazone counter has increased tremendously. Current technology is able to investigate the shape and the duration of a series of sequential pulses, thus enabling the technologist to obtain information on the shape and size characteristics of a fineparticle system [4–7]. Research publications dealing with the applications of resistazone counters number many hundreds. Summaries of this technical literature are available from the manufacturers of the instruments [1, 2, 8]. In this presentation we do not discuss specialist applications of the instruments, but confine the discussion to a review of the evolution of significant instrument features.

Consider the ideal case of a single smooth sphere present in the center of a cylindrical orifice. It is possible to deduce a simple relationship relating resistance changes in the orifice to the dimensions of the fineparticle and the orifice. Early workers with resistazone counters established that this relationship held for spherical fineparticles provided that the diameter of the fineparticle did not exceed 40% of the diameter of the orifice. Other experimental studies show that, in general, using the simple orifice available in the early instruments, fineparticles with dimensions of less than 3% of the diameter of the orifice could not be characterized easily. This lower limit has been reduced by modern techniques such as hydrodynamic focusing and the use of contoured orifice equipped with flow straighteners of the type discussed later in this section. Because of the success in correlating the volume of spheres with the electrical signals generated by their passage through the orifice, it is often claimed that resistazone counters measure the volume of a fineparticle. It must be stressed that this is an unwarranted generalization. It has been shown that the signal generated by such counters is dependent on the shape of the fineparticle and hence cannot be related in a simple manner to the volume of an irregularly shaped fineparticle [4]. In practice, resistazone counters are calibrated by using standard fineparticles such as the latex spheres available from the Dow Chemical Corporation. Often the assumption that the pulse height is proportional to the volume of the fineparticle is used to extend a one-point calibration by using spheres of known size to calibrate the dial of the instrument in equivalent spherical diameters. Therefore, data from most resistazone characterization studies are not absolute, and the sizes quoted are for equated spheres. Any report of data from a resistazone counter study should clearly state any use of a calibration standard.

Early workers using resistazone counters anticipated that the behavior of metal fineparticles in the counter would differ from that of dielectric fineparticles. In practice, it was found that all fineparticles investigated behaved as if they had insulating properties. This has been variously attributed to the fact that many fineparticles have oxide layers on them or chemisorbed layers of nonconducting species that in the electrolyte serve to give the fineparticles insulating properties under the operating conditions of the resistazone counter.

The earlier models of the resistazone counters were found to be particularly sensitive to the presence of any chips or irregularities on the periphery of the orifice delineating the interrogation zone. Studies showed that this arose from the fact that the Reynolds numbers of the flow dynamics through the orifice were on the order of 100–1000. This is sufficiently close to the Reynolds number for the changeover from laminar to turbulent flow that the presence of a chip or an irregularity was likely to initiate turbulent flow. Eddies in the turbulent flow passing through the sensitive zone of the instrument, which to the instrument looked like fineparticles, formed islands of "unavailable" electrolyte. It has been shown that a contoured orifice leads to a smoother flowline configuration at the entry into the orifice, with a consequent lessening of the effect of any problems due to chipping or any slight protuberance. Also, in general, the contoured orifice decreases noise signals in a characterization study and improves the resolution achievable with the instrument [5–7]. Since the operative size range of a characterization study is 3–40% of the orifice diameter, the characterization of a wide-size-distribution fineparticle system requires the use of a series of orifices of different diameters.

The presence of fineparticles larger than the orifice clogs the orifice. An instrument innovation that prevents problems arising from a blocked orifice is to protect the entrance to the orifice with an electrode made of metal screen, with the apertures of the screen 40% of the diameter of the orifice [5]. If this type of electrode is combined with a contoured orifice plus a flow-straightening collar, the resolution of the measurements is increased and background noise decreased. The orifice system incorporating these various modifications, all developed at IITRI, is illustrated in Figure 8.2.

Early discussions of the performance of resistazone counters assumed that the electric field in the orifice of the resistazone counter was isotropic and that the exact location of the fineparticle within the zone was unimportant. However, later investigations showed that the shape of the pulses generated by fineparticles in the orifice off center were different from those generated by those exactly in the center of the interrogation zone. Thus Grover in Israel [9] described the occurrence of four different types of pulse generated by fineparticles following different paths through the

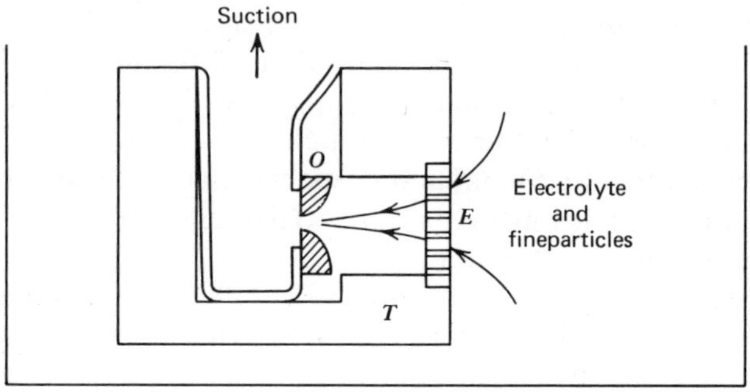

Fig. 8.2 The IITRI improved resistazone counter system. Symbols: *O* contoured orifice; *E*, electrode made from electroformed metal mesh; *T*, Teflon block. The tube bored in the Teflon block between the electrode and the orifice acts as a flow-straightening throat.

orifice. Similar results were reported by Davies et al. of IITRI [6]. At about the same time, Thom [10] at the medical clinic of the University of Berlin published studies on the pulse shape obtained when spheres were drawn through an orifice on nylon threads [10]. They showed that when the spheres moved along the axis of the orifice, one obtained smooth peaked pulses. Different-shaped pulses were found to be associated with the passage of the sphere on an off-axis trajectory, with the shape depending on the proximity of the sphere to the wall of the orifice. It was found empirically that by cutting the edges of the orifice to create a 45°-angle plane to the axis of the orifice, the distortion of the pulses could be reduced, but that the pulse magnitude was greater for peaks created by off-axis fineparticles than for those for on-axis fineparticles. To avoid distorted pulses that result in error in the measured size, various approaches have been adopted. Thom used a capillary tube to fire the fineparticles down the axis of the orifice. This approach is effective with systems such as blood cells, where the range of sizes present is relatively small, but is difficult with a wide-size-range powder. Davies et al. [6] concluded that their flow straightener plus contoured orifice in conjunction with the protective mesh electrode in effect had structured the flow into the orifice, thus enabling the fineparticles to move down the axis of the system.

The system developed by Thom and co-workers was exploited commercially by the German company Telefunken. The physical principles employed in the Telefunken system are illustrated in Figure 8.3. When suction is applied on the downside of the orifice, clean electrolyte is drawn through the orifice along with a stream of suspension from the capillary *N*. The clean electrolyte is said to focus the suspension stream,

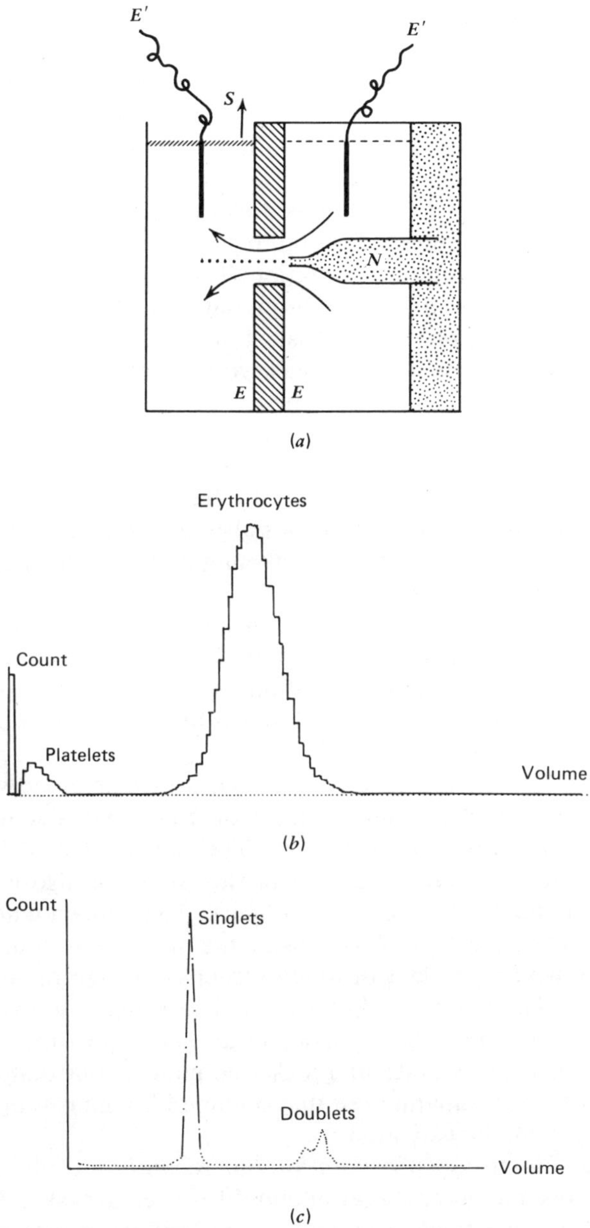

Fig. 8.3 The Coulter counter TF® system uses hydrodynamic focusing: (*a*) measurement system, (*b*) size analysis of whole human blood and (*c*) size analysis of 2.02-μm latex showing singlets and doublets. Symbols: *S*, applied suction; *E*, electrolyte; *N*, suspension of fineparticles to be characterized; *E'*, electrodes.

and the technique is described as *hydrodynamic focusing* [12]. The Telefunken system has been licensed by the Coulter Electronics Organization and sold under the name Coulter counter TF®. In the commercial literature provided with the Coulter counter TF system, it is stated that the clean flow of electrolyte narrows down the stream of suspension leaving the capillary throat by a factor of 10. The high-resolution data obtainable with the Coulter counter TF system is illustrated in Figure 8.3*b*. The data for the latex shown in Figure 8.3*b* is interesting in that the doublets appear to be resolved with respect to orientation. The trace shown in Figure 8.3*b* is remarkably like the high-resolution data for latex fineparticles as generated by the IITRI photocentridisk shown in Figure 6.10. The slightly smaller component of the doublet peak would seem to correspond to the signal for doublets traveling with their axes along the axis of the orifice, whereas the slightly sharper component represents doublets traveling with their axes perpendicular to the orifice.

The Coulter Electronics Organization has developed an electronic editor that looks at the structure of a pulse and rejects all types except those characteristic of fineparticles traveling down the axis of the orifice [11]. Karuhn and Berg [13] described the use of a combination of hydrodynamic focusing plus electronic pulse inspection to measure fineparticle shape. The system used by these workers is shown in Figure 8.4. These workers claim that the leading edge of the pulse is indicative of shape and that the length of the pulse gives further information on such things as fiber length.

Davies et al. [14] have used a resistazone orifice shielded with a mesh electrode to measure fiber length. They protected the orifice with an electrode having apertures of a given size. The registration of fineparticles having equivalent volumes greater than this aperture size is then interpreted as being due to the presence of fibers. To enable a whole range of fiber lengths to be investigated, a resistazone orifice was equipped with a steel wheel on which six electroformed screens of different aperture sizes were mounted. The wheel could be rotated to bring any given disk into position in front of the orifice. Davies et al. [14] report that this system was rather difficult to operate in a leak-free manner but that fiber length data obtained were comparable to that obtained by microscope examination for the systems investigated.

The Coulter Electronics Corporation has recently introduced two versions of their resistazone counter adapted for special tasks. The Coulter on-line monitor consists of three resistazone orifices mounted in parallel in a measurement cell fed with suspension to be characterized. Initially, measurements are carried out with the first orifice until it is blocked. Then a monitor system immediately switches data collection to the second

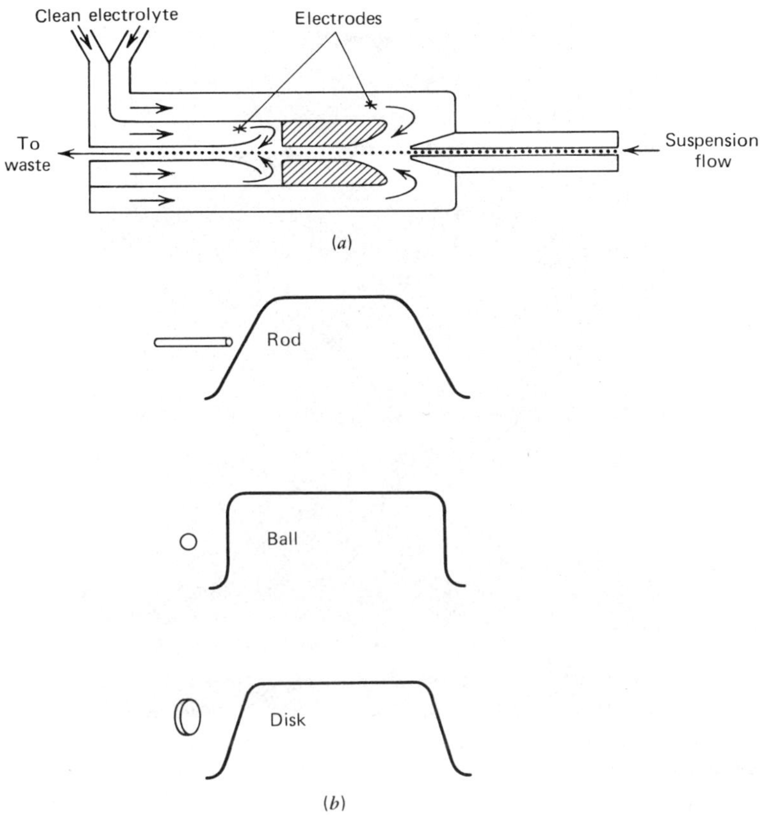

Fig. 8.4 Kahrun and Berg [13] used a combination of contoured orifice hydrodynamic focusing and pulse editing to obtain shape information from a resistazone counter: (*a*) orifice system; (*b*) leading edge shape and duration of pulse yields information in the magnitude and shape of a fineparticle.

orifice while initiating a cleaning procedure at the first orifice. To accommodate the very rare occasion when two orifices are blocked, a third orifice is provided that collects data until orifices 1 and 2 are cleared.

In the second special Coulter system, a large orifice mounted vertically is used to size fineparticles normally sized with the aid of sieves. The instrument is calibrated so that the printout of the data is in size intervals that match the corresponding size intervals in standard test sieve series [1].

DeBlois and Wesley [3] used a Nuclepore filter to create a resistazone counter of very small aperture that could be used to measure the size of small objects such as bacteria and viruses. Typical pulses obtained by

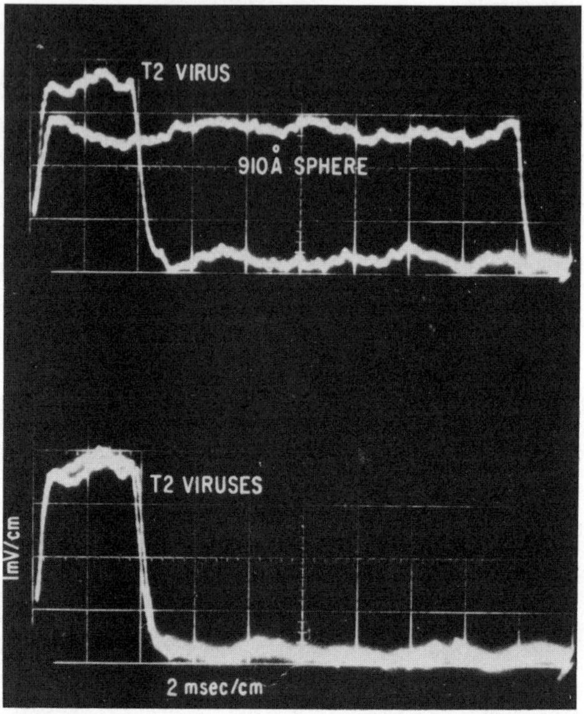

Fig. 8.5 DeBlois and co-workers employed a Nuclepore® filter to create a resistazone orifice small enough to measure the size of a virus [3].

DeBlois are shown in Figure 8.5. The equipment is available commercially as the Nanopar analyzer® [3].

8.3 SONIZONE STREAM COUNTERS

A stream counter in which the size of a fineparticle present in an interrogation zone is measured with the use of ultrasonic waves was reported by Albertson [15] and evaluated by Gayle and co-workers [16]. In this instrument, shown in Figure 8.6, a piezoelectric crystal mounted in the base of a fluid column functioned alternately as an emitter and receiver of ultrasonic waves. At the start of a measurement cycle, the piezoelectric crystal emitted a short, converging pulse of ultrasonic waves. The crystal was then placed in a neutral mode for a time α, after which it became a receiving crystal for a very short period, $\Delta\alpha$. As the pulse of ultrasonic waves moved up the column whenever a fineparticle was encountered, a

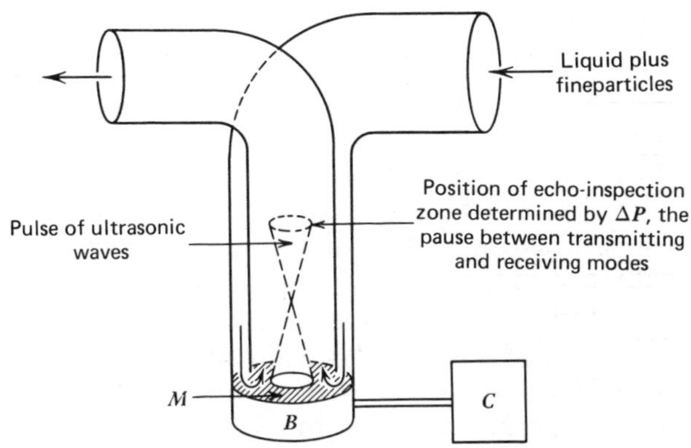

Fig. 8.6 Sonizone stream counter. Symbols: *C*, control operating crystal in transmitting and receiving mode as required; *M*, mask; *B*, base containing piezoelectric crystal.

signal was reflected back to the crystal. However, only pulses reflected from a height equal to $\alpha/2V$, where V is the velocity of the ultrasonic waves in the fluid containing the fineparticles to be characterized, are recorded. Thus the location and the dimensions of the interrogation zone and its height above the crystal were varied by altering the periods α and $\Delta\alpha$. The resolution of the technique was also limited by the duration of the burst of ultrasonic waves emitted from the crystal. In theory, the reflected signal is proportional to the area of the fineparticle; in practice, extensive calibration procedures had to be employed in order to make the instrument functional. It was found that turbulence and bubbles in the fluid moving through the instrument gave false signals during the measurement procedure. The instrument was originally developed for monitoring the cleanliness of opaque fluids since the procedure did not require that the fluid be either conducting or transparent. However, the instrument did not receive further development, presumably because of the difficulty of calibrating it. Future developments in ultrasonic technology may perhaps make it possible to reexamine this type of sonizone stream counter. The developers of the instrument indicated that they thought that it could be developed to monitor fineparticles as small as 10 μm, but that at the stage of development reported in the literature, a more realistic limit was 25 μm.

The Autometrics Corporation has developed a system for determining both solids concentration in a suspension and the average size of the fineparticles in a stream counter that uses ultrasonic sound waves [17]. A beam of ultrasonics passing through a slurry of fineparticles suffers power loss from two sources [17, 18]. The first source of power loss occurs because the fineparticles attempt to move with the wavefront of the ultrasonic wave. As the fineparticles attempt to move with the wavefront they experience viscous drag in the fluid, which leads to power dissipation. The second type of power loss occurs when sonic energy is diffracted out of the ultrasonic beam. It has been shown that for solids that are less than 20% by volume of fineparticles, it is possible to find two different ultrasonic frequencies such that at one frequency the change in power is basically due to the mass of fineparticles present in the suspension, whereas at a different frequency, small changes in size distribution function of the suspended fineparticles will cause a large power loss. For any given slurry of fineparticles, these two different frequencies have to be determined by experimental investigation. The system used in the Autometrics device is illustrated in Figure 8.7*a*. Typical performance data are summarized in Figure 8.7*b* [19]. Within the current state of the art, the suitability of this device for monitoring any given industrial slurry must be explored empirically.

An interesting sonizone stream counter is the acoustic fineparticle counter. This instrument was originally developed by Langer [20], who observed that fineparticles moving down a vacuum hose created audible clicks as they moved through a constriction made by a clamp.

Langer built and tested various-shaped capillary tubes and established that the sound made by a fineparticle could be heard only upstream of the capillary constriction and that the walls of the tube connecting the capillary to the aerosol stream had to be smooth and gradually contoured down to the capillary orifice size. He also showed that the shape of the exit from the capillary was immaterial and that no sound could be heard downstream of the capillary. He showed that the performance of the instrument could be greatly improved by making the entrance to the capillaries section bell shaped. Later work by various investigators has established that the sound produced in this type of instrument is generated in the following manner. The Reynolds number of the airflow through the capillary has to be greater than or equal to 2000, and for this type of flow one would normally anticipate turbulent flow. Indeed, if rough surfaces are present in the throat leading to the capillary tube or in the capillary itself, the flow appears to break down into turbulent motion with consequent continuous noise generation. In the absence of a fineparticle, the contoured throat leading to the capillary section maintains laminar flow conditions (see discussion of Reynolds number in Chapter 3). Thus Scarlett showed by using Doppler shift measurements in laser light (see discussion of Doppler shift in Chapter 5) that the air jet emerging from the capillary tube maintained its integrity for a considerable distance down the tube and that the conditions inside the jet were lamina [22]. By

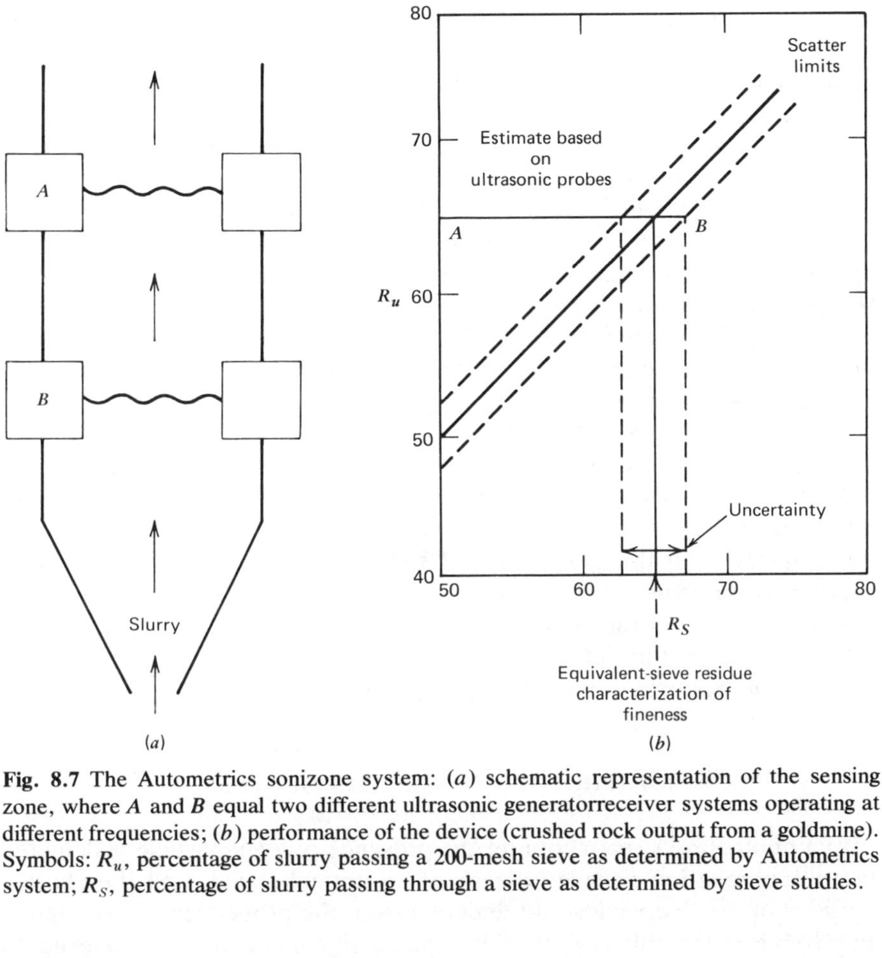

Fig. 8.7 The Autometrics sonizone system: (*a*) schematic representation of the sensing zone, where *A* and *B* equal two different ultrasonic generatorreceiver systems operating at different frequencies; (*b*) performance of the device (crushed rock output from a goldmine). Symbols: R_u, percentage of slurry passing a 200-mesh sieve as determined by Autometrics system; R_S, percentage of slurry passing through a sieve as determined by sieve studies.

measuring the pressure fluctuations down the capillary pipe section of the instrument, Karuhn [21] postulated that the mechanism producing the sound was as follows. The fineparticle entering the capillary section of the instrument penetrates the boundary layer of the laminar flow initiating the shedding of a vortex. This vortex will travel at a much higher speed than the fineparticle and would be reflected by the end of the pipe to create a sound wave in the capillary [21]. Scarlett and Sinclair showed that the frequency of the sound generated by the fineparticle was dependent on the length of the capillary tube, and the length was related to half the wavelength of the emitted sound wave. Thus the capillary tube acts as an organ pipe with one end effectively closed and the bell-shaped end open. This explains the effectiveness of the bell-shaped entrance in both contouring the streamline flows into the capillary to maintain laminar flow and at the same time acting as a resonator to enhance the note emitted by the capillary pipe when stimulated into acoustic emission by the vortex shed from the boundary layer by the fineparticle.

Until the acoustic signal has faded away and laminar flow conditions have been reestablished in the pipe, the presence of a subsequent fineparticle cannot be detected. It has been suggested that this type of acoustical counter could be very useful in measuring the gross level of dust fineparticles in industrial working areas. The extreme simplicity and robustness of the equipment makes it attractive for such situations.

It should be noted that Scarlett and Sinclair established that the fineparticles emerging from the capillary were still accelerating and were traveling at velocities well below those of the gas.

In general, the strength of the acoustic signal emitted from the counter does not seem to depend on the size of the fineparticle to within any achievable sensitivity at the present state of the art in acoustical engineering. However, Karuhn has pointed out that since the basic mechanism involved in the generation of the acoustical signal was an interruption in the type of flow present in the emergent jet issuing from the capillary section of the instrument, it should be possible to measure pressure changes in this jet and relate them to the size of the fineparticle causing the perturbation in the jet flow [21].

8.4 PHOTOZONE STREAM COUNTERS

In a photozone stream counter the presence of a fineparticle is detected in an interrogation zone by means of an optical signal, and the characteristics of the fineparticle are deduced from the properties of that signal. In passive photozone counters the optical signal is generated external to the interrogation zone and the fineparticle characteristics deduced from the magnitude and spatial distribution of the reflected signal. In an active photozone counter the fineparticles emit light when stimulated by means of some external energy beam. Thus fineparticles could fluoresce in the optical region when stimulated by ultraviolet light. Active photozone counters have received little attention; however, they do have the advantage of being species specific and could pick out certain types of fineparticles present in a general aerosol or slurry stream. The need to detect specific air pollutants could give stimulus to the development of this type of photozone counter.

There is a vast literature on the theory and design of photozone stream counters. In the space available, it is possible to give only a brief review of the various instruments that are available commercially and to indicate trends in current design that may lead to the evolution of new photozone counters. The light-scattering properties of fineparticles depend in a complex way on their refractive index, physical dimensions, and the wavelength of light incident on the fineparticles. Entire books have been written on the theory of light scattering; however, only in rare cases is one able to relate the size of the light-scattered signal to the magnitude of the fineparticle by theoretical considerations. In practice, most widely used photozone counters are calibrated empirically by using standard fineparticles of known dimensions. Variations between the different optical designs of the various instruments makes it difficult to correlate their performance except on an empirical basis. When the fineparticles are considerably larger than the wavelength of light, the fineparticles can be considered as generating a geometric shadow so that the energy removed from the light beam is a simple function of the diameter of the fineparticle. In such situations it has become commercial practice to refer to the instrument as being based on the measurement of "light blockage."

As the dimensions of a fineparticle start to approach the wavelength of light, the complex scattering behavior is studied by use of scattering theory developed by a German scientist, Mie, whereas when the fineparticles are much smaller than the wavelength of light, the appropriate light-scattering theory is called the *Rayleigh light-scattering theory*. Other specialized light-scattering theories are known as Rayleigh–Gans and Lorentz–Mie theories [23–25].

The first photozone stream counters were developed in the United States in the late 1950s and early 1960s. They were used to measure the size distributions of aerosols [25, 26]. The stream of aerosol to be characterized was injected into an illuminated area by using hydrodynamic focusing to provide a well-defined stream of aerosol passing down the middle of the optically defined interrogation zone. To minimize problems that could arise from using gases of different constitution and/or different temperature to that of the aerosol stream, the stream of gas used in the hydrodynamic focusing is usually obtained by filtering a diverted portion of the aerosol stream. For the same reason, any dilution of the aerosol required to achieve single occupancy of the interrogation zone should be carried out with a portion of the filtered aerosol stream.

The Royco Instrument Corporation manufactures a whole range of photozone stream counters [28]. The optical system used in the different instruments depends on the information requirements of the particular study. For example, in Royco Model 203, shown in Figure 8.8*b*, right-angle scattered light is used to measure the fineparticles. This instrument is similar to a family of photozone counters developed at IITRI known by

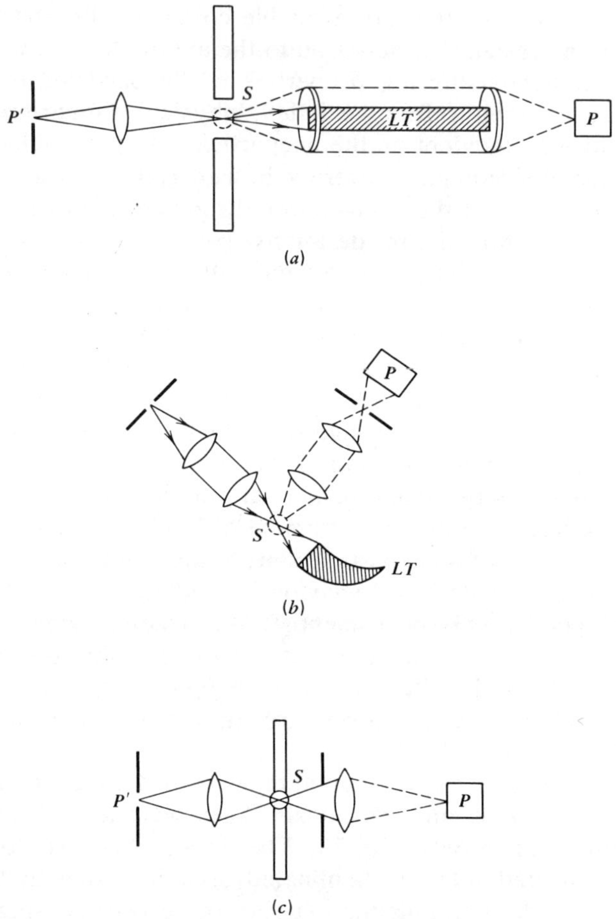

Fig. 8.8 Various photozone stream counters manufactured by Royco Corporation [28]: (*a*) Royco Photozone system (model 218); (*b*) Royco Photozone system (model 203); (*c*) Royco (model 345) for fineparticles in liquid suspension. Symbols: *P'*, white light "pinhole" source of light; *S*, inspection zone; *P*, photoelectric device; *LT*, light trap.

the family name *aerosoloscope* [27]. In Royco Model No. 218, shown in Figure 8.8*a*, a near-forward scattered light is used. The company recommends that the near-forward scattering instrument be used when one must monitor large volumes of gases in which the suspended fineparticles can vary widely in composition size and optical properties. Royco Model 345, shown in Figure 8.8*c*, is used to characterize fineparticles in a liquid suspension. In this instrument the energy removed from a forward-traveling beam is measured. Royco instruments are calibrated using fineparticles of known characteristics. However, if one is interested in research studies of specific aerosols and/or fineparticles in a liquid suspension, the optics of the various instruments are well defined, and one can carry out more complex interpretation of experimental data by using the appropriate light-scattering theory.

The Bausch and Lomb Corporation's photozone stream counter for the characterization of aerosols is shown in Figure 8.9 [29].

From the calibration curve in Figure 8.9, it can be seen that the slope of log I against log D has a slope of 2 for spheres greater than 1.5 μm. This indicates that for spheres greater than this size, the power removed from a white-light beam and scattered into the reception solid angle used in this

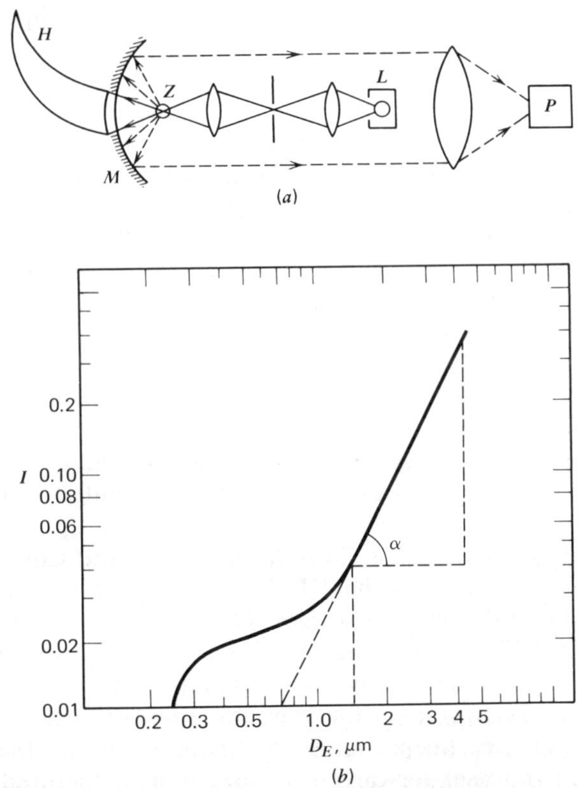

Fig. 8.9 The Bausch and Lomb photozone stream counter: (*a*) optical system (Z, inspection zone for aerosol moving at right angles to the plane of the diagram; H, absorption horn for forward-traveling beam; L, white light source; P, photoelectric device; M, concave mirror); (*b*) calibration curve [I, current in photoelectric device (in microamperes; D_e, diameter of latex spheres (α is slope of geometric optic equivalent region of response curve)].

counter is proportional to the geometric shadow of the sphere. For a region of 1.5 μm to approximately 0.3 μm, however, the fineparticles scatter far more light into the forward direction than would be anticipated from simple diffraction theory. This result is in accordance with the predictions of Mie scattering theory.

The Climet photozone stream counter, shown in Figure 8.10, characterizes the fineparticles by collecting and measuring most of the forward-scattered light and a high proportion of the back-scattered light with the aid of an elliptical mirror [30]. The interrogation beam enters through a small aperture at the front end of the elliptical mirror and is focused onto the interrogation zone. This zone is at one focus of the elliptical mirror. Light not scattered from the interrogation beam is absorbed by a light trap. If a source of radiation is placed at one focus of an ellipse, the energy is focused at the other. Therefore, all the back-scattered and forward-scattered light within the confines of the elliptical mirror are reflected to the other focus of the elliptical mirror where the photoelectric device is placed. It can be seen from the calibration curve shown in Figure 8.10 that by measuring the scattered light over a much larger solid angle, the calibration curve is more sensitive to changes in size below 1.5 μm than in the Bausch and Lomb system shown in Figure 8.9.

In the HIAC (an acronym for high accuracy) photozone counter, fineparticles in a liquid stream are characterized by the power drop they cause in a forward-traveling beam of white light [31]. The system along with calibration data are shown in Figure 8.11. It can be seen that the calibration curve deviates slightly from the square relationship predicted for straightforward geometric shadow power removal. The deviation is not unreasonable in the sense that the signals for less than 7.3 μm are slightly higher than they should be, which would indicate that for these fineparticles, the instrument is beginning to pick up some forward-scattered light (see curves for Bausch and Lomb and Climet counters). Beringer et al. [32] adapted the HIAC system to study relatively large fineparticles suspended in a gas stream. They showed that for fineparticles in the size range 20–100 μm, the geometric shadow power loss relationship can be used to interpret the experimental data.

A photozone stream counter for characterizing fineparticles in the range 2–100 μm, where the fineparticles are characterized by the white light they scatter in the near-forward direction, is manufactured by Particle Technology, Inc. [33].

An instrument developed for the characterization of submicron fineparticles is shown in Figure 8.12. Fineparticles too small to be resolved as images can still be seen in dark-field-illuminated microscope systems by the scintillations of light that they create as they move into the

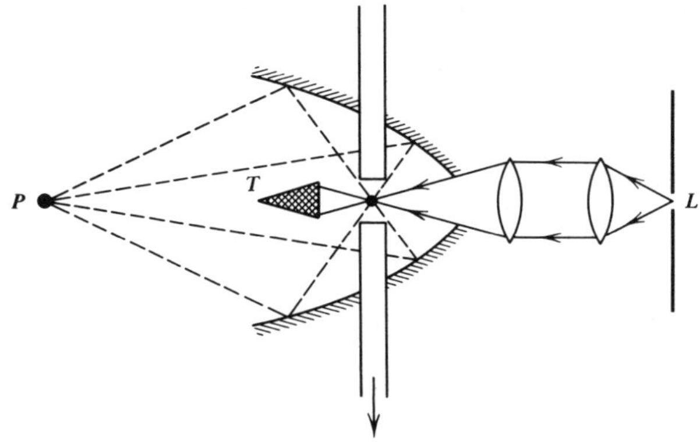

(a)

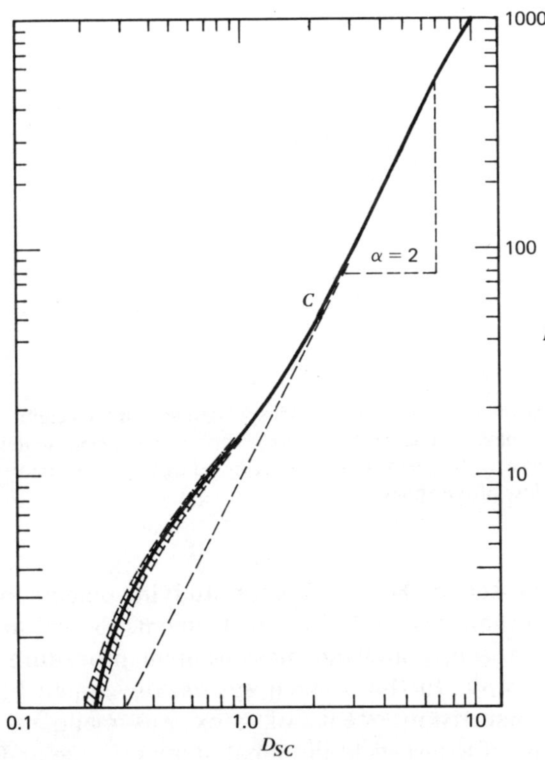

(b)

Fig. 8.10 The Climet photozone stream counter: (*a*) optical system; (*b*) calibration curve; (*c*) measured relationship. Symbols: *P*, photoelectric device; *T*, light trap; *L*, point white-light source; P_H, height of the pulse generated in the photoelectric device; D_{SC} sphere size in microns.

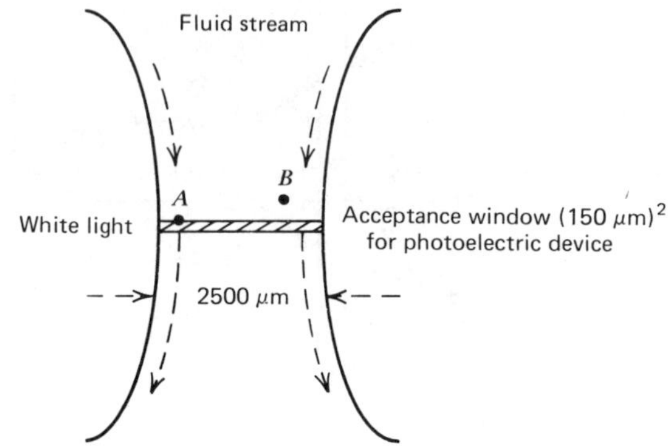

(a)

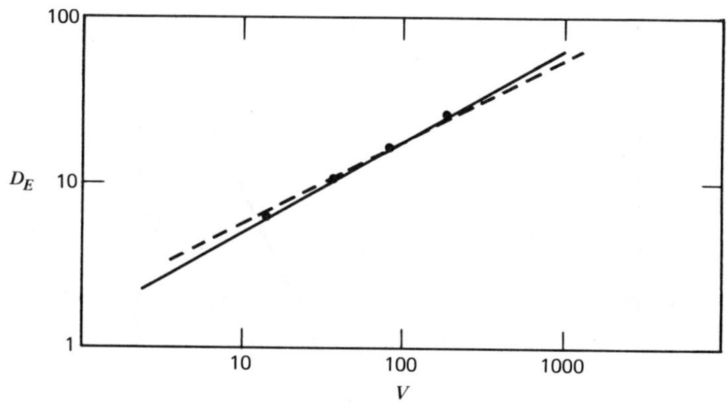

(b)

Fig. 8.11 The HIAC photozone counter: (*a*) inspection system; (*b*) calibration curve. Symbols: D_E, sphere diameter in microns; *V*, voltage drop of photoelectric device in millivolts; —— calibration curve; ---- curve of slope 2 corresponding to power removal proportional to the geometric shadow of the sphere.

plane of the microscope. Techniques for studying objects in this way are described as ultramicroscopy. The instrument shown in Figure 8.12 exploits ultramicroscopy and the measurement procedure is known as *flow ultramicroscopy*. In the original system developed by Collins and co-workers, a suspension stream of latex was made to flow up and through the plane of a dark-field-illuminated microscope and the scintillations created by the presence of the latex fineparticles counted. In the automated version of their equipment, shown in Figure 8.12, the microscope is replaced by a high-amplification photoelectric device linked to data processing equipment. A laser light shines through the latex suspen-

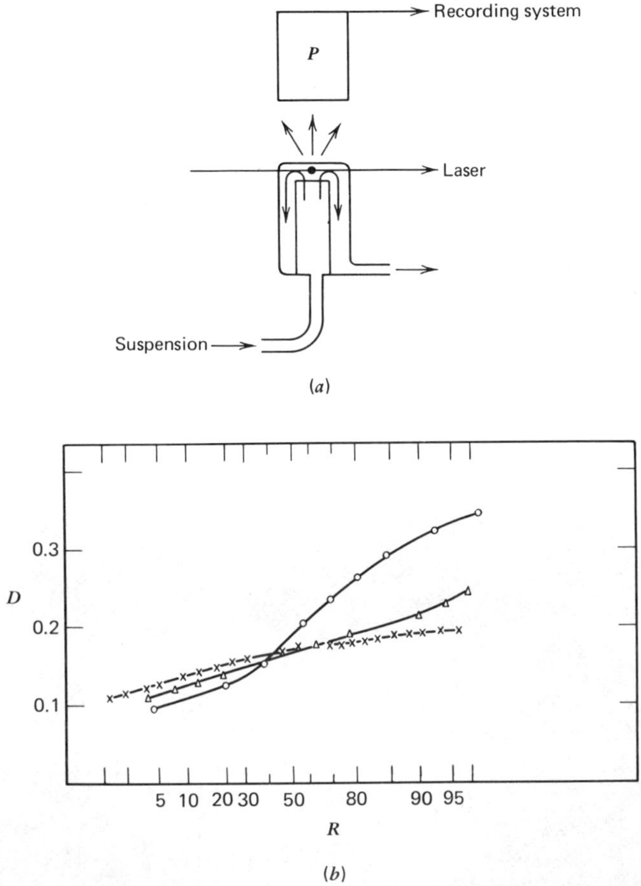

Fig. 8.12 The flow ultramicroscope, for the characterization of submicron latex suspensions: (*a*) optical system; (*b*) typical data. Symbols: *P*, photoelectric device; *R*, percentage by weight less than stated size; *D*, diameter in microns; ⊙, photocentridisc data; Δ, electron micrograph data; × flow ultramicroscope data.

sion welling up from a feed system and flowing over into a discharge container. As the latex spheres move into the laser light, they scintillate and generate a signal at the photomultiplier. In the experiments reported by Collins and co-workers, the magnitude of the spheres was related to the energy scattered into the receiving angle of the photoelectric device using the Lorentz–Mie theory of light scattering of monochromatic radiation [34]. From the data shown in Figure 8.12, it can be seen that measurements made by this technique compare well with those obtained from electromicrographs. However, the sizes estimated by these procedures

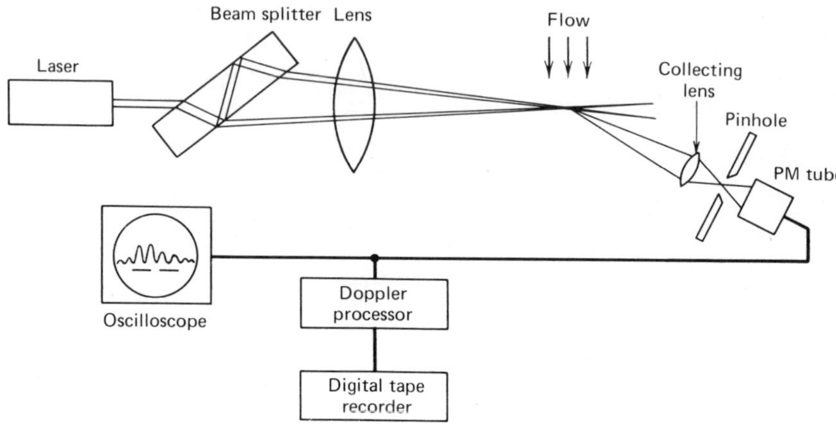

(*a*)

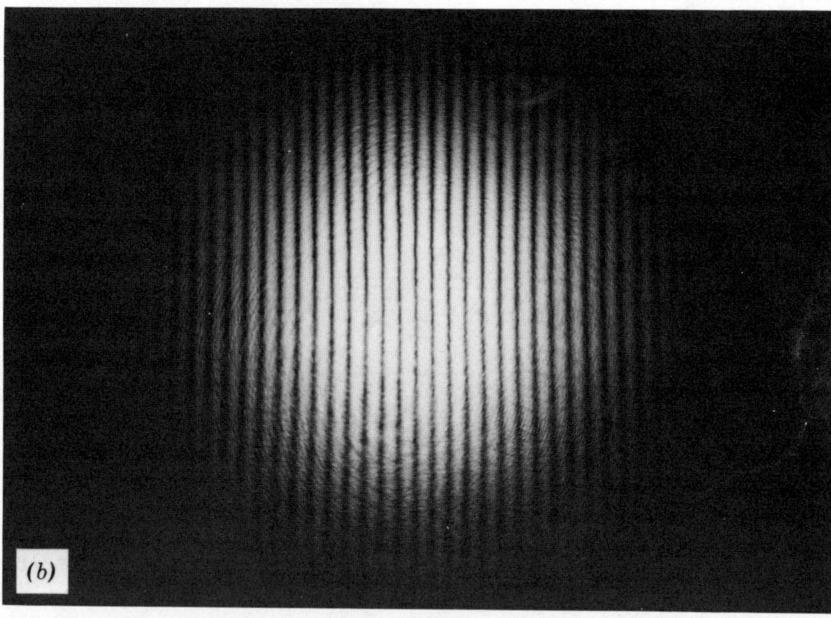

Fig. 8.13 Yanta's stream method, based on Doppler shift measurements [36]: (*a*) experimental system; (*b*) fringes imaged in the interrogation zone; (*c*) photomultiplier tube output showing fineparticle moving through interrogation zone.

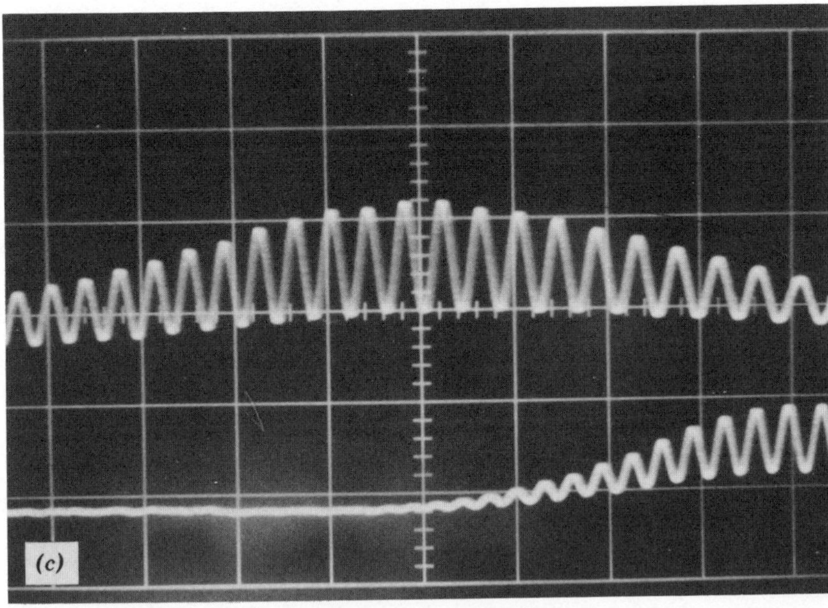

Fig. 8.13 (*Continued*)

are smaller than the estimates obtained by photocentridisk characterization. It may be that for such very small spheres in the centrifugal disk equipment, colloidal drag forces are creating Stokes diameters physically different from those of the physical diameters as viewed under the electromicrographs or as determined from radiation cross-sectional measurements.

Wu [35] has described a dual beam laser interrogation photozone stream counter for measuring the size and the velocity of fineparticles in the range 20–800 μm moving in a fluid stream. A large-area laser beam is used to intercept the track of a fineparticle moving in a fluid stream. Downstream of the first beam and slightly displaced in the vertical direction is a second laser beam. From a comparison of the pulses generated by the fineparticles crossing these two beams, one can deduce both the size and the velocity of the fineparticles.

Yanta [36] has used the Doppler effect to measure the mass of the fineparticle in a stream counter. In his system the fineparticles are accelerated through a constriction leading to the interrogation zone. The Doppler shift generated by the interference between the light reflected by the fineparticle from the crossed light beams is measured. The system used by Yanta and a typical pulse of reflected light are shown in Figure 8.13. From the known forces in the feed throat and the measured velocity of the fineparticle, the mass of the fineparticle can be measured.

It is quite possible that further development of Yanta's system, in which the power of the reflected signal and the Doppler shift were measured, would enable simultaneous information on the cross-sectional area and the mass of the fineparticle to be obtained [37].

The Prototron fineparticle counter is a specialized photozone system used to inspect parenteral solutions for tramp fineparticles [38]. Parenteral solutions are pharmaceutical preparations injected into the body [39].

In the Prototron a laser beam is used to scan the contents of a sealed vial containing liquid in a circular pattern. The unscattered laser light is absorbed by a trap. Any fineparticle lying within the scan pattern will scatter light. The light scattered in the forward direction is picked up by a lens and sent to a photoelectric device. It is claimed by the manufacturer that the instrument can size fineparticles in the range 1–100 μm. Presumably the instrument is calibrated empirically by using spherical fineparticles.

8.5 "HOT FILAMENT" STREAM COUNTER

Vonnegut and Neubauer [40] have described an instrument for sizing droplets. The basis of their instrument is the fact that when a droplet strikes a hot wire, the energy taken from the wire as the droplet evaporates causes a temperature drop in the wire, thus increasing the electrical resistance of the wire. Monitoring the electrical resistance of the wire results in a series of pulses that can be related to the size of droplets hitting the wire [40].

NOTES AND REFERENCES

1. Information on the Coulter Counter is available from Coulter Electronics, Inc., 590 West 20th St., Hialeah, Florida 33010.
2. Information on Celloscope and Electrozone resistazone counters is obtainable from Particle Data, Inc., P. O. Box 265, Elmhurst, Illinois 60126.
3. R. W. DeBlois and K. A. Wesley, "Size Concentration and Electrophoretic Mobilities of Several C-Type Encornaviruses, and T-even Bacteriophages by the Resistive-pulse Technique," reprint available from the General Electric Corporation, Research and Development Distribution, P. O. Box 43, Building 5, Connecticut, Schenectady, New York 12301. Commercial information on the Nanopar analyzer is available from the same address.

4. R. Davies, R. Karuhn, and L. Townsend, "The Effect of Particle Shape on the Response of an Electrical Resistance Zone Counter," in *Proceedings of the Symposium on Particle Size Analysis, Nuremberg, Germany, 1975*; published as *Dechema Monographien*, 79 (1589–1615), Teil B (1975).

5. For an introduction to the literature on the Coulter counter and other resistazone counters, see R. Karuhn, R. Davies, B. H. Kaye, and M. J. Clinch, "Studies on the Coulter Counter, Part One, 'Investigations into the Effect of Orifice Geometry and Flow Direction on the Measurement of Particle Volume,'" *Powder Technol.*, 11 (1975), 157–171; see also references 6 and 7.

6. R. Davies, R. Karuhn, and J. Graf, "Studies on the Coulter Counter, Part Two, 'Investigations into the Effect of Flow Direction and Angle of Entry of a Particle on Both Possible Volume and Pulse Shape,'" *Powder Technol.*, 12 (1975), 157–166.

7. R. Davies, R. Karuhn, J. Graf, and J. Stockham, "Studies on the Coulter Counter, Part Three, 'Applications of the IITRI Sensing System,'" *Powder Technol.*, 3 (1976), 193–201.

8. T. Allen and K. Marshall, "The Electrical Sensing Zone Method of Particle Size Measurement," bibliography published by University of Bradford, England, 1972.

9. N. J. Grover, *Biophys. J.*, 12 (1972), 1099.

10. R. Thom, *Zeitschrift Gesamte Exp. Med.*, 151 (1969), 331.

11. R. W. Lines, "A Short Contribution to the Proceedings of the First European Particle Size Analysis Conference," Nuremberg, September 1975, pp. 424–431. Also see comments on this contribution by R. Davies, pp. 432 and 433 in reference 4.

12. L. Spielman and S. I. Goren, "Improving Resolution in Coulter Counting by Hydrodynamic Focussing," *J. Colloid Interface Sci.*, 26 (1968), 175–182.

13. R. F. Karuhn and R. H. Berg, "Particle Shape Analysis Via Electrolytic Sensing Zone," paper presented at the International Symposium on Powder and Bulk Solids, Chicago, May 1978; proceedings published by the International Powder Institute, Chicago, Illinois.

14. R. Davies, R. Karuhn, J. Graf, and J. Stockham, *Bull. Parenteral Drug Assoc.*, 29 (2) (March/April 1975), 98–109.

15. C. Albertson, "Detection of Particles in Liquids by the Use of Ultrasonics," in *Proceedings of the First National Symposium on Nondestructive Testing of Aircraft and Missile Components, San Antonio, Texas*, February 16, 1960.

16. J. B. Gayle, "An Evaluation of the Sperry Ultrasonic Particle Counter," *J. Am. Assoc. Contamination Control*, 2 (10) (October 1963), 16–21, 30.

17. Commercial information is available from the Autometrics Corporation, 4946 North 63rd St., Boulder, Colorado 80301.

18. The basic physics of the attenuation of ultrasonic beams passing through a slurry of fineparticles was studied in depth by G. H. Flammer, "The Use of Ultrasonics in the Measurement of Suspended Sediment Size Distribution and Concentration," Ph.D. thesis, University of Minnesota, 1958.

19. A. L. Hinde and P. J. D. Lloyd, "Realtime Particle Size Analysis in Wet Closed Circuit Milling," *Powder Technol.*, 12 (1975), 37–50.

20. G. Langer, "An Acoustic Particle Counter Preliminary Results," *J. Colloid Sci.*, 20 (1965), 602–609.

21. R. F. Karuhn, "The Development of a New Acoustic Particle Counter for Particle Size Analysis," in *Proceedings of the Conference on Powders, IIT Research Institute, September 1974*; proceedings published by IITRI, 10 West 35th St., Chicago, Illinois.

22. B. Scarlett and I. Sinclair, "Operation of the Langer Acoustic Counter," contribution to the proceedings of the Nuremberg Conference on Fineparticle Measurements, September 1975 (see reference 4).

23. M. Kerker, *The Scattering of Light and Other Electromagnetic Radiation*, Academic, New York, 1969.

24. H. C. Van der Hulst, *Light Scattering by Small Particles*, Wiley, New York, 1957.

25. D. Sinclair, "A New Photometer for Aerosol Particle Size Analysis," *J. Air Pollution Control Assoc.*, 7 (2) (February 1967), 105–108.

26. J. K. Channell and J. Hanna, "Experience with Light Scattering Particle Counters," *Arch. Env. Health*, 6 (March 1963), 386–400.

27. Information on the various versions of the aerosoloscope is available from Fineparticle Section, IITRI, 10 West 35th St., Chicago, Illinois.

28. Information is available from Royco Instruments, Inc., 141 Jefferson Drive, Menlo Park, California 94025.

29. A. E. Martens, "An Electro-optical Dust Counter," paper presented to the American Association for Contamination Control, Albany, New York, February 1966. Commercial information is available from Bausch and Lomb, Rochester, New York.

30. Information is available from the Climet Corporation, 1320 Colton Avenue, Redlands, California 92373.

31. Information is available from HIAC Instrument Division, P. O. Box 3007, 4719 West Brook Street, Mont Clair, California 91763.

32. A. J. Behringer, T. S. Mika, D. H. Wood, and T. W. Nash, "Use of a Modified HIAC Criterion PC 320 as an Airborne Particle Size Analyzer," paper presented at the Conference on Particle Size Analysis, September 1977, Salford, England; proceedings to be published by the Analytical Chemistry Section of the Royal Institute of Chemistry, Great Britain.

33. Particle Technology, Inc., 734 North Pastoria, Sunnyvale, California 94086.

34. J. A. Davidson, E. A. Collins, and H. S. Haller, "Latex Particle Size Analysis, Part 3, 'Particle Size Distribution by Low Ultra Microscopy,'" *J. Polymer Sci.*, Part C (34) (1971), 235–255.

35. J. Wu, "Fast-Moving Suspended Particles—Measurements of Their Size and Velocity," *Appl. Opt.*, 16 (3) (March 1977), 596–600.

36. W. J. Yanta, "Measurement of Aerosol Size Distributions with a Laser Doppler Velocimeter," *National Bureau of Standards Special Publication 412, Aerosol Measurement*, proceedings of a seminar on aerosol measurement held at National Bureau of Standards, Gaithersburg, Maryland, May 7, 1974.

37. R. J. Adrian and K. L. Orloff, "Laser Anemometer Signals Visibility Characteristics and Applications to Particle Sizing," *Appl. Opt.*, 16 (3) (March 1977), 677–684.

38. Commercial information is available from Prototron, Inc., 1020 Corporation Way, Palo Alto, California 94303.

39. M. J. Groves, "Parenteral Products," William Heineman, Medical Books Limited, London 1973.

40. B. Vonnegut and R. Neubauer, *Anal. Chem.*, 24 (1954), 100.

CHAPTER

9

DIRECT MEASUREMENT OF FINEPARTICLE DIMENSIONS

9.1 IMAGING SYSTEMS

Many procedures for making direct measurements of fineparticle dimensions from a study of the enlarged image of a fineparticle have been reported in the literature. The image generated by an optical microscope system can be viewed directly through the eyepiece or projected onto a ground-glass screen. Direct measurements can then be carried out on the image by using measurement grids placed on the screen. The image can also be recorded photographically and measurements made on the negative or the photographic print. An electronic image can be generated by focusing a television camera on the image plane of the optical microscope. Most current television cameras use *linescan* procedures for generating the image on a television screen, where the overall image is created by the persistence of fluorescent light generated on the screen of the television monitor by the scanning beam of electrons. The stream of signals coming from the television camera can be diverted to a computer where logic operations to evaluate various parameters of the fineparticle profile can be performed. Such image analysis procedures are referred to as *linescan logic procedures*. In the late 1970s, development of alternative television camera design employing arrays of electrooptic elements to generate digital mosaic images became available [1, 2]. Processing of image mosaics generated by these new imaging procedures offers many advantages over linescan logic procedures.

Electron microscopes are used to generate images of fineparticles that have dimensions smaller than 1 μm. In the transmission electron microscope (TEM) principles similar to those employed in generating a picture by optical means are employed. Electromagnetic lenses analogous to optical lenses are employed to produce the final picture. A technique known as *shadow casting* can be used when generating a TEM image to obtain three-dimensional information. In this procedure, metal vapor in a vacuum is directed at the fineparticles being imaged to create a shadow. From the dimensions of this shadow, it is possible to measure the height of the fineparticle as well as its projected parameters. In the scanning electron microscope (SEM) a primary beam of electrons, which can be moved in a scanning pattern, is directed onto the specimen to be imaged. The electron beam–specimen system are in a vacuum. As the primary beam hits the specimen, there is a burst of secondary electrons generated by the interaction of the primary beam and the surface. These secondary electrons are collected in a detector and used to modulate the linescan pattern of a slave television monitor moving in synchronous motion with the scanning pattern of the primary electron beam. A SEM picture is referred to as a *SEM image*. A SEM image has a three-dimensional appearance. Since the image of the scanning electron microscope is generated by a linescan procedure, the signals from the scanning electron microscope can be fed directly into a computer for the evaluation of fineparticle characteristics by linescan logic manipulation procedures [3, 4].

The electromagnetic energy incident on the fineparticles during the imaging process may change its dimensions. For example, latex emulsion fineparticles can be enlarged by the electron beam energy in an electron microscope. In optical microscopy the problem can be particularly acute with systems such as pollen grains, which can swell under the combined effects of heat from the imaging beam and the presence of fluid used to disperse the system.

In another important imaging technique, holography, coherent light is used to generate a hologram that, in turn, can be illuminated by coherent light to produce a three-dimensional image of a fineparticle system. The invention of the laser has made this imaging system important in fineparticle science for two reasons: (1) it enables one to image a system with greater depth of focus; and (2) the short pulse of light generated by a laser enables one to take very rapid images of transient and/or hostile systems. Thus holographic imaging has been used to image aerosol sprays from cans and the exhaust from rockets.

In another technique, off-axis holography, the primary laser beam is first split into two beams. The first beam falls on the object to be imaged, and the diffracted light from this object is then allowed to mingle with the second undiffracted beam, which follows a route different from that of the photographic plate. The interference pattern produced between the diffracted light and the undiffracted light is recorded in the photographic image. The resultant negative constitutes the hologram. When suitably illuminated, it enables one to generate a three-dimensional image of the object.

Another procedure, in-line holography, can be used in imaging aerosol fineparticles. In such systems in which the fineparticles occupy little space in the laser beam there is no need to provide a reference beam. The undiffracted forward-traveling portion of the beam serves as the reference beam. This technique has been used by Thompson and co-workers to study fineparticle systems [5, 6]. Van Ligten of the American Optical Corporation has used a similar system to image systems viewed through the microscope so that a greater depth of focus can be achieved [7, 8].

9.2 PREPARATION OF FINEPARTICLES

Fineparticles to be imaged in a microscope system are often dispersed on a glass slide with the aid of a drop of viscous fluid. The technologist should avoid excessive shear forces when spreading out a set of fineparticles in a drop of fluid by using a glass rod moved over the surface of the glass slide. If a drop of slurry to be examined is allowed to fall onto a slowly spinning disk at some distance from the center of the disk, and if the surface of the disk has been initially treated with the clear fluid used in preparing the slurry so that there are no problems associated with the wetting of the disk surface, the mild centrifugal action of the slowly spinning disk can spread the various fineparticles in the disk so that they are spread out from each other in the field of view. This technique can be used with a thermosetting resin to give a permanent film of fineparticles trapped in a resin matrix. Alternatively, a quick-drying transparent household cement can be used to give a similar film of fineparticles.

When making measurements by direct viewing with the human eye on an array of fineparticles, it must be remembered that the eye is part of the measurement system. Its idiosyncrasies must be considered as part of the measurement procedure. In several widely used measurement procedures a human operator visually compares the size of an image with a set of standard circles. In such procedures the size characteristics of the immediate neighbors of a fineparticle influence this judgment, as illustrated by the two systems shown in Figure 9.1. In the arrays of Figure 9.1, the two center circles are of the same size but look as if they differ in size because of the optical illusion created by their near neighbors. To avoid such optical illusions, very low density of fineparticles per unit area should be used. A special viscous medium, which can be easily evaporated after it has set, for use in preparing fineparticles for examination has been described by Thaulow and White [9]. It may be possible to use selective stains and/or polarized light to isolate particular fineparticle species against a general background of inconsequential fineparticles [10]. Optical computing procedures for finding specific types of fineparticle in a general field of view have been described by Kaye and Naylor [11].

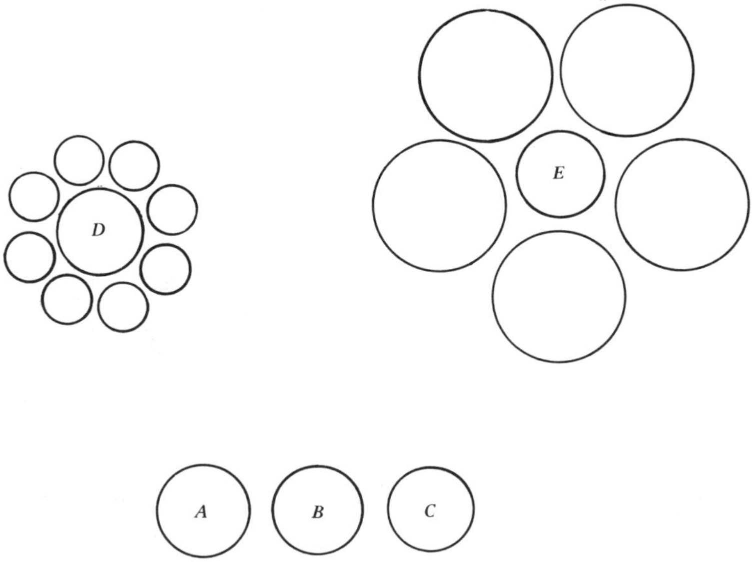

Fig. 9.1 The population environment affects the judgment of sizes by visual comparison.

9.3 DIRECT MEASUREMENT OF SINGLE FINEPARTICLE PARAMETERS

Very often in a characterization study, two- or one-dimensional measurements are made on a projected image of the fineparticle. Thus consider the two-dimensional image of an irregular-shaped fineparticle shown in Figure 9.2. The length of the profile is usually taken to be the magnitude of the maximum projected length as illustrated. The breadth is then taken to be the projected width of the profile at right angles to the direction in which its length is measured. It should be noted that Ridgeway has pointed out that the concept of length as defined previously causes problems when applied to images of some regular geometric shapes [12].

For historic reasons it has become common practice to describe the length of a profile projected in any given direction as the Feret diameter of the profile. Thus the maximum projected length of a profile is its maximum Feret diameter. The minimum width is its minimum Feret diameter, and the breadth is the Feret diameter at right angles to the maximum Feret diameter. A useful parameter of an irregular profile is its *convex hull*, which is the minimum line envelope free of concave sections that can be drawn around the profile. It can be shown that the average value of the

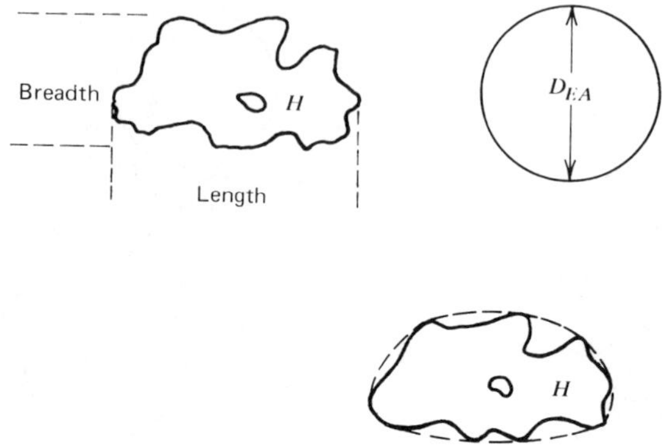

Fig. 9.2 Various parameters used to describe the size of a fineparticle (........ defines the convex hull of the profile; H denotes a hole in the profile).

Feret diameter of an irregular profile taken over many different directions is related to the perimeter of the convex hull by the relationship

$$F_{av} = P_{ch/\pi} \tag{9.1}$$

where P_{ch} is the perimeter of the convex hull [13]. The diameter of the circle of equal projected area is often used to describe an irregular fineparticle.

When discussing the structure of an irregular fineparticle, it is useful to describe four different features of the morphology. Any obvious regular substructure of the surface is described in this book as the texture of the profile. The general "ins" and "outs" of the profile in this book are referred to as the *structure* or the *topography* of the profile. The best way to distinguish between the texture and the topography of the profile is to imagine that the profile is an island as seen from the air. The topography is the bays and the peninsulas of the island, whereas the texture is the individual rocky structure of a beach on the island. The external morphology in this book is taken to be the shape of the convex hull. The fourth descriptive feature of the profile is its topology. In mathematics, topology is a study of relationships that remain the same in distorted space. If we take the profile shown in Figure 9.2 and stretch it on a rubber membrane, no matter how we stretch it, it remains basically a doughnut with a single hole included in the island area. In topology this is described as being of genus 1. The topology of a profile is important in the mathematics of automatic pattern recognition used in automated systems such as the Bausch and Lomb Omnicon® image analyzer. Its definition is given at this point for the sake of completeness, but it does not arise again in the text of this book [14].

Linear dimensions of a profile can be measured readily when examining the profile through a microscope by placing a linear scale in the eyepiece of the microscope. Any scale placed in an eyepiece of a microscope is known either as a reticule or a graticule; the latter name is more common in Great Britain.

When looking at a set of fineparticles through a microscope or on a photographic print, the first problem is to decide what is a separate fineparticle. Often a profile looks as if it is made from two or more overlapping subunits. The technologist must decide from a knowledge of the system if the components of the profile should be treated as separate entities or whether the group should be treated as a single fineparticle.

Figure 9.3 shows several reticules used in fineparticle science [15]. In the simplest grid, the Whipple grid, originally developed for counting the number of stars per unit area on photographs taken with celestial telescopes, the number of stars was determined within subsquares of the grid. To illustrate the way in which the Whipple grid is used to determine the number of fineparticles per area collected on a filter paper, consider the following set of data. Fineparticles settled onto a filter paper are examined through a Whipple grid. The number of fineparticles per square of the grid is counted. After 100 squares had been examined, the population of the squares was as summarized in Table 9.1. The measured average population per square could be calculated from these data, but a better way to treat the data is to calculate the expected population per square. Since the overall population per square is low, the appropriate statistical distribution for describing the probable occurrence of the various recorded populations is the Poisson distribution. Graph paper for processing data describable by the Poisson distribution is shown in Figure 9.4 [16]. The abscissa of the graph paper is a logarithmic scale indicating expected populations of a square; the ordinate is a normal–probability scale indicating the fraction of recorded populations greater than or equal to a stated population. The curved lines crossing the main field of the graph are population density curves [17].

To plot a point on this type of graph paper, one first calculates the fraction of populations greater than or equal to a stated population and then moves down the curve of that population until it intercepts the appropriate line linked to the population frequency axis. For example, in the data in Table 9.1, 19 of the recorded populations had populations of one or more dust fineparticles; that is, 90% or .90 of all the populations had fineparticle populations greater than this value. The lowest curve in the chart in Figure 9.12 is population = 1. One moves down this curve until it intercepts the .9 line from the population frequency line. The experimental data point is located at this intersection. Again, from Table 9.1 it is

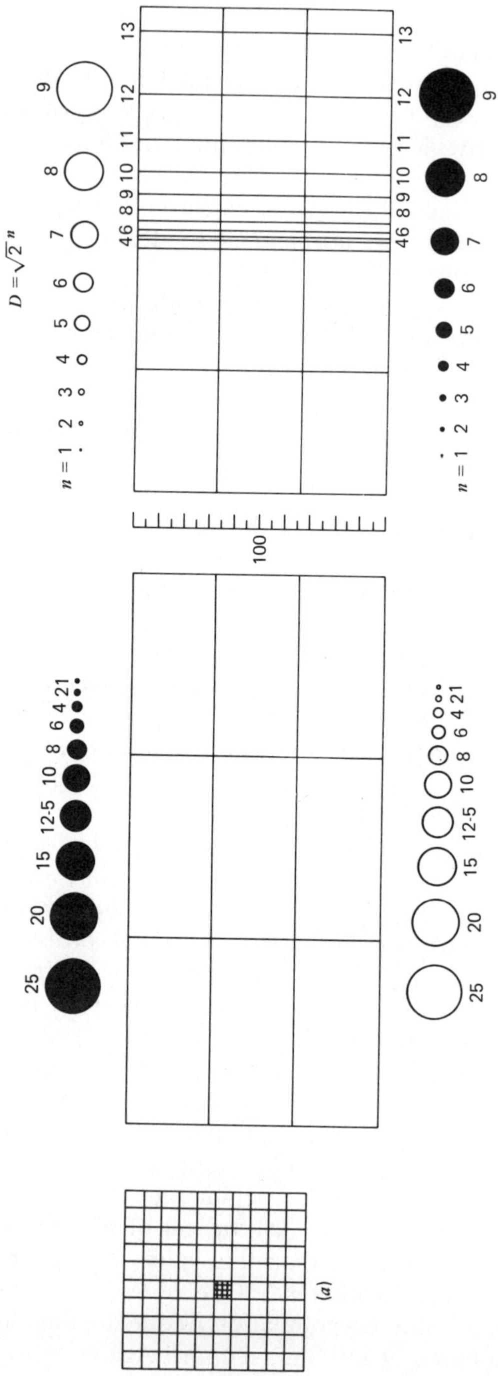
D = √2 n
n = 1 2 3 4 5 6 7 8 9
4 6 8 9 10 11 12 13
100
25 20 15 12·5 10 8 6 4 2 1
(a)
(b)
(c)

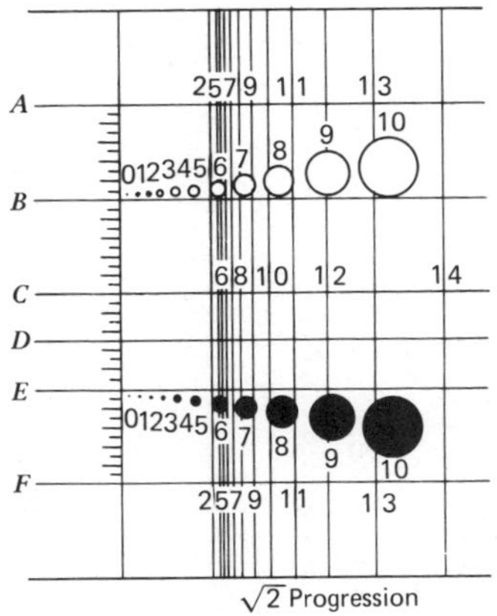

(d)

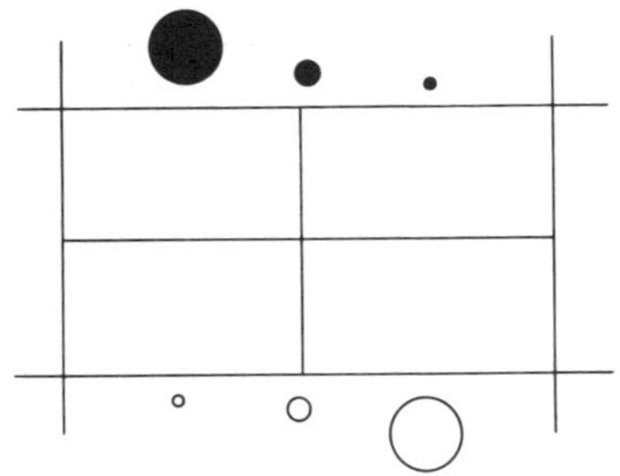

(e)

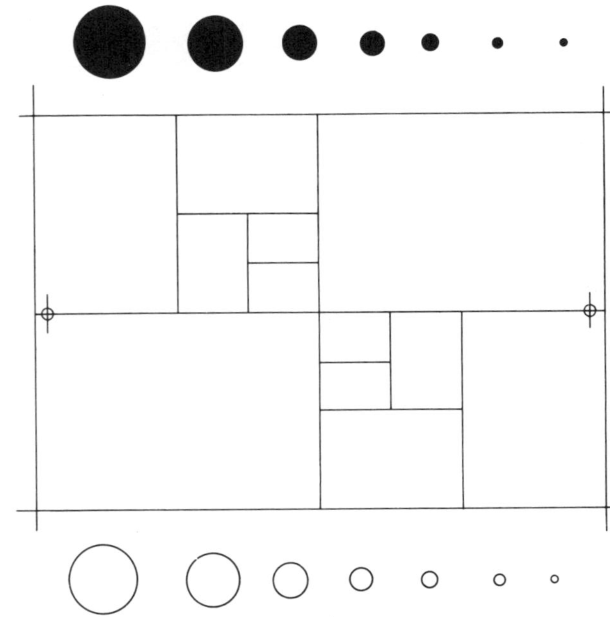

(f)

Fig. 9.3 Grids for placing in the eyepieces of a microscope for use in the evaluation of fineparticle images: (*a*) Whipple grid; (*b*) Patterson Cawood graticule; (*c*) globe–circle reticule developed by May; (*d*) "improved May graticule"; (*e*) Watson graticule; (*f*) BSS 3406 graticule.

TABLE 9.1. Dust Content of Air Samples[a]

Number of dust fineparticles contained in square	0	1	2	3	4	5	6
Number of squares having stated dust content	10	19	24	22	12	10	3
Fraction of samples with concentration greater or equal to stated size	1.00	.90	.71	.47	.25	.13	.03

[a]Expected population is 2.5 per unit square (see Figure 9.4).

noted that .71 of all the examined populations had fineparticle contents greater than or equal to two fineparticles. Moving down the population line $C = 2$, the data point is located at the intersection of this line with the population line .71. When the totality of experimental points from Table 9.1 are plotted on the graph paper it can be seen that they define a straight line parallel to the ordinate and crossing the expected population axis at the best estimate of the population of fineparticles per unit square. Any departure of the experimental data from a clustering about a vertical straight line indicates that the populations of various areas examined are not varying at random.

When measuring the dimensions of a set of fineparticles in an imaged field of view, it is essential to have a nonambiguous strategy for accepting or rejecting the fineparticles that intersect the boundary of the selection square. Consider, for example, the array of fineparticles shown in Figure 9.5. The rectangle *LMON* is taken to define the selection area. Unbiased selection rules for accepting or rejecting fineparticles such as those labeled *A* through *J* that fall on the boundary must be used in a characterization procedure. One unbiased selection procedure is to locate the approximate centroid of the fineparticles and only include in the measurement study those fineparticles whose centroid lies within the selection area. This is an unbiased and adequate procedure but unnecessarily difficult. If all profiles of a given size are considered, and then if they are randomly scattered in the two-dimensional space being examined, it is reasonable to assume that half of the centroids for the profiles in doubt will lie outside the boundary line and half of those will lie inside. Therefore, on average, it is just as adequate to accept every other fineparticle that crosses the boundary as an unbiased method of estimating the true

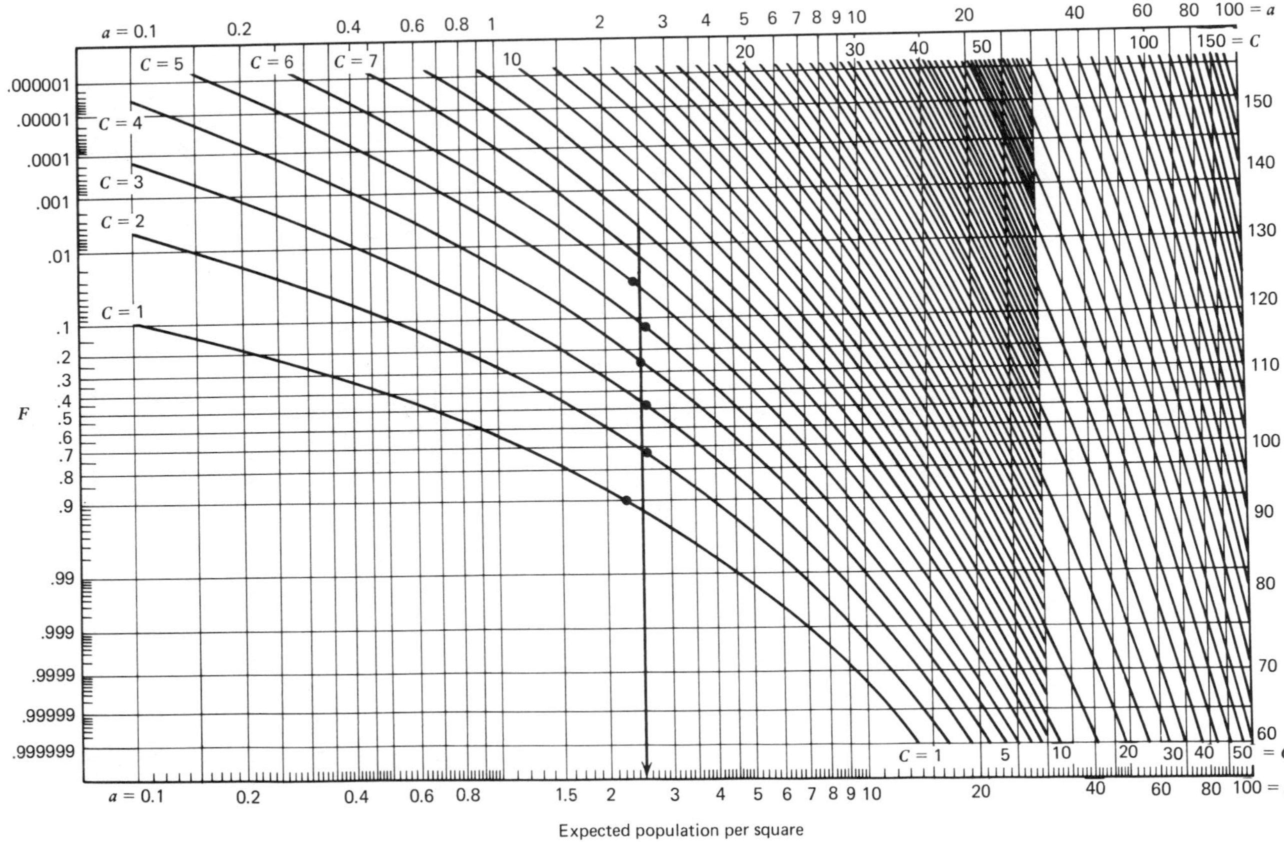

Fig. 9.4 Poisson population graph paper used to make the best estimate of the expected population of fineparticles per unit area of search (F, experimentally determined frequency of fineparticle population greater than or equal to C; C, population of fineparticles).

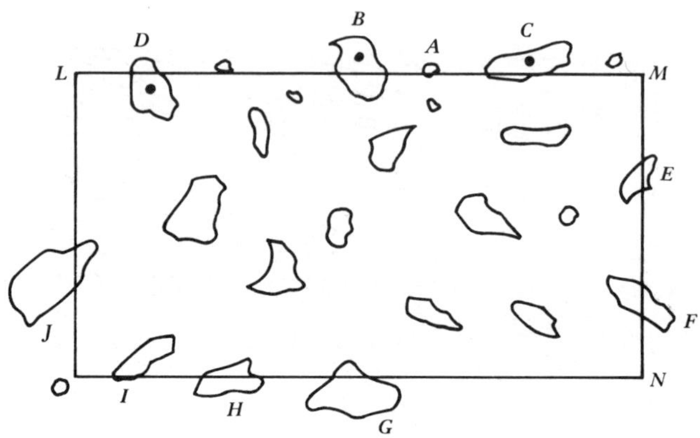

Fig. 9.5 Selecting a proportion of the fineparticles that are on the boundary of the selection grid must be carried out without bias.

population of fineparticles to be found within the search area. Again, this is an adequate and unbiased procedure for selecting the fineparticles to be included in the count; however, it is not easy to implement this logic in linescan logic iconometric systems. For such systems, one can use the equivalent selection procedure that all fineparticles that lie along line *LMN* be included in the count whereas all fineparticles on line *LON* be rejected, provided that the two lines are of equal length [18].

The Patterson–Cawood globe and circle graticule (see Figure 9.3) has been widely used to estimate the projected areas of an array of fineparticles [19]. The various squares of this reticule are used to select the fineparticles to be measured. The profiles are sized by comparing the projected area of the profile with the circles drawn on each side of the main rectangle. Early in the development of iconometric procedures it was noticed that operators would assign different sizes to the projected areas of a fineparticle if they used a set of solid circles for comparative purposes as compared to the sizes assigned if they used ordinary circles for comparative purposes. The solid circles by common usage are described as "globes," and hence the name of the reticule. Operators should be trained to use reticules of this kind using standard profiles [20]. They should be allowed to use either circles or globes according to their own preference. Initially many operators tend to oversize fineparticles, but they correct for this when their attention is drawn to this aspect of measurement technique.

When measuring the size distribution function of a set of fineparticles imaged in a microscope, measuring the dimensions of every profile in an array leads to data such as those shown in Table 9.2*a*. At first sight it

TABLE 9.2*a*. Distribution Data Collected in an Indiscriminate Manner by Measuring All Fineparticles in 10 Fields of View Selected at Random

Size Group, μm	0–10	10–20	20–30	30–40
N^a	1112	321	20	2

[a]Number of fineparticles.

TABLE 9.2*b*. Distribution Function Data for a Field of Fineparticles Collected by Stratified Count

Size Group, μm	0–10	10–20	20–30	30–40
$N_a{}^a$	1	4	350	780
Po^b	10.000	2.500	0.028	0.013
N^c	10.000	12.500	12.528	12.541
P^d	79.7%	99.6%	99.9%	100.0%

[a]Number of areas examined to find 10 fineparticles in size group.
[b]Population frequency per unit area of stated size.
[c]Cumulative number of fineparticles equal to or less than stated size.
[d]Percentage number of fineparticles equal to or less than stated size.

looks impressive that the distribution function calculated from the data is based on information from 1455 fineparticles. However, it must be appreciated that the average mass of the fineparticle in the largest size group in Table 9.2*a* is 343 times the average mass of fineparticles in the smallest size group. This means that the two fineparticles of Table 9.2*a* represent almost as much mass as the total 1112 fineparticles in the smallest group. A chance encounter with one more fineparticle in the largest size group would significantly shift the median size calculated from the data in Table 9.2*a*. For this reason, when measuring the size distribution function of fineparticles having a relatively wide range of sizes, a stratified count procedure should be used. In such a strategy, one sets out to look for the number of selection areas that contain, say, 10 fineparticles of each size. For example, if we were to examine the same population as that for which data are summarized in Table 9.2*a* we would probably generate data such as that shown in Table 9.2*b*. We would perhaps find that 10 of the smallest-size fineparticles, in the range 0–10 μm, occur within one selection area. We would now select another field of view at random and might have to look for four fields of view before finding 10 of the fineparticles in the range 10–20 μm. With a reasonable set of data based on this approach, we would probably find that it was necessary to examine approximately 350 fields of view before locating 10 of the next size group and 800 fields of view to locate 10 of the largest size group. To minimize labor in a stratified measurement procedure, we could change the optical magnification when looking for the larger fineparticles. For example, if we changed the magnification by a factor of 10 for the two largest size groups, we would only have to inspect 80 and 40 areas, respectively.

The two reticules shown in Figures 9.3*c* and 9.3*d* were evolved by May and co-workers at the Porton Down Research Centre in Great Britain. They are sometimes variously referred to as *Porton graticules*. These reticules use circles and globes having diameters that increase in $\sqrt{2}$ progression. This $\sqrt{2}$ grading of the reference circles arose historically from the fact that many of the aerosol systems studied by the Porton group obeyed the so-called lognormal size distribution functions. For such distribution functions, the use of the $\sqrt{2}$ progression gave equally spaced data points on logarithmic graph paper. The use of $\sqrt{2}$ gradations in the comparison circles was also useful in the early days of size characterization studies for scientists working with relatively coarse powders who wished to characterize by microscope examination fineparticles that represented the finer portions of a distribution, the coarser portion of which was evaluated with the use of sieves. The ratio of sieve aperture sizes in the BSS sieves is in a $\sqrt{2}$ progression, and hence it was only natural to carry over the same grouping into the microscope method to evaluate the fines of the distribution. It is unfortunate that the use of $\sqrt{2}$ reticules has become so widely accepted since these reticules are insensitive if one attempts to use a graticule to evaluate a population of fineparticles with only a small range of sizes present in the population. Their use can also distort measured distribution functions that are bimodal or do not follow a lognormal distribution function.

The major difference between reticules *d* and *c* in Figure 9.3 is that the reticule of the former has more flexibility built into it for selecting the areas of a field of view to be evaluated. Thus if a relatively heavily populated fineparticle distribution is being studied, only fineparticles lying between the lines in reticules *c* and *d* in Figure 9.3 are accepted, whereas those between reticules *a* and *f* are accepted for a less-populated field of view.

The set of straight lines to the left of the median line in both reticules *c* and *d* in Figure 9.3 are used to estimate the size of fineparticles larger than the largest circle shown in the reticule. Thus the distance from the median line to the line labeled 9 is the same as the diameter of the circle given the number 9 in the reticule. The sizes of the lengths from the median line to the lines labeled 10, 11, 12, and 13 continue the $\sqrt{2}$ progression size increases of the circles in the reticule.

Reticule *e* in Figure 9.3 is a special reticule developed by Watson for studying fineparticles that are considered to represent respirable hazards [21]. Reticule *f* in Figure 9.3 is that recommended in the British Standards procedure for evaluating images of fineparticles.

Reticules should always be calibrated at the particular microscope magnification being used with a stage micrometer such as a linear scale of known dimensions accurately etched or printed onto a microscope slide that can be viewed through the microscope.

9.4 IMAGE SHEARING PROCEDURES

In image shearing procedures for characterizing the dimensions of a profile, two images of the profile are moved with respect to each other. Usually the two images of the profile are initially coincident. One image is then moved away from the other. This movement of the image is described as *image shearing*. The overlap of the sheared images is usually clearly visible in the field of view. The images are moved with respect to each other until just separated. The shearing control can be directly calibrated to give a readout of the linear dimension equivalent to the image separation. As illustrated in Figure 9.6*b*, the magnitude of the dimension measured in this way for an irregular profile is dependent on the orientation of the major axis of the profile with respect to the direction of image separation. Image shearing equipment is provided with a system for rotating the image with respect to the direction of image separation so that the various dimensions of the profile can be measured.

The first image shearing device was described by Timbrell, who described his system as *double-image microscopy* [22]. He used a vibrating mirror inserted into the optical system of the microscope to provide two images of the field of view. The frequency of vibration in the mirror was so high that persistence of vision created the impression of two images in the field of view. The amount of separation between the two images was controlled by the amplitude of vibration. The image separation could be increased by increasing the amplitude of the mirror. A commercial version of Timbrell's instrument, manufactured by Flemming Instruments of Great Britain, is called "particle size micrometer and analyzer, type 526" [23]. An all-electronic implementation of the Timbrell principle of double-image microscopy, called the *Shearicon*®, has been developed by Coulter Electronics [24]. Watson used a dichroic beam splitter to obtain

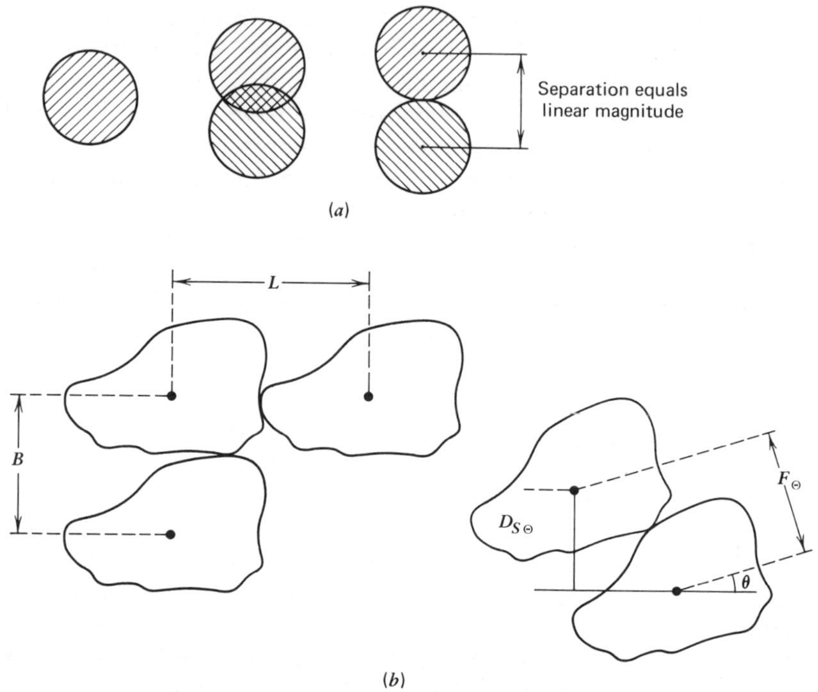

Fig. 9.6 Image shearing can be used to characterize image profiles: (*a*) basic concept used in dimension estimation by image shearing; (*b*) various image-shear measurements. Symbols: F_θ = Feret's diameter at angle θ to direction of shear; $D_{S\theta}$ = separation diameter.

the two images. In his instrument the two images are red and green with a black overlap [25]. The Vickers Instrument Company of York (England) manufactures an image separation device in which the two images are separated by using special prisms that can be rotated relative to each other [26]. A small television camera can be placed directly above a Watson or a Vickers eyepiece. Using this system, the image separation can be judged comfortably on a large screen without the eyestrain associated with extensive use of a microscope. Such a system is useful when training new operators [28]. One of the major advantages of double-image procedures is that the sizing of a profile is no longer a subjective decision on the part of the operator.

It is common practice to define three different types of average diameter for use in double-image microscopy: The first double-image diameter is defined as the average of the sizes estimated from image separation, with respect to a fixed direction, of an array of fineparticles in random array. This average diameter came into use in fineparticle science because it minimized the amount of work that had to be done when characterizing a field of view because it obviated the need to keep rotating the field of view for each fineparticle characterized to measure its "length." Physically, the first double-image diameter corresponds to the expected average of maximum chords drawn in all possible directions on all profiles of an array.

The second double-image diameter is defined in the published literature as that magnitude of individual profiles determined by separating the images when the fineparticle has been first aligned with its maximum dimension along the direction of image shear.

The third double-image diameter is defined as the magnitude of the image separation when the maximum dimension of the fineparticle is aligned initially to be perpendicular to the direction of shear.

9.5 THE ZEISS–ENDTER FINEPARTICLE CHARACTERIZATION INSTRUMENT

In the early 1950s Endter and Gebauer [27] devised a system in which the area of a profile was measured by superimposing on the profile a circle of light, the diameter of which could be varied by means of an iris diaphragm. This system was made available by the Zeiss Corporation, and the common name of this instrument is *the Zeiss–Endter size analyzer* [29].

To use the instrument, a photograph of the fineparticle to be characterized is placed on a special table. The particular profile to be evaluated is positioned over the center of a circular beam of light. Using a handwheel, the diameter of the transparent portion of the iris diaphragm is altered until the area of the light beam equals the area of the profile. The movement of the handwheel to alter the iris diaphragm also moves a wheel connected to a set of counters. When the operator judges the light beam to have the same area as the profile, he or she operates a foot switch that causes the circuit connected through the wheel to add a count in the appropriate register. It is important to realize that the recording wheel of the instrument is quantized. It has been found that the human eye cannot be relied on to remember which fineparticles in a field of view have been evaluated, and so the equipment has a spring-loaded pointer at the end of a lever that makes a hole in the center of the profile at the same time as the foot that operates the counter. The Zeiss–Endter system has enjoyed wide use in spite of its slow operation because of its relatively low cost and the objectivity of the measurement.

Several systems have been suggested for automating the basic principle employed in the Zeiss–Endter size analyzer; for example, see the work by

White et al. [30]. Kaye has suggested that the recording of the energy flux through a transparent profile outline placed on the iris diaphragm to obtain a graph such as that shown in Figure 9.7 could provide useful data [31]. Over the first region of this graph, the beam of light is sufficiently small that the whole of the beam lies within the fineparticle profile. For this range of iris diameters, the graph of power against diameter on a log–log scale is linear and of slope 2. As the light beam starts to move outside of the profile, there is a departure from linearity. The curve in this region will follow a path such as that shown in Figure 9.7 until all the fineparticle profile lies within the enlarged beam. From this graph, it is possible to deduce the circle of equal area as the intersection of the extrapolated horizontal curve with the extrapolated linear relationship for when the circle of thc iris is entirely within the profile. When locating the profile to be characterized in the beam of light, one can estimate the centroid of the profile. The largest circle that can be drawn around this point that lies within the profile can be deduced from the graph of power against iris diameter since it is the diameter at which the linear relationship starts to fail. On the other hand, the largest circle that can be drawn about the centroid and that just touches the outermost extremity of the profile is the value of the iris diameter beyond which there is no increase in transmitted power (it is assumed that the background around the profile is perfectly opaque—in practice, it may be necessary to make some small experimental adjustments for finite transmission through the background of the profile). Kaye has defined this larger circle as the embracing circle. Kaye has recently shown that several compactness ratios can be defined by using diameters of these inscribed and embracing circles in conjunction with the diameter of the circle of equal area and the diameter of the circle equal in area to the profile of the convex hull. These compactness ratios are useful in describing the shape of the profile. Their uses are discussed

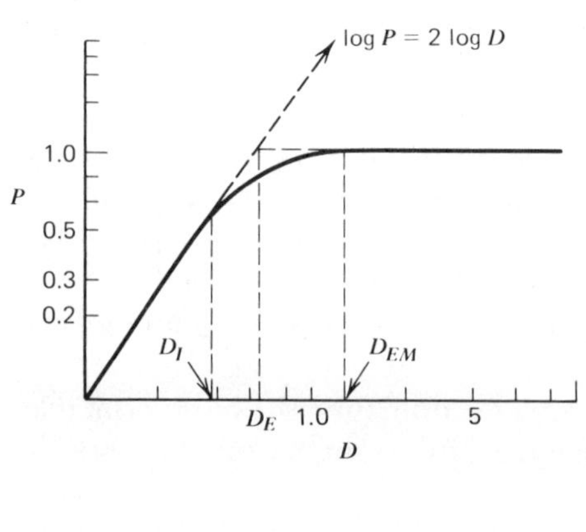

Fig. 9.7 Measurement of the power transmitted through the iris of a Zeiss–Endter instrument could be used to evaluate fineparticles profiles (D, iris diameter; P, power of light beam passing through the instrument).

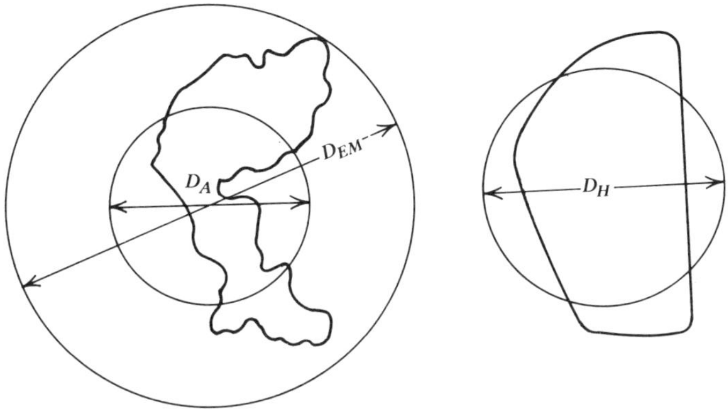

Fig. 9.8 Compactness ratios:

$$\text{Internal compactness} = \frac{\text{area of profile}}{\text{area of convex hull}} = \frac{D_E^2}{D_H^2} = IC$$

$$\text{External compactness} = \frac{\text{area of profile}}{\text{area of embracing circle}} = \frac{D_A^2}{D_{EM}^2} = EC$$

$$\text{Hull compactness} = \frac{\text{area of convex hull}}{\text{area of embracing circle}} = \frac{D_H^2}{D_E^2} = HC$$

For profile shown, $IC = 0.90$; $EC = 0.23$; $HC = 0.26$.

in Chapter 10, and their definitions are given in Figure 9.8. It should be noted that there will be slight variations in the values of D_I and D_{em} between operators depending on their ability to judge the location of the centroid of the profile but that D_E as deduced from the graph in Figure 9.7 will be precise and operator independent.

9.6 CHARACTERIZATION OF FINEPARTICLE PROFILES BY CHORD POPULATIONS

The expected values of various types of chord drawn through fineparticle profiles have been used to characterize fineparticles. Historically, the first type of chord measurement used to characterize fineparticles was the mean linear intercept, which is the value of chords generated by a line drawn at random across a field of view of fineparticles. Such a set of chords L_1, L_2, L_3, L_4, and L_5 is shown in Figure 9.9. Measurements of this kind are often made in petrography (a study of the structure of rocks), where the field of view in Figure 9.9 represents a section through a rock and the fineparticles are sections through ore grains embedded in the rock

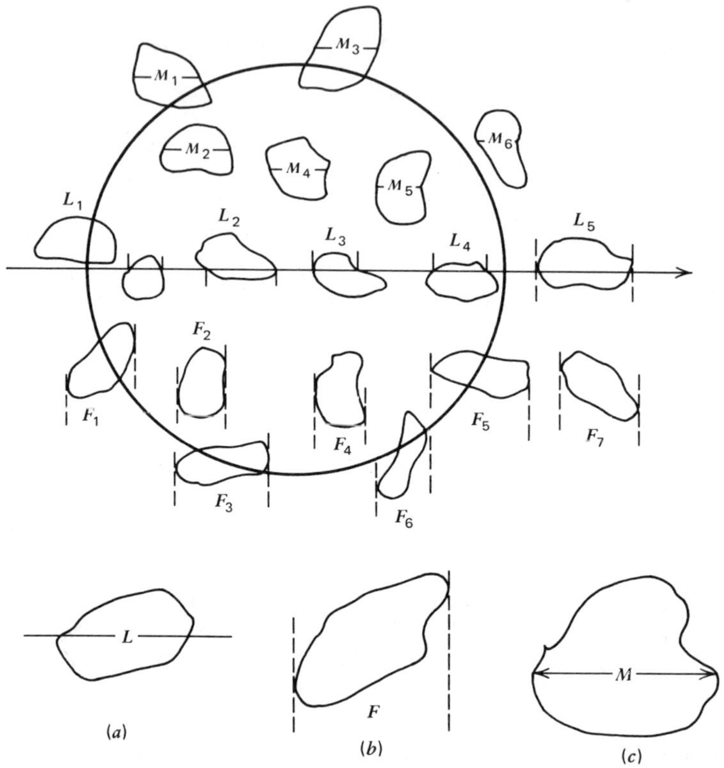

Fig. 9.9 Chord populations used to characterize fineparticle arrays: (*a*) "random chord" expected value is the mean linear intercept ($= \sum_{i=1}^{i=n} L_i/n = \beta$); (*b*) Feret diameter (projected length with respect to a reference direction); (*c*) Martin diameter (divides profile into two equal areas with respect to a fixed direction).

matrix. Assuming that the section is a representative random section and that the grains of the ore species are not in preferred orientation, Rosiwal [32] showed that

$$\alpha = \sum_{i=1}^{i=n} \frac{L_i}{L}$$

where α is the volume fraction of ore in the specimen and n is the number of chords generated by a linescan search of length L. It can also be shown that if the sectioned dispersed fineparticles exposed by the act of sectioning are convex, the following equation is true:

$$\frac{\sum_{i=1}^{i=n} L_i}{n} \rightarrow \beta \quad \text{as} \quad n \rightarrow \infty \tag{9.2}$$

where $\beta = 4\,\bar{V}/\bar{S}$, β is defined as the *mean intercept length* (it is also sometimes called the *grain size*), and $\bar{V}$ is the total volume of all the dispersed fineparticles.

Rosiwal's intercept method is based on theorems in a subject known as *geometric probability* [13, 32]. Many problems in geometric probability have been studied by metallurgists and petrographers because of their need to relate the exposed appearance and properties of two-dimensional sections through rocks and metal specimens to the three-dimensional structure of their specimens [33, 34]. Biologists have also been very active in geometric probability because of their study of sections of such three-dimensional systems as the brain, nerve cells, and cancerous tumors. In this chapter, geometric probability relationships are stated without proof. The interested student should consult advanced texts for the mathematical basis of the quoted relationships [13, 34, 35, 41].

It has been suggested that populations of chords drawn through three-dimensional bodies could be used to characterize a fineparticle system [38]. The drawings of chords on a two-dimensional section of a fineparticle system or on the projected images of a set of fineparticles is a special case of this general theory; however, the general characterization of fineparticle arrays by their chord populations suffers from the problem that the amount of work required to differentiate relatively similar systems from each other is very high [39].

An instrument has been described in which the population of random chords has been empirically correlated with the size distribution function of fineparticles generated in a crushing process. In this instrument, known as the "Procedyne model PA-110 on-line particle analyzer" [40], the slurry of fineparticles to be characterized is diluted to a volume concentration of 2%. It flows down through an inspection cell at the rate of 3 m/second. It is claimed that under these flow conditions the fineparticles to be inspected are in random array with no preferred orientation. It is also claimed that the speed with which the fineparticles are scanned with a laser beam is such that no significant movement of the fineparticles occurs during the measurement time. The instrument can be used to size fineparticles with dimensions ranging from 2000 to 10 μm. A spot of light having an Airy diameter of approximately 10 μm is generated by a laser. This beam scans the inspection window in a direction perpendicular to the movement of the slurry. The on–off signals from the scanning laser beam are interpreted as if they were measured chord lengths across a fineparticle silhouette. It has been demonstrated empirically by the manufacturers of the instrument that changes in the chord population for the slurry passing through the inspection zone can be related to shifts in the size distribution function of the slurry.

The set of maximum length chords perpendicular to a fixed direction that can be drawn within a set of profiles in random orientation has been used to characterize a set of fineparticles. This chord population shown in Figure 9.10 was used in the "Cassela–Vickers" scanning microscope [41]. This instrument, developed in the late 1950s, was commercially available for several years. Although the instrument is no longer commercially available, it is worth reviewing the logic used in it in some detail since modern electronic devices such as the linear array of photodiodes available from the Reticon Corporation could exploit the same type of logic [42]. In the Cassela–Vickers system, measurements were made directly on the image of the field of fineparticles imaged through a microscope and projected onto a photomultiplier. A slit of width W was placed on the eyepiece of the microscope, and the energy flux coming

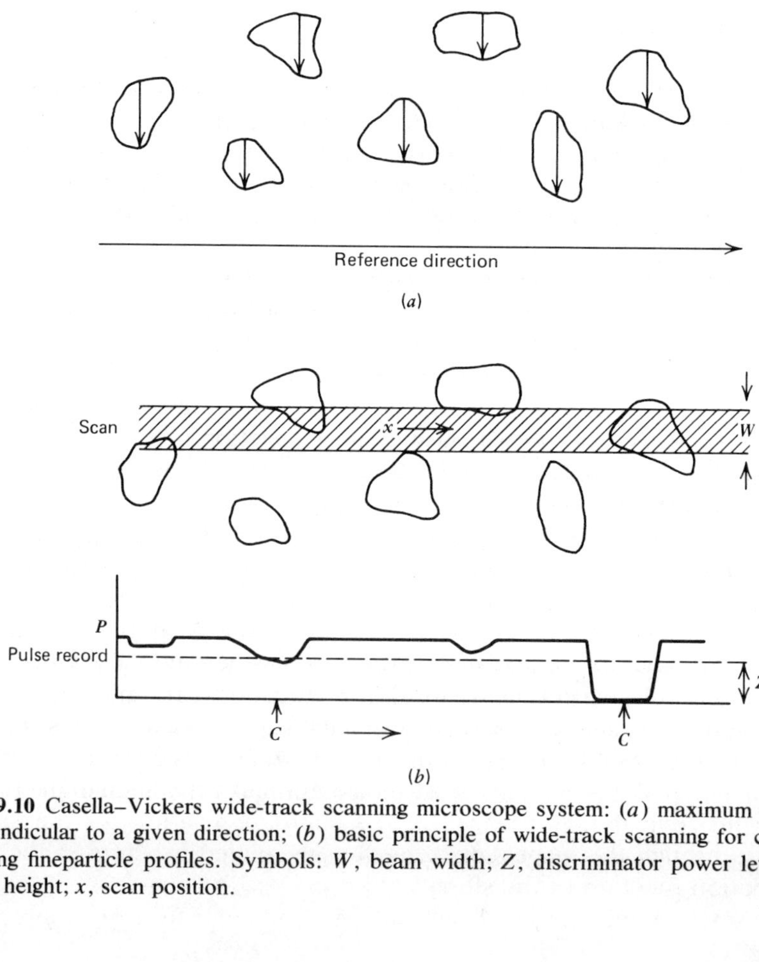

Fig. 9.10 Casella–Vickers wide-track scanning microscope system: (*a*) maximum chords perpendicular to a given direction; (*b*) basic principle of wide-track scanning for characterizing fineparticle profiles. Symbols: W, beam width; Z, discriminator power level; P, pulse height; x, scan position.

through this interrogation slit was measured by a photoelectric device. The stage of the microscope was mechanically scanned so that as a fineparticle passed into the field of view, it would cut down the energy reaching the photomultiplier. A record of the energy leaving the photo multiplier as the slide was scanned in direction X would be as shown in Figure 9.10*b*. The energy coming through the system was monitored and a counter operated whenever the energy dropped below a level Z, where Z was a known fraction of W. It is interesting to note that in this logic, one does not need to know where the fineparticle is, nor the exact point at which one moves over the boundary. One is only interested in the maximum signal generated by the presence of the fineparticle. Measurement of the maximum chord perpendicular to the direction of scan is generated by carrying out a series of measurements at various values of W and Z, and the data are transformed by using an equation developed by Hawkesly [41].

An average chord used to characterize fineparticle profiles is the Martin diameter [42]. The Martin diameter for a single profile is defined as the magnitude of the chord that divides the profile into two equal areas with respect to a fixed direction. The Martin diameter is used to facilitate the characterization of an array of fineparticles, as viewed through the microscope, by using statistical averaging to achieve a significant descriptor without constantly aligning a measurement scale, or the fineparticles, in the field of view to measure specific dimensions of the fineparticle. It is argued that if the fineparticles are in "random array" (which usually means that the orientation of the fineparticle is independent of its position within the field of view, with all orientations being equally probable), that the average value of the Martin diameter will be a characteristic diameter for a given array of fineparticles.

Another linear measure of the magnitude of a fineparticle profile that has been widely used in fineparticle science is the Feret diameter. The Feret diameter of a profile is defined as the length of the profile as projected with respect to a specific reference direction [43]. Again, the intent originally was to increase the speed with which an array of fineparticles could be characterized by using statistical averaging to measure an average Feret diameter for an array of fineparticles, thus obviating the necessity to align the measuring scale as each fineparticle is encountered in a field of view to measure a specific dimension. It can be shown that the expected value of the Feret diameter of an array of fineparticles is related to the sum of the perimeters of the convex hulls of the fineparticles by the relationship

$$F_{\text{av}} = \frac{P_{\text{ch}}}{\pi} \tag{9.3}$$

where P_{ch} denotes the sum of the perimeters of the convex hulls and π is the ratio of the perimeter to the diameter of the circle [37]. Originally, the Feret diameter was significant only as an average for an array of fineparticles. Automated image analysis makes it possible to study the variations in the Feret diameter of an individual profile and use it to characterize individual profiles (see discussion of the Feret diameter signature waveform as a shape descriptor in Chapter 10).

To focus attention on the fact that the Feret diameter and the Martin diameter were originally intended to be properties of an array rather than of an individual fineparticle, they are sometimes referred to as *statistical diameters*. The use of statistical diameters in the characterization of fineparticle profiles to minimize the amount of characterization work by using statistical averaging is an attractive idea. However, in practice, two problems usually cause difficulties: (1) to obtain a useful degree of certainty in the measured statistical diameter, it is usually necessary to carry out measurements on many profiles; and (2) any slight departure from true randomness in the field of view being characterized can cause bias in the data being generated.

There is considerable confusion in the literature of fineparticle characterization on the efficiency and the nature of the Martin and Feret diameters as useful parameters in fineparticle science. Thus Herdan stated that the use of the Feret statistical diameter in particle size analysis by microscopic methods of analysis should be avoided because of the large positive error introduced for elongated particles [32]. It appears that this statement has its origin in the results of a study conducted into the physical significance of various diameters used in the characterization of fineparticles described by Heywood [45]. At the time Heywood carried out his investigation, the physical significance of the average value of the Feret diameter was not generally known, and the aim of Heywood's studies was to investigate how the Martin and Feret diameters compared with equivalent-area diameters. The fineparticles used by Heywood had various shapes, but the minimum diameter of the profiles did not vary significantly since all the fineparticles had passed through a square sieve aperture of size 1.13 in. and retained on a square aperture of 1.00 in. Heywood classified the profiles according to their elongation ratios. This ratio is defined and illustrated in Table 9.3. The fineparticles were dense, chunky rocks, so that when viewed in a stable position on a flat surface, their profiles tended to be rectangular in shape. The information for the various-shaped profiles reported by Heywood is summarized in Table 9.3. As the fineparticles become more elongated, the average value of the Feret diameter appears to be approximately 10% larger than the other measures of fineparticle profile, and hence the statement of Herdan

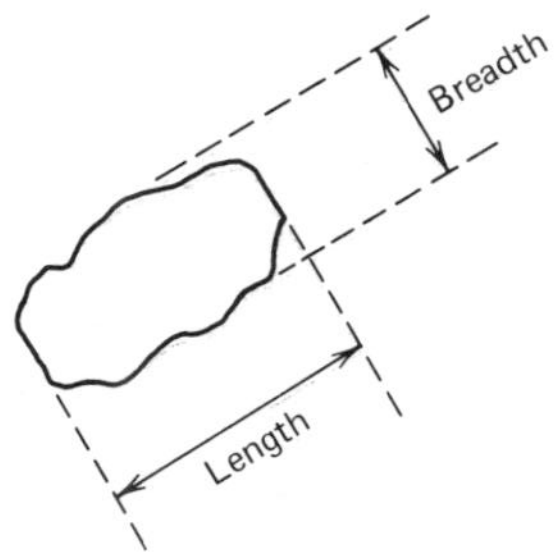

The elongation ratio is written length/breadth = $n = L/B$ and the estimate of perimeter, $2(B + nB)$. In Heywood's data B is expressed relative to unit size for the square particle. Therefore, a dimensionless estimate of the perimeter is $P = 2(1 + n)$.

TABLE 9.3. Comparative Characterization Data for a Set of Pebbles Reported by Heywood [45]

Parameter	Magnitude				
Mean projected diameter	3.465	3.640	3.850	5.180	4.540
Reticule Comparison — Opaque Circle	3.655	3.841	4.116	4.404	4.720
Reticule Comparison — Transparent Circle	3.502	3.656	3.953	4.264	4.720
Martin statistical diameter	3.442	3.594	3.800	3.960	4.060
Feret statistical diameter	3.783	3.961	4.370	4.810	5.450
For pebbles having elongation ratios of	1.00–1.20	1.21–1.40	1.41–1.70	1.71–2.00	2.01–2.75
Perimeter estimates for each group of pebbles[a]	2.10	2.30	2.56	2.85	3.38

[a]Linear average of n within the range used in the calculation of this value.

quoted earlier. However, this difference is not an error but an important piece of information because the Feret diameter is not a measure of the area of the fineparticle but, as indicated earlier, of the perimeter of the convex hull of the profile. This fact can be illustrated by using Heywood's original data; thus, since the profiles used by Heywood approximated to rectangles, we can estimate the perimeter of the profiles as being $2(L + B)$. The estimated perimeters of the various sets of fineparticles, in dimensionless units, calculated this way are summarized in Table 9.3. The value of the perimeter is plotted against the average Feret diameter for each shape group of fineparticles in Figure 9.11. It can be seen that a straight line can be drawn through the data points, demonstrating the direct relationship between the Feret diameter and the perimeter of the profiles. It should be noted that in a discussion of Heywood data, published in 1948, Walton [46] drew attention to the fact that for nonreentrant fineparticles the average value of the Feret diameter was related to the perimeter. At that time Walton commented that this relationship between the average value of the Feret diameter and the perimeter "seems to have been overlooked hitherto in particle size analysis." Herdan's influence on the statistics of fineparticle systems has been great, and his discouragement with regard to the utility of the Feret diameter is a lingering influence in modern characterization literature.

The Martin diameter does not appear to have a simple relationship to any obvious fineparticle behavioral parameter. Its usefulness in current

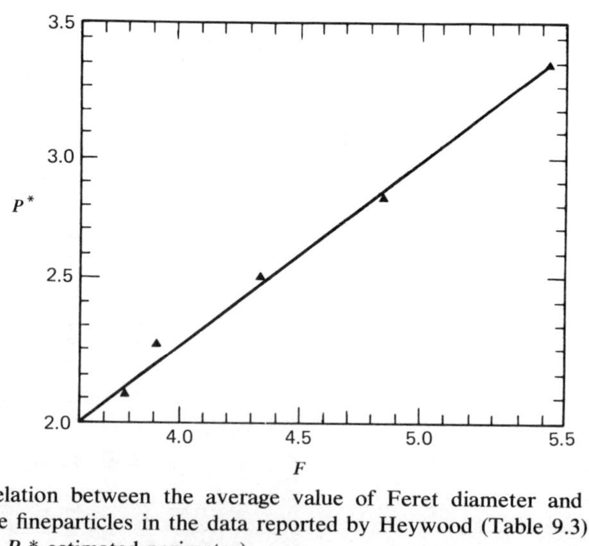

Fig. 9.11 Correlation between the average value of Feret diameter and the estimated perimeter of the fineparticles in the data reported by Heywood (Table 9.3) (F, measured Feret diameter; P,* estimated perimeter).

particle size techniques is strictly limited, although this fact is obscured because of some confusion of the Martin diameter with other statistical diameters that have been used to describe fineparticle systems. Thus Herdan states that "Martin's statistical diameter is sometimes referred to as the mean linear intercept or mean chord" [32]. Herdan does not give a justification for this statement, but from the subsequent equation he gives for the "mean chord," it is obvious that he is equating the Martin diameter to the quantity used by metallurgists to measure what they call the "mean grain size" and that is identical to the quantity defined in equation 9.3. It is obvious that this quantity is not the same as the Martin diameter and that Herdan's statement is incorrect.

Another unfortunate obfuscation of the physical significance of the Martin statistical diameter is to be found in an otherwise excellent paper on the measurement of significant dimensions of fineparticle profiles by Davies [47], where the Martin diameter is equated to the average of diameters drawn through the centroid of the fineparticle profile. If we imagine any profile treated as a lamina and broken down into small elemental areas, then if x is the distance of the elemental area from a line through the centroid, the condition for the reference line to pass through the centroid is that $\Sigma(\Delta A \cdot x)$, where ΔA denotes the elemental area, is equal on both sides of the line. On the other hand, the condition for the line to be a Martin diameter is that $\Sigma(\Delta A)$ is equal. Stated mathematically, the condition for a line to be a Martin diameter is that the zero moments of the areas about the line are equal whereas a line through the centroid diameter required that the first order moments of the elemental areas be equal. The two criteria will be equivalent only for certain symmetrical profiles.

9.7 CHORD LOGIC CHARACTERIZATION OF FINEPARTICLE PROFILES

Profile parameters can be deduced from the statistics of sets of parallel chords generated by using a television–scan interface to a computer. The steps in the estimation of the Feret diameter and the area of the profile from the statistics of parallel chord sets are illustrated in Figure 9.12. The profile shown in Figure 9.12 is intercepted by 10 scanlines. Therefore, the Feret diameter in the direction at right angles to the scan pattern is 9δ, where δ is the separation of the scanlines plus the fragments α and β indicated in the diagram. Neither α nor β are known, nor can they be measured without increasing the resolution of the scan pattern. The profile has been placed at random on the scan pattern; therefore, if the chords generated by the pattern were to be studied for several random locations on the scan pattern, the value of α could be anywhere between 0

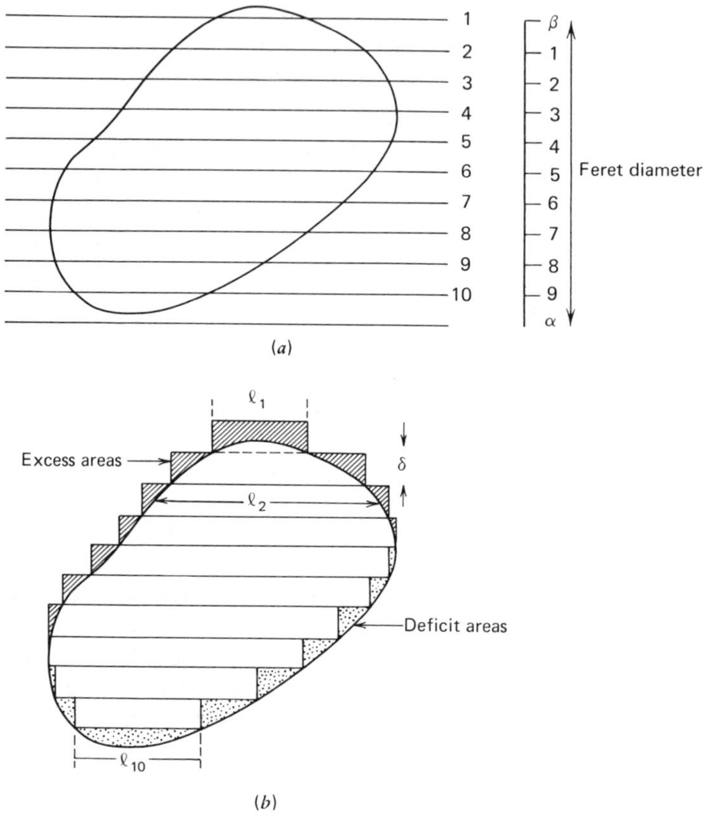

Fig. 9.12 Estimates of profile area and Feret diameter made from the statistics of sets of parallel chords: (*a*) estimate of Feret diameter; (*b*) estimate of area.

and δ, with all values being equally probable—thus the expected value of the lower fragment in a series of measurements is 0.5δ. In the same way, the expected value of the fragment β is 0.5δ. The magnitude of the two end fragments vary independently of each other. Therefore, the expected value of the sum of the fragments of the Feret diameter is δ. Thus one can allow for the effect of the end fragments on the estimate of the Feret diameter in an unbiased manner by adding one unit to the scan element number that lies entirely within the projection. Thus in the case of Figure 9.12*a*, the estimate would be 9 + 1. Note that 9 + 1 equals the number of lines intercepted by the profile. In general, the number of lines crossing the profile, times their separation δ, is an unbiased estimate of the Feret diameter of the profile (see discussion in reference 37). The fact that this procedure is an unbiased estimate of the Feret diameter is important when considering the confidence to be placed in the estimate of the overall perimeter of the profile based on the average of the Feret diameter measured over many orientations. The uncertainty in the average decreases rapidly because the uncertainties in the individual measurements are self-canceling as the average is accumulated. Operations such as the estimate of the Feret diameter from the number of parallel chords intercepted on a television scan pattern are known as linescan logic operations.

The area of the profile is approximated by taking the sum of the elemental areas defined by $l_i\delta$, where δ is the separation of the chords and l_i is the length of the ith chord. It can be seen from Figure 9.12 that this operation will result in an overestimate of the area of the profile in certain regions and an underestimate in others. Since the profile is a random line meandering across the pattern of parallel lines used to generate the chords, there is no inherent bias in the sum of the difference between excess elemental areas and the deficit areas. If the same profile is subjected to a sequence of measurements with random variations in its orientation with respect to the set of parallel lines, the accumulated uncertainty in the sequence of measurements will decrease without bias to give a good estimate of the area. The uncertainty in the estimated area increases as the profile decreases in absolute size with respect to the separation of the scanlines. If the estimate of the area is occurring in a sequence of measurements, the purpose of which is the estimation of the distribution function of an array of fineparticles, the uncertainty per profile is a random variable that tends to be a self-compensating uncertainty in the allocation of the individual profile area to a set of size ranges again. As in the case of the average Feret diameter calculated by linescan logic procedures, the confidence to be placed in the distribution function is much higher than the confidence to be placed in the measurement of individual areas [37].

9.8 DIGITAL EVALUATION OF FINEPARTICLE PROFILES

If a profile of A_1 is located in a field of view of area A_2 and if points are scattered at random on the field of view, it can be shown that the proportion of dots falling within the profile is given by the relationship

$$n = \frac{A_1}{A_2} \times N$$

where n is the number falling within the profile and N is the total number of dots. It follows that an unbiased estimate of the area of the profile is given by

$$A_1 = A_2 \times \frac{n}{N}$$

It is also true that if an irregular profile of area A_1 were dropped at random onto a regular array of dots contained in an area A_2, the number of dots α contained within the profile boundaries would tend to

$$\alpha = \frac{n}{N} \times A_2$$

where n represents the total of dots falling within the various profiles. This latter statistical relationship is the basis of a method known as *Chayes's point count method*. He developed this technique for modal analysis, which is the name given by geologists to techniques for evaluating the proportions of different rock species exposed in a section through a composite rock. The technique is also known as Chayes's method of modal analysis. Consider the system shown in Figure 9.13. The field of view *ABCD* can be considered to be a field of view of a sectioned rock. The white area is the matrix rock. The irregular outlines of different color would be sections through dispersed elements in the rock. For the system shown in Figure 9.13, it is assumed that there are three different dispersed species in the matrix rock. An overlay is made on the field of view of a grid containing 100 points which it is intended would be scanned in the sequence illustrated (grids such as those shown in Figure 9.13, with connecting lines joining up the dots to enable the geologist to proceed in an orderly manner through the scan sequence, are available commercially). They are known as *integration eyepieces* [29]. When dealing with points that lie on the profile boundary, it is statistically correct and unbiased to take half the points in doubt to include in the estimate of the points that lie within the profile. For the system shown in Figure 9.13, the point count estimate of the areas of the three different types of profile are 5, 2, and 20%. A direct automation of Chayes's point count method is available in the instrument known as the *AMEDA* [50].

Kaye and Murphy [36] have shown that point count estimates of areas at relatively low resolution can be used to obtain a good estimate of the size distribution function by using compensating errors in the statistical strategy employed in the measurement procedure. The basic principle employed in their strategy is that the variations in the point count content of a profile placed at random in various orientations on a regular grid will vary according to the normal probability distribution. Therefore, because the error in allocating fineparticles to a given group is a random error, one

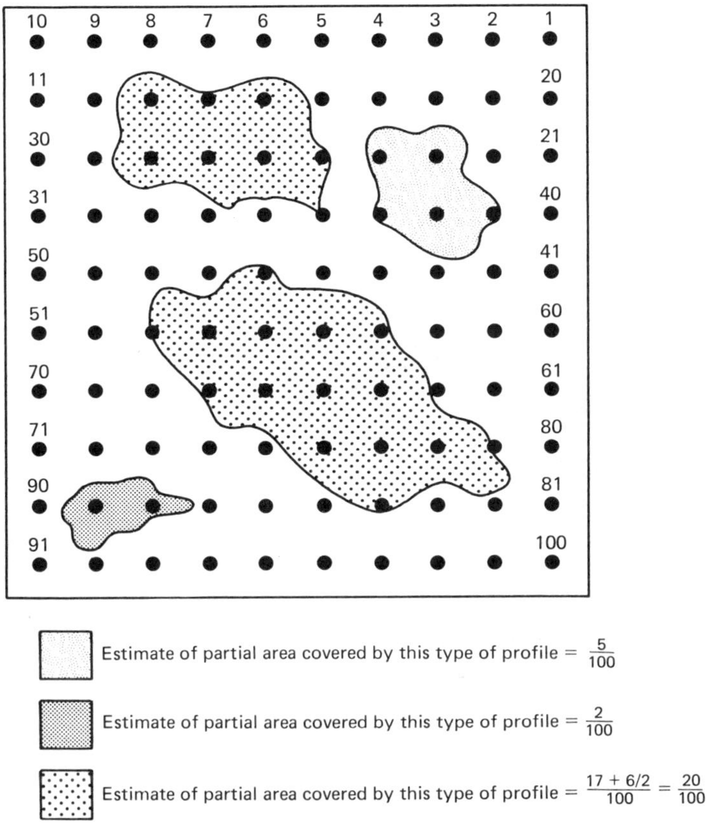

Fig. 9.13 In Chayes's point count method for the modal analysis of rocks, the number of points of a regular grid falling within the fineparticle profiles is used to characterize the profiles [33].

will wrongly classify large particles as small as often as one will make the opposite error.

It has been suggested that for low-cost characterization studies, the digitization of the image could be carried out efficiently using a fiber optic array as shown in Figures 9.14 and 9.15 [51–53]. More sophisticated fiber optic systems for image analysis have been described by Kaye and co-workers; however, the falling costs of digital image transforming equipment has caused a shift in image analysis technology that has tended to make it uneconomical to employ fiber optics in high-speed, high-resolution systems.

Several commercial instruments are available for transforming a photo-

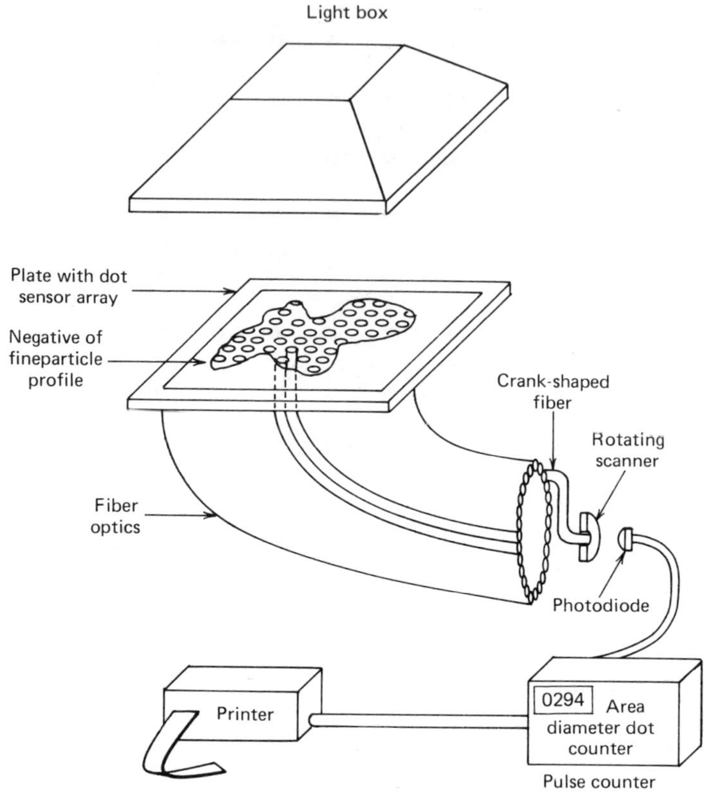

Fig. 9.14 Low-cost fiber-optic device for point sampling transformation and evaluation of a fineparticle image [53].

graph or an image into a highly resolved mosaic of points [54–57, 60]. Various logic operations can be performed on such matrix representation of the fineparticle profiles.

9.9 AUTOMATED EVALUATION OF FINEPARTICLE PROFILES

Since the early 1950s there have been many instruments developed for automated parameter evaluation from fineparticle images. In the first generation of instruments, developed in the early 1950s, various scanning procedures were used, but they were all essentially linescan logic systems. Several of these instruments, such as the Mullard L188 automated particle size analyzer, the "flying spot particle resolver," and the

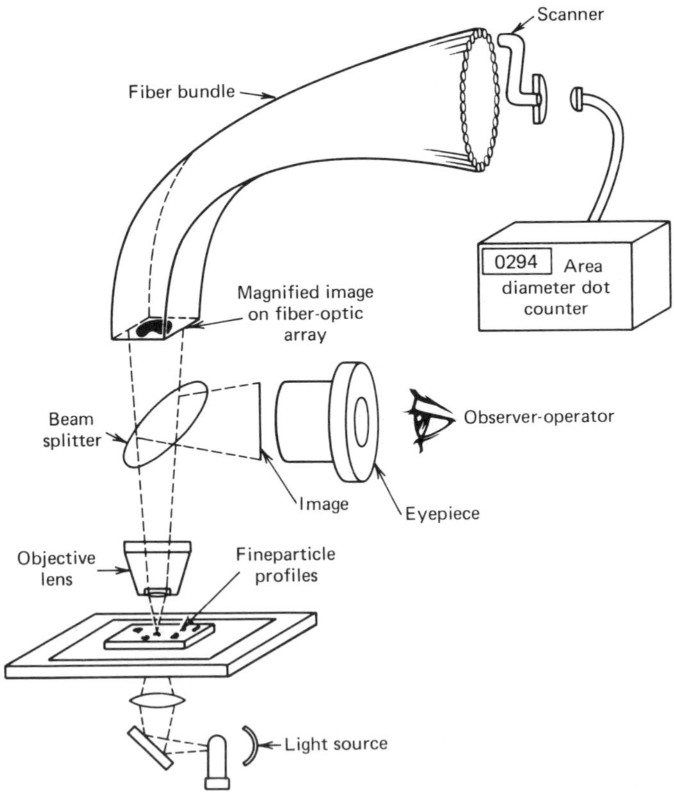

Fig. 9.15 Alternative fiber-optic implementation of point sampling image characterization system [53].

Casella–Vickers instrument described in Section 9.6 were commercially available for a short period [58–60]. The availability of lower-cost television cameras in the early 1960s led to the development of a second generation of automated image analysis systems based on the interfacing of television cameras to digital computers. Typical of image characterization equipment developed at this time were the early models of the Quantimet image analyzer and the πMC image analyzer. This second generation of image analyzers, based on linescan logic and television cameras, has continued to evolve, and one of the difficulties of giving a coherent account of automated image analysis systems is the fact that many manufacturers implement different logic systems at different times in sequential models of their brand name image analyzers. A third generation of image analyzers based on the new solid-state low-cost image digitizers began to emerge in the middle 1970s. Typical of this third generation are the analyzers developed by Knollenberg and the Omnicon image analyzer [49, 61, 63]. In the third-generation equipment descriptive parameters of fineparticle profiles are generally derived by mathematical operations performed on a digital representation of the profile stored in the memory of a computer.

The Bausch and Lomb Corporation began to manufacture automated iconometric devices, based initially on linescan logic systems, in the early 1960s. One of their systems which was marketed by the Millipore Corporation became known as the *πMC image analyzer* [62]. A different version of this equipment, sold directly by Bausch and Lomb, was known as the *QMS system,* which was originally intended for use in the quantitative microscopy of metals (hence the name) [34]. The most sophisticated image analyzer marketed by the Bausch and Lomb is known as the *Omnicon,* which employs software manipulation to derive parameters of the fineparticles profiles. It can be programmed for extra operations by the operator [49].

The *Quantimet* is an image analyzer based on television camera interfacing to a computer. Its name derives from the fact that it was originally developed for quantitative metallurgy [48]. In the mid 1960s a corporation was established to market the system, and this corporation was known as Image Analyzing Computers Ltd., which was contracted to Imanco [48]. Through the merger of several instrument companies in Great Britain, the company marketing the Quantimet image system has recently become known as the Imanco Division of Cambridge Instruments [48], and the latest version of the equipment is known as the Imanco image analysis system.

The Joyce–Loebel organization has developed an automated image analysis system in cooperation with the department of medical physics at the University of Manchester in England. This system is a third-generation device employing software manipulations to derive information from a digital representation of the image. It is commercially available as the *Magiscan system* [64].

A semiautomatic image analyzer, which can have various levels of sophistication built into the digital processing elements, is manufactured by the Kontron Corporation of Germany. Their system is known as the *MOP modular system for image analysis* [65].

An image analysis system developed by White and co-workers at the Materials Research Laboratory of Pennsylvania State University for the automatic processing of scanning electron microscope images is available commercially as the Computer Evaluation of Scanning Electron Microscope Images (CESEMI) system for image analysis [66, 67].

The Leitz Organization developed an image analyzer system using chord logic and television interfacing that was made available commercially under the name of the Leitz *Classimat* [68]. A more sophisticated image analyzing system using electronic and software manipulation of a matrix representation of the image is known as the *Leitz texture analyzer system* [68]. A software-based system of image evaluation is the *OPTOMAX modular image analyzer* [69]. Another image analyzing system is known as the *Luzex 450 particle size analyzer* [70].

In the automated iconometric systems available from Particle Measuring Systems, Inc. moving fineparticles are imaged by a laser beam onto a highly resolved array of photoelectric elements. Optical criteria are used to decide when a fineparticle is in the focal plans of the system. A high-speed electronic data processing system dumps the information from the linear array of photoelectric devices in a sequential pattern into a computer memory. The speed of the system is such that the computer contains an image array of the system built up of image slices generated by the linear array of photoelectric devices. Various logic operations are available and the instruments can develop size distribution functions of an evaporating aerosol in real time. The instruments have been used extensively in cloud physics, where they have been mounted on aircraft flying through clouds and snow storms. The systems obviously have a wider application to the characterization of other fineparticle systems, and recently other workers have applied the same basic logic to the characterization of fineparticles in a liquid [61, 71, 72].

The Zeiss Corporation manufactures a television interface linescan logic image analyzer known by the name of *Microvideomat,* or as it is sometimes referred to, the *Videomat* [29].

Perhaps one of the most sophisticated image analysis systems developed so far is that known as the *COSMOS image analyzing system* for studying the brightness and the shape of galaxies as photographed through astronomical telescopes. The system is able to locate elliptical galaxies and then use a sophisticated search pattern to zero in on the axes of the ellipse and measure its eccentricity by evaluating the major and minor axes of the ellipse. When originally built, the system was too expensive for consideration as a basic system for fineparticle characterization; however, the falling cost of mainframe digital computers may make it possible to employ similar logic in the characterization of the shape of fineparticles [73, 74].

Shellman and co-workers at the Argonne National Laboratory in Chicago have developed several sophisticated systems, known by such acronyms as ALICE and POLLY, for detailed analysis of photographs. The POLLY system was originally developed for measuring particle tracks in bubble chambers as used in atomic physics. These systems have also been used for such complicated tasks as the evaluation of fingerprint systems and the analysis of pictures obtained from space satellites [75].

Another program that was originally developed for evaluation of bubble tracks in atomic physics but that could have applications in fineparticle characterization is known as SWEEPNIK and was developed by Frisch at the Cavendish Laboratory in Cambridge, England [76].

The Perkin–Elmer Corporation has developed a system for characterizing blood cells by using automated iconometrics [81]. This could obviously be also used to size fineparticles.

Tipton [77] recently described the use of television interface in conjunction with a xenon laser light source to image fineparticles passing up an effluent stack. Another television computer system for evaluating for spray droplets has been described by Simmons and Lapera [79].

Cheng et al. [80, 78] have developed a system for evaluating the size and the structure of chromosomes that could be used to characterize fineparticle profiles.

NOTES AND REFERENCES

1. See commercial literature of the Fairchild Corporation, Palo Alto, California.
2. R. Price, "Packets of Charge Instead of Currents," *New Scientist* (June 19, 1975 article on new solid-state technology).
3. O. Johari and S. Bhattacharyya, "The Application of Scanning Electron Microscopy for the Characterization of Powders," *Powder Technol.*, 2 (1968–69), 335–348.
4. E. W. White, H. Görz, G. G. Johnson, Jr., and R. E. McMillan, "Particle Size Distributions of Particulate Aluminum from Computer Process SEM Images," in *Proceedings of the SEM Symposium*, IITRI, Chicago, Illinois, 1970, pp. 57–64.
5. B. J. Thompson, "Holographic Particle Sizing Techniques," *J. Phys. (Series E, Sci. Instrum.)*, 7 (1974), 781–788.
6. M. E. Fourney, J. H. Matkin, and A. P. Waggoner, "Aerosol Size and Velocity Determination via Holography," *Rev. Sci. Instrum.*, 14 (2) (1969), 205–213.
7. R. F. Van Ligten, *Holographic Microscopy and Small Particles*, American Optical Corporation Research Group, P. O. Box 187, Framingham Center, Massachusetts.
8. Image Analyzing Computers, Inc. has developed a system for the automatic evaluation of aerosols, as recorded in a hologram. The regenerated image is inspected with the television camera focusing automatically on each fineparticle and carrying out the necessary measurements. Information is available

from Image Analyzing Computers, Inc., 40 Robert Pitt Drive, Monsey, New York 10952.

9. N. Thaulow and E. W. White, "General Method for Dispersing and Disaggregating Particulate Samples for Quantitative SEM and Optical Microscopic Studies," *Powder Technol.*, 5 (1971–72), 377–379.
10. A wealth of information on the optical characteristics of various types of fineparticle can be found in the *Particle Atlas*, available from McCrone Associates, 2820 South Michigan Avenue, Chicago, Illinois 60616.
11. B. H. Kaye and A. G. Naylor, "Spatial Filtering for Detection of Particles of a Particular Shape," *J. Colloid Interface Sci.*, 39 (1) (1972); G. M. Hidy, Ed., *Aerosols and Atmospheric Chemistry,* Academic, 1972, pp. 105–110.
12. K. Ridgeway and J. B. Scotten, "The Effect of Particle Shape on the Properties of Powders," paper presented at the Heywood Memorial Symposium, held at Loughborough University, September 1973.
13. M. J. Kendall and P. A. P. Moran, *Geometric Probability,* No. 10 of *Griffith's Statistical Monographs in Courses*, Charles Griffith, London, 1963.
14. Those unfamiliar with topology will find a readable introduction in I. Stewart, *Concepts of Modern Mathematics,* Penguin Books, Harmondsworth, Middlesex, England, 1975; also, 41 Steelcase Road West, Markham, Ontario, Canada, 1975.
15. Reticules are available from BGI Incorporated, 58 Guinan Street, Waltham, Massachusetts 02154.
16. G. Herdan, *Small Particle Statistics*, Academic, New York, 1960.
17. This special graph paper is available from Chartwell Company of Great Britain.
18. British Standards 3406 1961, "Methods for the Determination of Particle Size of Powders," information available from the British Standards Institute, 2 Park Street, London W1.
19. H. S. Patterson and W. Cawood, *Transact. Faraday Soc.,* 32 (1936), 1084.
20. B. H. Kaye, "A Simulated Array of Particle Profiles for Use in Microscope Methods of Particle Size Analysis," *Powder Technol.*, 4 (1970–71), 275–279.
21. H. H. Watson, *Br. J. Ind. Med.*, 19 (1952), 80.
22. V. Timbrell, *Nature*, 170 (1952), 318, 319.
23. Commercial information is available from Flemming Instruments, Caxton Way, Stevenage, Hertfordshire, England.
24. Information is available from Coulter Electronics Ltd., Coldharbour Lane, Harpenden, Hertfordshire AL5 4UN, England.
25. Commercial information is available from W. Watson & Sons, Ltd., Barnet, Hertfordshire, England.
26. Information on the Vickers Image Shearing Eyepiece is available from Vickers Instruments Ltd., Bhax By Road, Yorkshire, England.

27. F. Endter and H. Gebauer, *Optik,* 13 (1956), 97.
28. J. Donato, B. H. Kaye, and R. Murphy, "Television Display System for Use with the Timbrell Double Image Microscope," *Powder Technol.*, 2 (1969–70), 249–250.
29. Commercial information is available from Carl Zeiss Corporation, Oberkochen, Württ, West Germany.
30. E. T. White, R. J. Batstone, J. Day, and P. G. Wright, "Modifications to Zeiss-Endter Particle Size Analyzer for Data Logging," *Powder Technol.*, 5 (1971–72), 127–130.
31. B. H. Kaye, "Possible Automation Procedures for the Zeiss Endter Particle Size Analyzer," in *Assessment of Airborne Particles*, T. T. Mercer, P. E. Morrow, and W. Stober, Eds., 1972, Springfield, Ill., Chapter 12.
32. For a readily accessible discussion of the Rosiwal intercept method, see G. Herdan, *Small Particle Statistics*, 2nd ed., Butterworth's, London, 1960.
33. F. Chayes, *Petrographic Modal Analysis*, Wiley, New York, 1956.
34. R. T. DeHoff and F. N. Rhines, *Quantitative Microscopy,* McGraw-Hill, New York, 1968.
35. E. E. Underwood, *Quantitative Stereology*, Addison Wesley, Reading, Mass., 1970.
36. B. H. Kaye and R. Murphy, "Size Distribution Estimation Using Dot-Counting Logic," *Powder Technol.*, 4 (1970–71), 203–312.
37. B. H. Kaye, "Some Aspects of the Efficiency of Statistical Methods of Particle Size Analysis," *Powder Technol.*, (2) (1968–69), 97–110.
38. B. Scarlett, "A Statistical Description of Particulate Systems," in *Particle Size Analysis*, M. J. Groves and J. L. Wyatt-Sargent, Eds., Society for Analytical Chemistry, London, 1970.
39. G. Mason, "Random Chord Distributions from Triangles," *Powder Technol.*, 12 (1975), 277–281.
40. Information is available from the Procedyne Corporation, 221 Somerset Street, New Brunswick, New Jersey 08903.
41. B. B. Morgan and E. W. Meyer, *J. Sci. Instrum.,* 36 (1959), 492–501.
42. Commercial information is available from Reticon, Inc., 910 Benicia Avenue, Sunnyvale, California 94086.
43. G. Martin, C. Blythe, and H. Tongue, *Transact. Ceramics Soc.*, 23 (1923), 60.
44. R. L. Feret, (1931) Association International pour l'Essai des Mnt., 2 Group D, Zurich, Switzerland.
45. H. Heywood, "A Comparison of Methods of Measuring Microscopical Particles," *Transact. AIME*, 4 (1945), 391.
46. W. H. Walton, "Feret's Statistical Diameter as a Measure of Particle Size," *Nature*, 162 (1948), 329.
47. C. N. Davies, "Measurement of Particles," *Nature*, 195 (1962), 768.

48. C. Fisher, "The Metals Research Image Analyzing Computer," in *Proceedings of the Particle Size Analysis Conference,* Loughborough, England, 1966, Society for Analytical Chemistry, London, England, 1967.

49. See, for example, the commercial literature on the Omnicon system marketed by Bausch and Lomb, available from Analytical Systems Division, Bausch and Lomb, 820 Linden Avenue, Rochester, New York 14625.

50. Commercial information on the AMEDA system is available from Femco Incorporated, Irwin, Pennsylvania.

51. B. H. Kaye, A. G. Naylor, and N. R. Jones, "Mosaic Fibreoptic Reading Head," in *Proceedings of the International Conference on Particle Technology*, IIT Research Institute, Chicago, Illinois, August 1973; proceedings published by IIT Research Institute, 10 West 35th Street, Chicago.

52. A. G. Naylor and B. H. Kaye, "Fibreoptic Reading Head for Diffraction Pattern Analysis," in *Proceedings of Seminar in Depth on Fiber Optics*, San Francisco, proceedings of Society of Photo-optical and Instrument Engineers, 31 (October 1972), pp. 55–61.

53. B. H. Kaye, R. Davies, and R. Olhaber, "Fibreoptic Implementation of Dot Counting Methods for Particle Size Analysis," unpublished report of the IIT Research Institute, 10 West 35th St., Chicago, Illinois 60616.

54. Optronics International Incorporated, 7 Stewart Road, Chelmsford, Massachusetts 01824.

55. Quantex Corporation, 1011 Commercial Street, San Carlos, California 94070.

56. Spatial Data Systems Incorporated, Box 249, 508 South Fairview Avenue, Goleta, California 93017.

57. Photometrics, Incorporated, 442 Marrett Road, Lexington, Massachusetts 02173.

58. G. F. Collins, "Some Tests of the Mullard Type L188 Automatic Particle Analyzer," in *Proceedings of the Particle Size Analysis Conference*, Loughborough University of Technology, England, 1966; published by the Society for Analytical Chemistry, 9/10 Saville Road, London W1, England.

59. C. G. L. Furmidge, "Assessment of the Accuracy of the Flying Spot Scanning Microscope for the Measurement of Microscope Particles," *Br. J. Appl. Phys.*, 12 (1961), 268–274.

60. For a review of these various instruments describing the physical principles embodied in the various instruments, see B. H. Kaye, "Automatic Microscopes for the Paint Technologist," *Paint Oil Colour J.* 155, No. 3681 (1969), 870–872; 156, No. 3693 (1969), 285–287.

61. R. J. Knollenberg, "The Optical Array: An Alternative to Scattering or Extinction for Airborne Particle Size Determination," *J. Appl. Meteorol.*, 9 (1) (February 1970), 86–103. Commercial information on the systems developed by Knollenberg is available from Particle Measuring Systems, Inc., 1855 South 57th Court, Boulder, Colorado 80301.

62. Millipore Corporation, Ashby Road, Bedford, Massachusetts 01730.

63. Information is available from Cambridge Instruments, Melbourn, Royston, Hertfordshire, SG8 6EJ, Great Britain; see also reference 8.

64. Joyce-Loebel Ltd., Marquis Way, Team Valley, Gateshead, NE11 OQW, England.

65. Commercial information is available from Kontron Messgeräte, GMBH D-8057, Eching, West Germany. For a brief review of the MOP system, see *Powder Metallurgy International*, 10 (2) (May 1978), 95.

66. Lemont Scientific Company, 1359 East College Avenue, State College, Pennsylvania 16801.

67. E. W. White, H. A. McKinstry, and G. G. Johnson, Jr., "Computer Processing of SEM Images," paper presented at the Symposium on the Scanning Electron Microscope—the Instrument and its Applications, IITRI, Chicago, April 30–May 1, 1968; proceedings published by IIT Research Institute, Chicago 60616, pp. 97–129.

68. Commercial information is available from Ernst Leitz, GMBH B-633 Wetzlar, Postfach 2020, Germany. For a recent article leading into the literature on the performance of the Leitz "Classimat," see D. F. Alliet, T. A. Tietjen, and D. H. Wood, "Particle Size Analysis of Multi-component Toner Systems," paper presented at the Conference on Particle Size Analysis, University of Salford, England, September 12–15, 1977; proceedings to be published by the Analytical Division of the Chemical Society of Great Britain.

69. Commercial literature from Micromeasurements, Three "A" Street, Burlington, Massachusetts.

70. Commercial information is available from Nihon Regulator Company Ltd., Nichido Building, Sinbashi, 1-6-5 Minato-Ku, Tokyo, Japan. Commercial agent in Great Britain is the Joyce–Loebel Corporation; see reference 62.

71. D. L. Reichard, H. J. Retzer, L. A. Liljedahl, and F. R. Hall, *Transact. ASAE*, 20 (2) (1977), 232–237, 242.

72. C. A. Rhodes, L. F. Stowers, I. Hawkins, R. D. Bonnell, and W. Raines, "In line Dynamic Control Monitoring of Fluids for Space Systems," *Powder Technol.*, 14 (1976), 203–208.

73. Anonymous, "Automation Meets the Milky Way," *New Sci.*, 29 (January 29, 1970), 192.

74. Commercial information on the COSMOS system is available from Faul-Coradi, Scotland Ltd., Gateside, Harrington, East Lothian, Scotland.

75. Information is available from the Argonne National Laboratories, Chicago, Illinois.

76. D. G. M. Davies, O. R. Frisch, and G. S. B. Street, *Nucl. Instrum. Meth.*, 82 (1970), 54.

77. D. F. Tipton, "A Particle Analyzer for Stack Emissions," *Powder Technol.*, 14 (1976), 245–252.

78. R. S. Ledley and F. H. Ruddle, *Sci. Am.* (April 1966), pp. 40–46.

79. H. C. Simmons and D. J. Lapera, "A High-Speed Spray Analyzer for Gas Turbine Fuel Nozzles," Parker-Hannifin Corporation, Cleveland, Ohio; unpublished presentation made at the ASME Gas Turbine Conference, Cleveland, Ohio on March 12, 1969.

80. G. C. Cheng, R. S. Ledley, D. K. Pollock, and L. Rozenfeld, *Pattern Recognition*, Thompson Book Company, Washington, D.C., 1968.

81. Commercial information on the Perkin–Elmer "Cellscan" system for blood cell characterization is available from the Perkin–Elmer Corporation, Norwalk, Connecticut.

CHAPTER

10

SHAPE CHARACTERIZATION

10.1 DIMENSIONAL INDICES AND CORRELATION FACTORS

Davies, in a review published in 1975, discusses the fact that a literature survey carried out that year uncovered 160 references to publications dealing with the shape of fineparticles [1]. The simplest way to describe the shape of a fineparticle is to use a metaphor or to compare the fineparticle to a known object as in such terms as "egg-shaped" and "dendritic" (a word from the Greek meaning "tree-like"). Sometimes the recognition of the metaphor requires a knowledge of other languages; thus the word "scoriaceous" is used by geologists to describe lava-type rocks full of blow holes. This word is derived from a Greek word *skoria,* meaning "dung" [2]. The British Standards Institute has prepared a standard glossary of terms of use in the description of fineparticles.

Acicular	needle shaped
Angular	sharp edged or having roughly polyhedral shape (polyhedral derives from *poly,* meaning "many" and *hedra,* meaning "a base"; therefore, a polyhedral is understood to be a geometric shape having many faces, each of which can act as a base)
Crystalline	a geometric shape freely developed in a liquid
Dendritic	having a branched crystalline shape
Fibrous	regularly or irregularly thread-like
Flaky	no formal definition in the British Standards—it is assumed that one intuitively recognizes a flaky fineparticle
Lamellar	plate-like
Granular	having approximately equidimensional but irregular shape
Irregular	lacking any symmetry
Modular	having rounded, irregular shape
Spherical	globular shaped

Hausner has published a set of two-dimensional profiles with corresponding term to indicate morphology [7].

Various workers have attempted to achieve shape related descriptors from studies of such properties as powder packing, slurry packing, rheology, and group dynamics [8–11].

Shape descriptive parameters that relate, say, the average Stokes diameter of a set of fineparticles to their average resistazone diameters should be regarded as shape dependent correlation factors. (Such correlation factors have been called "shape factors," but they are seldom simply or directly related to fineparticle morphology, and the term "correlation factor" focuses attention on the physical significance of the parameter). A general discussion of the way in which shape-related correlation parameters are evaluated involves a somewhat lengthy discussion of the physical behavior of specific fineparticle systems and hence lies outside the scope of this present book and is not discussed further.

It should be noted that another quantity that has been termed a "shape factor" in the fineparticle literature is any parameter correlating a linear dimension of a fineparticle to some other property such as the surface area or to some functional property of the fineparticle, such as its hiding power in a paint film, or to describe the variation in behavior of a fineparticle with respect to the equivalent behavior of a spherical fineparticle.

One technique that has been used to separate the description of shape from the magnitude of a fineparticle is to use dimensionless ratios to define indices of shape. A great deal of the pioneering work in this field was done by mechanical engineers concerned with the efficiency of crushing and grinding machinery and by geologists concerned with the properties of fineparticles that were constituents in sedimentary rocks and/or beach sand. In both instances these scientists were concerned with entities that could usually be handled and characterized in three dimensions [3–5]. Thus Heywood studied relatively large rock fragments that were viewed in their position of maximum stability on a plane surface. He then defined the thickness T as the minimum distance between two parallel planes that are tangential to opposite surfaces of the fineparticle, with one plane as that of maximum stability. He then defined the breadth B as the minimum distance between two parallel planes that are perpendicular to the planes defining the thickness and are tangential to the opposite sides of the particle. The length L is the distance between two parallel planes that are perpendicular to the planes defining thickness and length and are tangential to the opposite sides of the particle. Heywood emphasized the minimum width (i.e., breadth) as an operative dimension because it corresponds to the minimum sieve aperture through which the fineparticle can move. Using these definitions, he then defined

$$\text{Elongation ratio} = \frac{\text{length}}{\text{breadth}}$$

$$\text{Flakiness ratio} = \frac{\text{breadth}}{\text{thickness}}$$

Since the Heywood-type indices are based on measurements made with respect to three mutually perpendicular axes, they are sometimes referred to as *triaxial shape indices*. Extensive reviews of procedures for deriving shape indices from geometric measurements have been prepared by Pahl et al. [12] and Loveland [13].

White and co-workers [14] have used dimensionless ratios to characterize a fineparticle population having a range of shapes. The measurements they carried out are illustrated in Figure 10.1.

The simplest presentation of data used by these workers, a graph of frequency of occurence against elongation ratio, is shown in Figure 10.2. This type of data summary ignores any dependence of shape on size. To bring out this latter aspect of their data, White and co-workers made use of the three-dimensional plot shown in Figure 10.3. Figure 10.3*b* presents the same information collapsed into a two-dimensional display by making use of contour density lines as shown. This graph brings out the fact that there is a range of shapes at all sizes but that the larger crystals are generally more elongated. A different type of three-dimensional space for the display of shape distribution information that could be used to present data such as those reported by White and co-workers is illustrated in Figure 10.4*a*. In this type of display, information on elongation, size, and frequency is plotted about three mutually perpendicular axes. (Note in this display that the cumulative frequency of fineparticles having a given elongation with size is being displayed. This differs from the information display in Figure 10.3, where frequency of occurrence, and not cumulative frequency of occurrence, is plotted.) Various two-dimensional graphs can be abstracted from the data display in Figure 10.4*a*. Thus the line representing the locus of mean shape for a given size of Figure 10.3*b* becomes a line in the elongation size plane denoted by 0.5 cumulative frequency. (Strictly speaking, this is the median line, and not the average line; however, for comparative interpretive purposes, we note that for the type of distribution functions shown in Figure 10.3, there will not be a great deal of difference between the median line and the mean line.) Lines representing the frequency of different elongations for a given-size fineparticle now become lines in space at right angles to the median line, see Figure 10.4*b*. In general, the lines showing the frequency of given elongations against size for a whole range of sizes determines a two-dimensional surface within the three-dimensional domains of Figures 10.4*a* and *b*.

Dimensionless indices used in the description of shape are not neces-

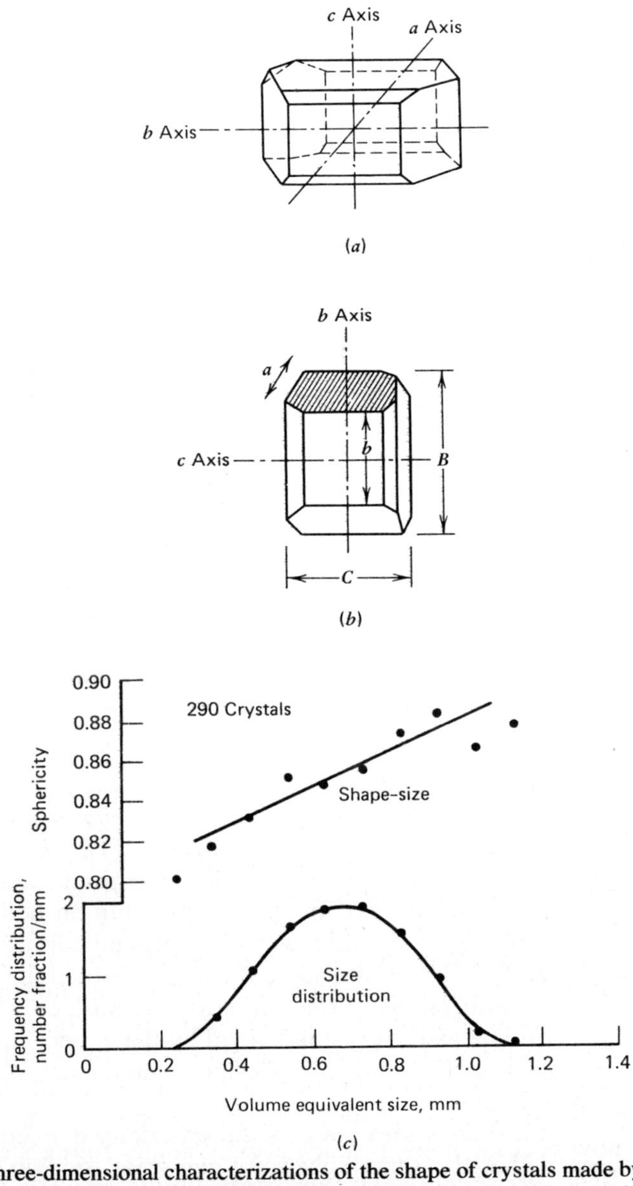

Fig. 10.1 Three-dimensional characterizations of the shape of crystals made by White et al. [12]: (*a*) three-dimensional representation of the sugar crystal; (*b*) with crystal in position of maximum stability B and C measured; elongation ratio $= B/c$; B taken as size; (*c*) variation of shape with size. To calculate size, measure facet dimensions a, b, and c: then (1) calculate the surface area of the crystal S_c; (2) calculate the volume of the crystal V_c; (3) calculate the surface area of the sphere of equivalent volume S_E. [Sphericity $= S_E/S_c$; "size" $= \sqrt{6V_c/\pi}$ (defined as the volume equivalent size)].

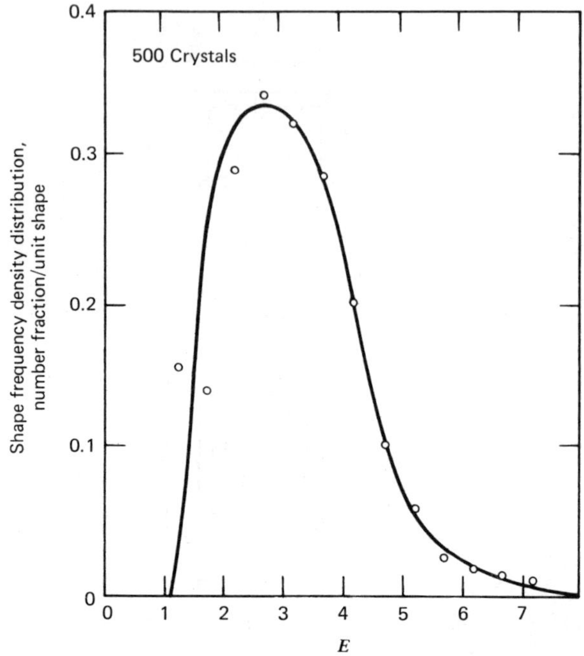

Fig. 10.2 A frequency–shape graph used to summarize shape distribution information for a fineparticle population [14] (E, elongation ratio = length/width).

sarily based on linear dimensions of the fineparticle. Thus White and co-workers also used a quantity they called *sphericity* to describe the shape of some of the crystals that they investigated. (The term "sphericity" has been used by different workers to describe similar but different dimensionless ratios. The context for the exact definition being used in any discussion of sphericity should always be checked carefully.) It can be seen from Figure 10.1 that White and co-workers defined the sphericity of a crystal as the ratio of the surface area of the surface of a sphere having the same volume as the crystal divided by the experimental determined area of the crystal. A plot of sphericity versus size reported by White is shown in Figure 10.3.

The difficulty of a shape index such as the sphericity defined by White and co-workers is that the same sphericity can arise for two obviously different shaped fineparticles. Thus if we had a sphere whose surface was covered by a series of spikes, it would have a high surface area and an apparently low sphericity when, to external circumstances, it appeared to be spherical. In the same way, a smooth rectangular crystal would have a high sphericity but, in fact, would look rather elongated from outside.

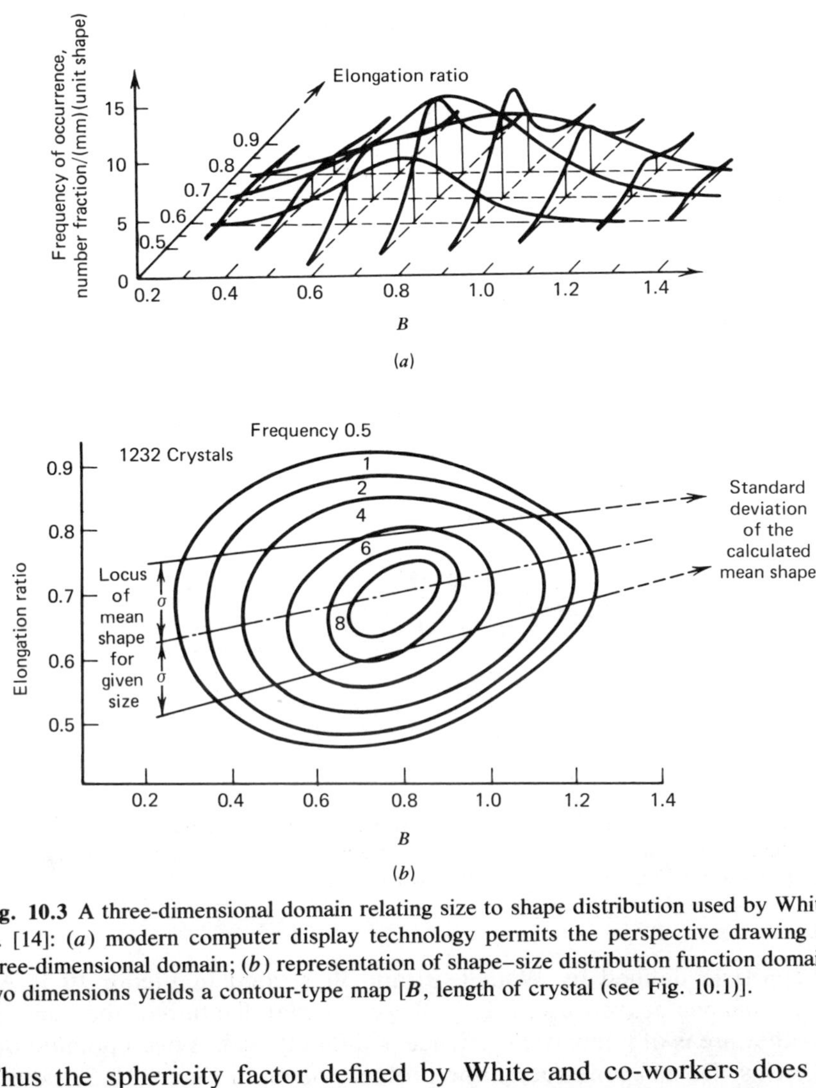

Fig. 10.3 A three-dimensional domain relating size to shape distribution used by White et al. [14]: (*a*) modern computer display technology permits the perspective drawing of a three-dimensional domain; (*b*) representation of shape–size distribution function domain in two dimensions yields a contour-type map [*B*, length of crystal (see Fig. 10.1)].

Thus the sphericity factor defined by White and co-workers does not correspond to one's intuitive appreciation of shape if the texture of the fineparticle and its structure and/or external morphology are varying. Stated in another way, sphericity does not independently measure texture, topography and external morphology, and these attributes can interact to give the same numerical index for fineparticles that look very different. This fact was appreciated by pioneers in the field who thus distinguished between what they called "form factors" and "roundness" [1, 2]. A good example of a "roundness descriptor" is the "roundness" of

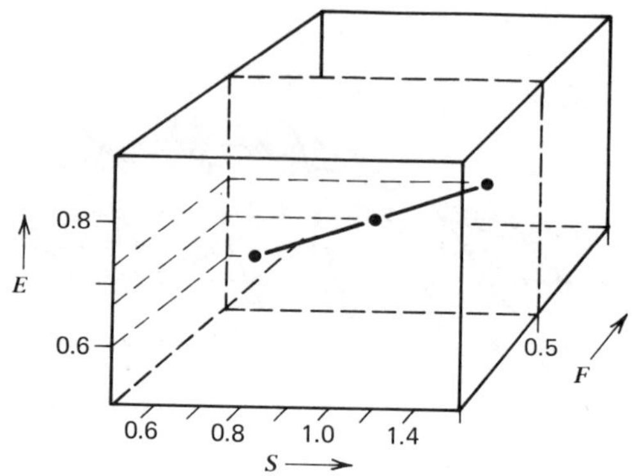

(a)

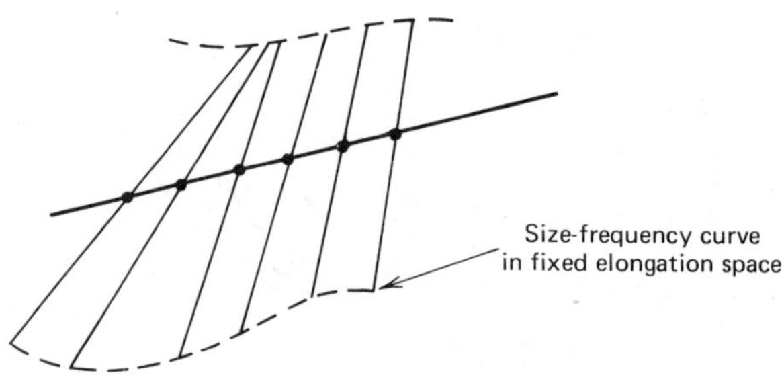

(b)

Fig. 10.4 The appearance of White's data as summarized in an alternate three-dimensional data space: (*a*) median line in three-dimensional space (*S*, size; *F*, frequency; *E*, elongation); (*b*) elongation–size curves pivot about the median line of *a*.

a fineparticle defined by Wadell [3] and illustrated in Figure 10.5. Although this has proved useful in geology, a generalization of the concept into other areas of fineparticle science is difficult, as has been pointed out by Davies. Davies constructed the profiles shown in Figure 10.5*b*, which are of the same roundness but that obviously have different structural details and external morphology.

Beddow [21] introduced the use of set theory into shape description terminology, and his concepts can be extended to provide a consistent set of descriptors to define the variation in texture, structural features (i.e., profile topography), and external morphology between fineparticle profiles. Thus, as illustrated in Figure 10.6, fineparticle profiles can belong to one set, all having the same external morphology, but to a different set on

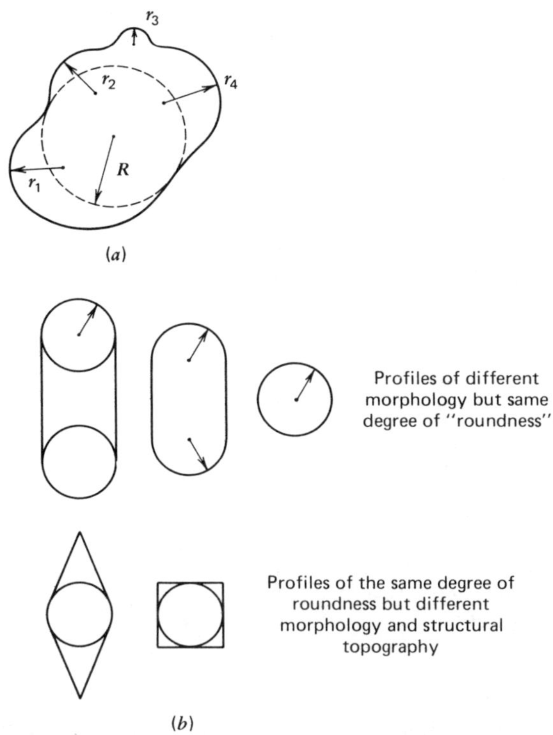

Fig. 10.5 Roundness has been suggested as a measure of profile structure–texture [1, 3]: (*a*) roundness as defined by Wadell as a measure of structure–texture (R, maximum inscribed circle; T, radius of curviture of a projection; roundness, $\sum_{i=1}^{i=n} r_n/nR$, (*b*) profiles drawn by Davies to illustrate the limitations of roundness as defined by Wadell.

the basis of texture or structural features. (Note that "chunkiness," which is discussed later in this section, is the reciprocal of elongation.)

Davies has developed a nondimensional graphical method of summarizing shape information based on triaxial measurements. The method is illustrated in Figure 10.7. The measurements for an individual fineparticle are rendered dimensionless and connected by dividing each individual measurement by the total of the three measurements for the individual profile. Thus for the fineparticle shown as A in Figure 10.7, the thickness and the breadth are one unit each and the length is 3. These measurements are normalized by using the total of the three measurements to give 0.2 as a dimensionless measure of the thickness and 0.6 for the breadth and the length. The data points that correspond to fineparticles A and B in Figure 10.7 are plotted on triangular graph paper in Figure 10.7. Shape data for

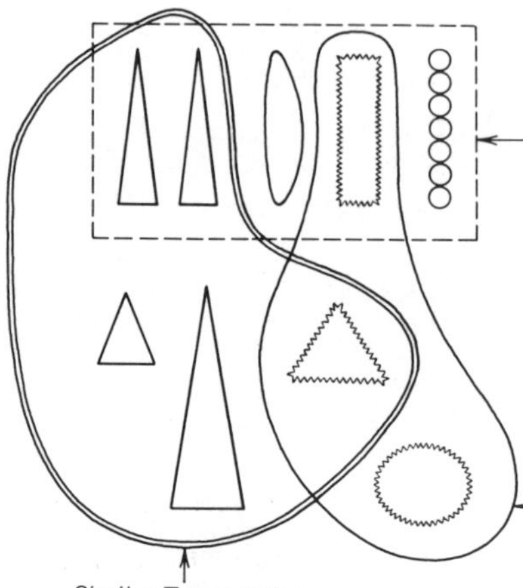

Fig. 10.6 Beddow and co-workers introduced the concepts of set theory to describe shape. Kaye has extended their original vocabulary to illustrate multiple membership by a given fineparticle in several sets.

two abrasive powders reported by Davies are summarized on triangulated mesh scales in Figure 10.8. Davies studied the abrasive fineparticles by gluing them along a 12-μm glass fiber, one end of which was held in a rotary mechanism. The other end of the fiber passed over a pulley and was attached to a tensioning weight. The image of an individual fineparticle was projected onto the screen of a Quantimet image analyzer. The magnification was adjusted to give one fineparticle profile in the measurement area of the screen at any one time. To characterize a fineparticle, the fineparticle was rotated slowly with measurements being taken of the area of the profile until the orientation of the profile corresponding to maximum projected surface area was located. With the fineparticle in this position, the length and the breadth of the profile were evaluated. The fineparticle was then rotated 90° and the thickness of the fineparticle measured. The data in Figure 10.8 are based on 500 sets of measurements obtained this way for the two abrasive powders—one designated "sharp" and the other "dull." The contour lines on these graphs represent data frequency limits of the normalized interconnected data. Davies has suggested that such contour lines could also be used to delineate surfaces in a three-dimensional data space as shown in Figure 10.9.

Hausner has pointed out that many of the fineparticles encountered in

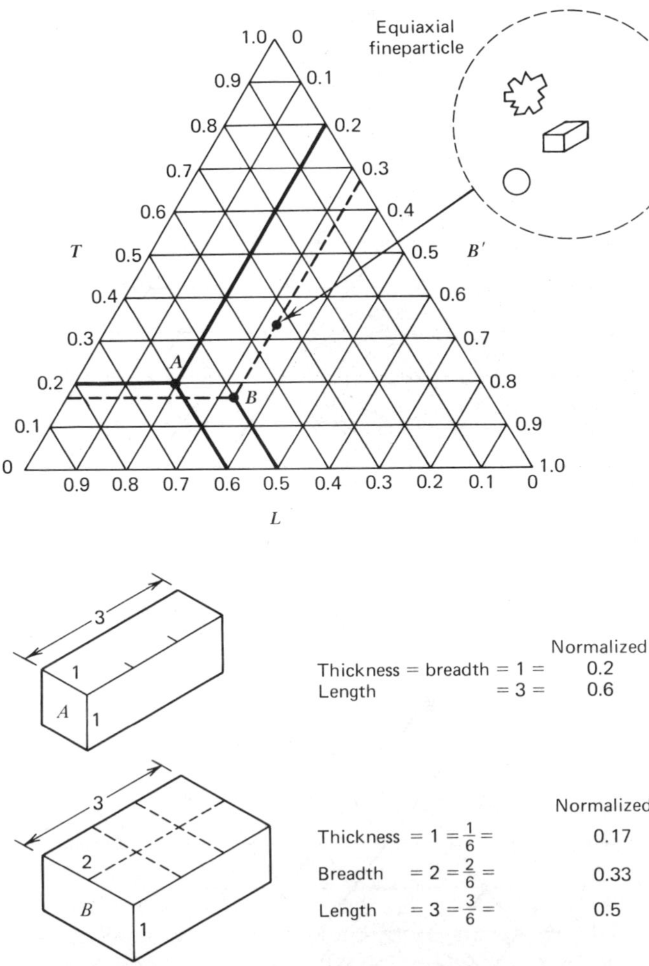

Fig. 10.7 "Triangular mesh" graph paper for plotting shape information [1] (*T*, normalized thickness; *B′*, normalized breadth; *L*, normalized length).

powder metallurgy have a certain amount of symmetry, thus implying that when the fineparticles are viewed in the position of maximum stability, the significant variations in two-dimensional space are directly displayed [7]. Hausner has suggested that a useful way of generating dimensionless indices of shape is to construct a rectangle around the profile so that its area is at a minimum when considering other rectangles that could be drawn about the profile. If A is the projected area of a profile, C is the actual perimeter of a profile, and a and b are the lengths of the sides of the

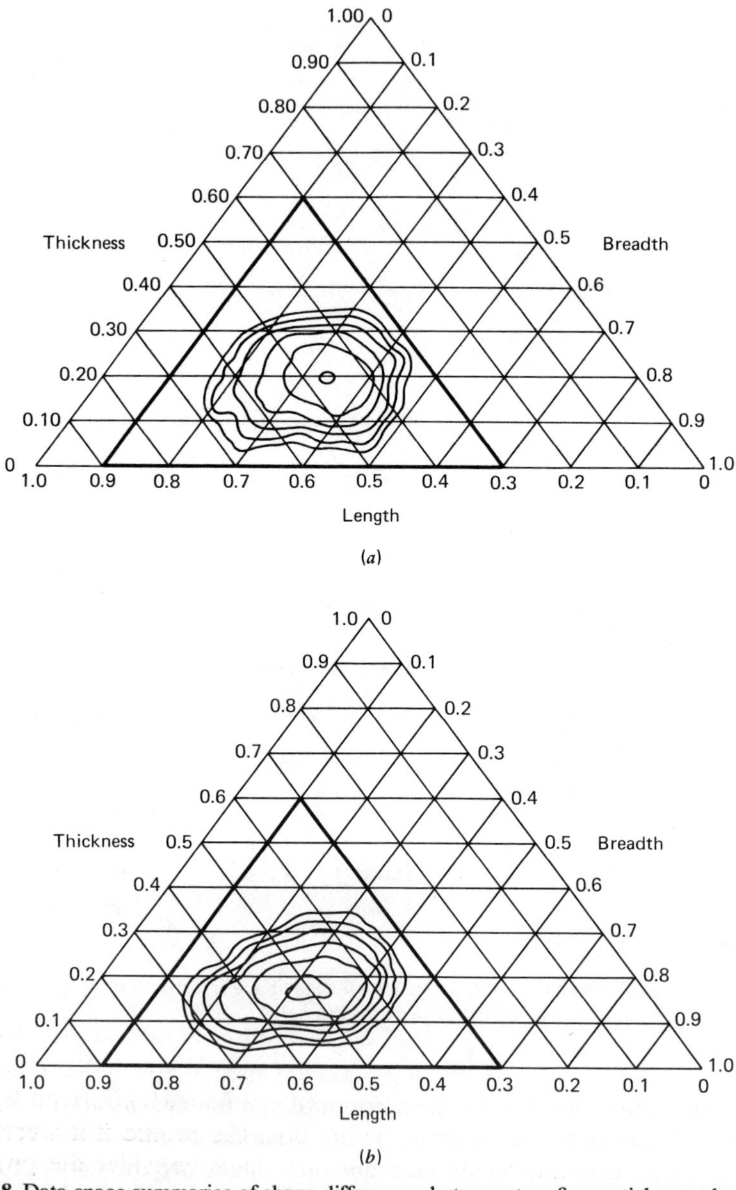

Fig. 10.8 Data-space summaries of shape differences between two fineparticle populations [1] (contour lines represent data frequency limits): (*a*) ''sharp'' silicon carbide; (*b*) ''dull silicon carbide.

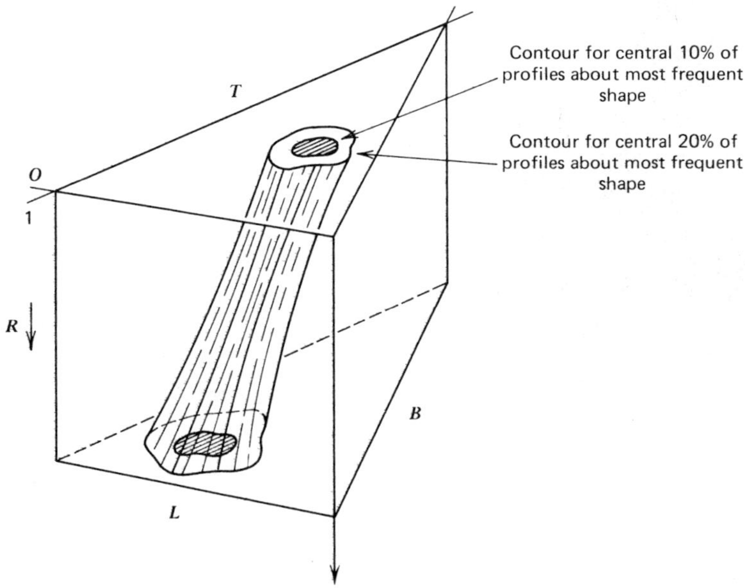

Fig. 10.9 Three-dimensional dataspace for representing the shape of a fineparticle [1] [*L*, length (normalized); *B*, breadth (normalized); *T*, thickness (normalized) *R*, roundness].

minimum area embracing rectangle, then the following shape description indices can be defined:

$$\text{Elongation factor} = \frac{a}{b}$$

$$\text{Bulkiness} = \frac{A}{a \times b}$$

$$\text{Surface factor} = \frac{c^2}{12.6A}$$

These are known as the *Hausner shape indices*. Medalia [16] has discussed the possibility of ambiguity in Hausner shape indices and has suggested that a more useful bulkiness measure can be derived by considering the distribution of mass elements in the profile if it were to be considered as defining a lamina mass [17]. Thus consider the profile in Figure 10.10; treating the profile as a thin lamina, one calculates the moments of elemental areas δA at a distance R and angle Θ from the centroid of the profile. Medalia defined the two-dimensional anisometry of the profile as the ratio of the large axis to the smaller one of the ellipse having the same radius of gyration as the lamina. For the profile in Figure

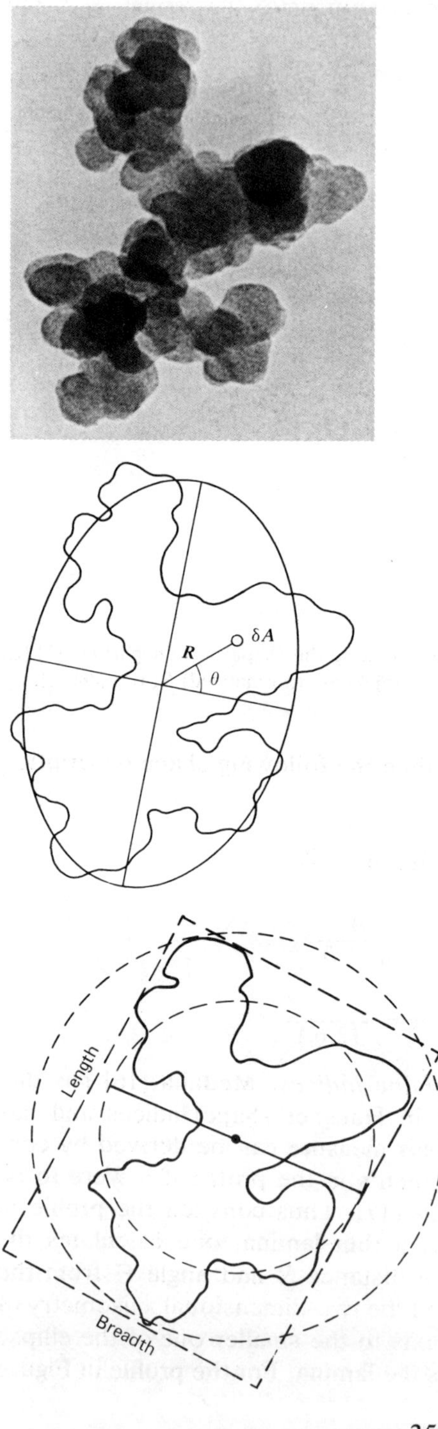

Fig. 10.10 Dimensionless shape indices of a carbonblack agglomerate (anisometry, 1.74; bulkiness, 1.54; Exc, external compactness = 0.43 for profile shown; C, chunkiness = 0.77 for profile shown).

10.10, the ellipse that would have the same radius of gyration as that of the profile is shown superimposed on the centroid of the profile. Medalia also defined the two-dimensional bulkiness of the profile as the quotient of the area of the ellipse divided by the area of the profile. For the profile in Figure 10.10, the anisometry was 1.74 and the bulkiness was 1.54.

Medalia and co-workers have also developed the use of the "circularity shape factor," which is defined as the ratio of the perimeter of the profile squared divided by 4π times the area of the profile [19].

Kaye [20] has investigated methods of describing the shape of profiles such as that in Figure 10.10 and has suggested the use of two other dimensionless ratios for the characterization of shape. These are the external compactness and the chunkiness [20]. The external compactness is defined as the ratio of the square of the diameter of the circle of equal area to that of the profile as divided by the square of the diameter of the embracing circle. The embracing circle is that circle drawn about the centroid of the profile, which just touches the extremity of the profile furthest out from the centroid. The chunkiness is the reciprocal of elongation as defined by Heywood. For the purposes of presenting graphical summaries of data, the chunkiness ratio has the advantage over the elongation factor that it runs from 0 to 1 with 1 corresponding to a square chunky fineparticle, whereas the elongation ratio goes to infinity for a very thin fiber and requires long-range logarithmic scales for adequate presentation of wide-ranging data. The external compactness is derived from two measurements, which can be made easily with the Zeiss–Endter equipment. For the profile shown in Figure 10.10, the external compactness, denoted by E_{xc}, has a value of 0.43 and the chunkiness, denoted by the letter C, equals 0.77. Using dimensionless indices such as the external compactness and the chunkiness, it is possible to construct data spaces that can be used to describe the range of shapes present in a population of fineparticles. The profiles in Figure 10.11 have been described by Beddow and co-workers in several studies of shape characterization methodology [21]. The sponge iron fineparticles have a rugged surface texture, whereas the atomized copper–lead fineparticles tend to be smooth. The external compactness ratio and the chunkiness for each of the sponge iron profiles in Figure 10.11 were measured, and the data are summarized in Figure 10.12. To help develop a vocabulary for describing multidimensional data space, it is useful to describe a graph such as that in Figure 10.12 as a two-dimensional data space. The shape dependent ratios defining a point in the data space can be considered to be coordinates defining fragments of a data point cluster. The centroid of the data point cluster in the two-dimensional data space can then be calculated. In the same way, one can calculate the radius of the second- and third-order moment equivalents of

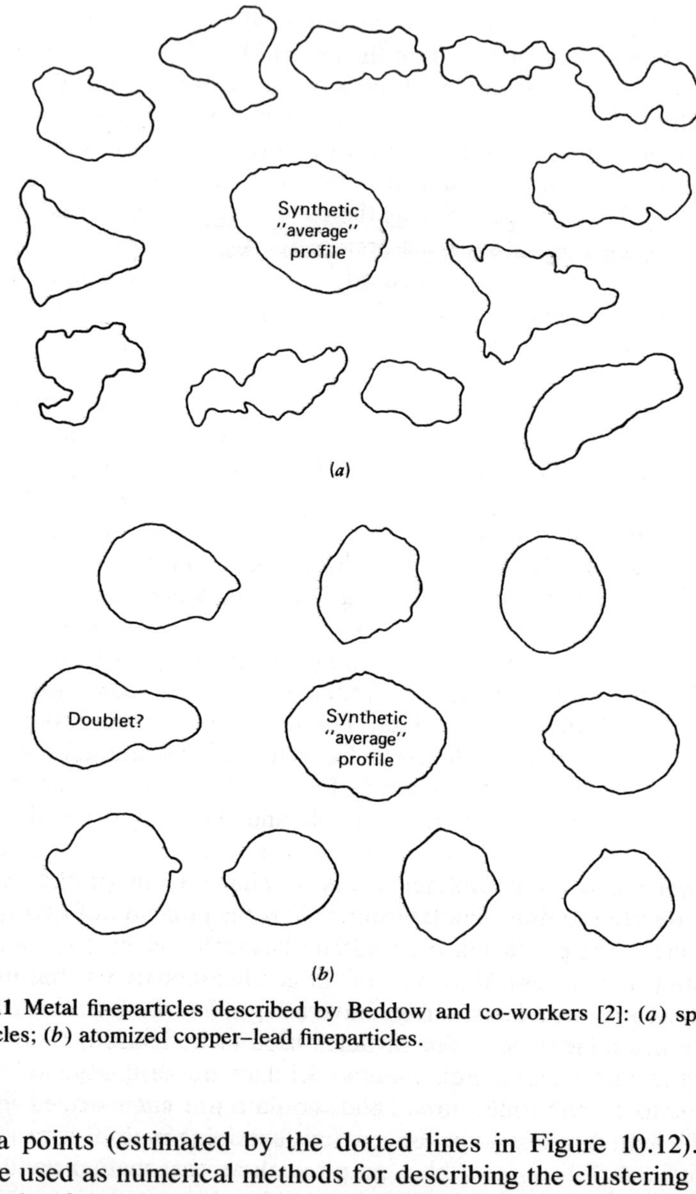

Fig. 10.11 Metal fineparticles described by Beddow and co-workers [2]: (*a*) sponge iron fineparticles; (*b*) atomized copper–lead fineparticles.

the data points (estimated by the dotted lines in Figure 10.12). These could be used as numerical methods for describing the clustering of various clouds of data points representing different fineparticle populations. Thus the external compactness and chunkiness data for the two sets of profiles are summarized in Figure 10.13. Figure 10.13 shows that the points cluster in different areas of the data space and that except for one point, the two sets of fineparticle profiles occupy completely different

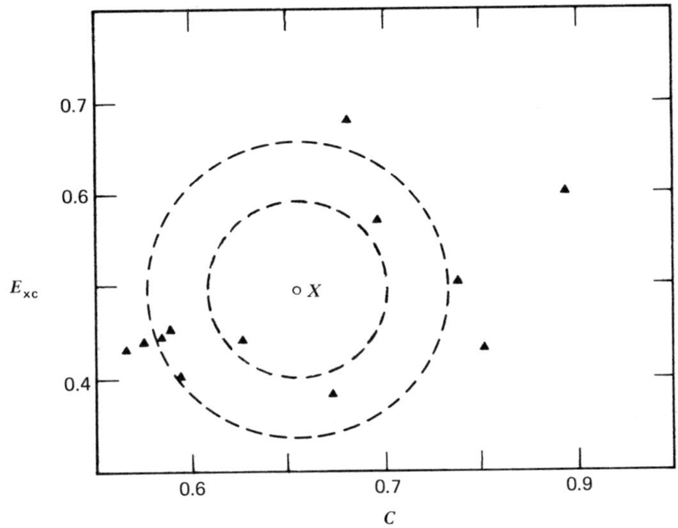

Fig. 10.12 Chunkiness–compactness domain for sponge iron fineparticles (X, centroid of data cluster; E_{xc}, external compactness; C, chunkiness.

regions. The one exceptional point is interesting in that it represents the profile labeled "doublet?" in Figure 10.11. When producing fineparticles by atomization, a liquid stream of the metal is blown into a spray, and the fineparticles created are allowed to solidify as they fall down through a cooling column of air. There is always a chance of collision producing a doublet just when the fineparticles are on the point of solidification so that they cannot flow together after joining. The doublet profile in Figure 10.11*b* resembles a fineparticle produced by the collision of two spherical fineparticles. Thus it would not be surprising if, when one plotted the information for an atomized powder in a domain such as that in Figure 10.20, there may be a substantial doublet cloud interpenetrating the space normally occupied by a different type of powder. Indeed, there may be other subclouds of points representing triplets, quadruplets, and so on. It is interesting to note that the doublet would represent a decrease in both the profile external compactness and chunkiness (hence its location in the lower left-hand corner of the domain), whereas the single atomized fineparticles are compact and chunky and hence occupy the upper right-hand corner of the data space.

If one were to construct a three-dimensional data space in which the three axes were external compactness, chunkiness, and a measure of the roughness of the surface, the data would appear somewhat as shown in

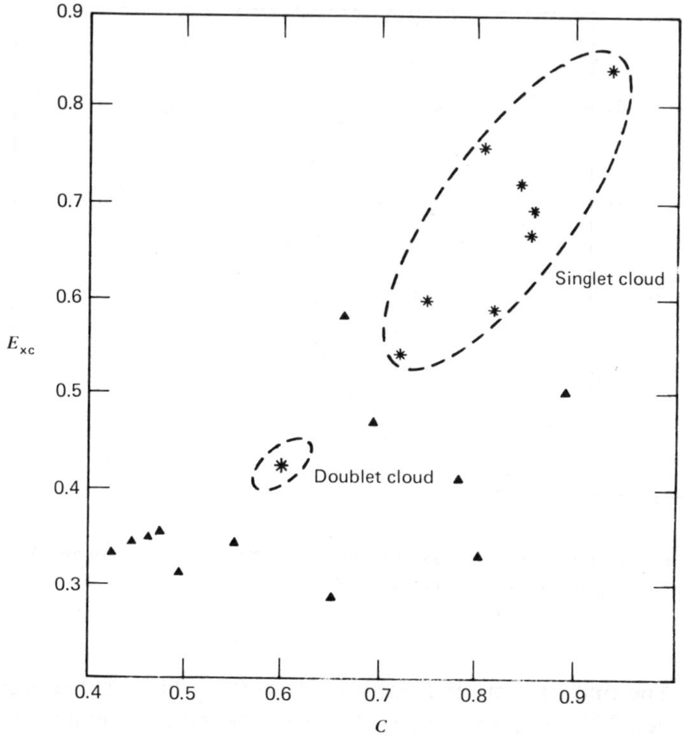

Fig. 10.13 The two types of powder plotted in a chunkiness–compactness domain (*, atomized powder; Δ, sponge iron powder; C, chunkiness; E_{xc}, external compactness).

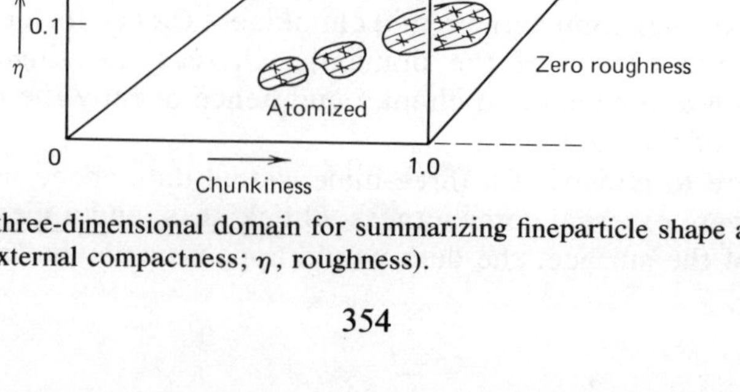

Fig. 10.14 A three-dimensional domain for summarizing fineparticle shape and size information (E_{xc}, external compactness; η, roughness).

Figure 10.14. The use of such data space would separate the two types of profile shown in Figure 10.11. The sponge profiles would lie in a data space plane defined by a roughness of 0.2, whereas the atomized powders would lie in a plane of zero roughness. When storing data in a computer memory, it is possible to manipulate the data in multidimensional space. Such multidimensional data space is hard to visualize specifically but increases the power of data group discrimination by numerical methods.

It is possible to conceive of dimensionless ratios based on the expected values of the so-called statistical diameters such as the Feret and Martin diameters discussed in Chapter 9. The possibility of using such statistical averaged dimensionless ratios was explored by Church [22], who pointed out that the achievement of useful confidence in the expected value of such a ratio, which itself is made up of two elements containing considerable uncertainty, requires very high numbers of experimental measurements.

10.2 SIGNATURE WAVEFORMS OF FINEPARTICLE PROFILES

To store information on a fineparticle boundary in a computer memory, the information on the location of the fineparticle boundary must be available as a dimensionless list. In the late 1960s and early 1970s several workers independently conceived of the idea that a dimensionless description of the profile of a fineparticle could be obtained from a Fourier analysis of a waveform representing the profile [15, 23, 24]. Thus Schwartz and Shane suggested in 1969 that one could generate a wave function characteristic of the fineparticle profile using the rotating vector transform illustrated in Figure 10.15. To locate a reference point within the fineparticle profile, they constructed the smallest circumscribing circle. The magnitude of a vector from the center of this circle to the periphery of the profile was then measured as a function of linearly distributed value of Θ, thus generating the wave-type curve illustrated in Figure 10.15. (It should be noted that in the waveform in Figure 10.15 the values of the vector producing the waveform have been normalized using the maximum value of the vector.) This wave-type curve can be treated as one complete cycle of a continuous periodic function so that it can be subjected to Fourier analysis. Fourier analysis is concerned with the fact that a continuous complex waveform can be broken down into a list of simple harmonic waves that can be combined to produce a complex waveform. The lowest frequency found to be a component of the composite wave is termed the *fundamental frequency*. Multiples of this frequency are termed *second, third, fourth harmonics,* and so on with the number before the term

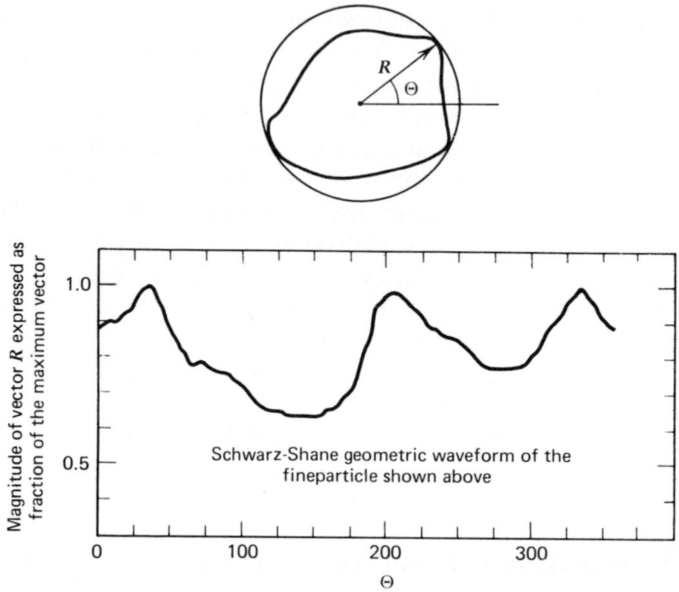

Fig. 10.15 A geometric signature waveform used to describe fineparticle profiles.

"harmonic" indicating the multiple of the fundamental frequency. The profile of a composite wave constructed from simple harmonic components depends on the phase relationship of the constituent waves and their amplitudes. However, in many situations the physicist is content to list the amplitude of the various harmonics in a wave to give a dimensionless list describing the complex waveform. It should, however, be appreciated that a reconstruction of any given waveform from a list of harmonics strengths requires phase information as well as amplitude information. When the waveform in Figure 10.15 is subjected to Fourier analysis, the list of the coefficients of the resultant harmonic series (i.e., of the relative amplitudes of the contributing simple harmonic waves that are multiples of the fundamental frequency) is a nondimensional list descriptive of the waveform.

The human ear is a good harmonic analyzer, and it has been suggested that a wave such as that shown in Figure 10.22 could be fed repetitively into an electronic system capable of producing a corresponding acoustic signal. Then the technologist could listen to the shape of his fineparticle and hear the difference between a smooth and a textured ellipsoid. (A textured ellipsoid signature waveform would have higher harmonics and would probably sound more violin-like than its smooth equivalent.) Prior to the availability of packaged Fourier analysis programs in a digital computer system, the mathematical labor involved in Fourier analysis would have made it impossible to consider using this approach in characterization of fineparticle profiles. With present-day computer technology, however, the preparation of a list description of a Fourier harmonic series as a shape descriptor can be considered as a routine operation [29, 30]. Erlich and Weinberg [24] chose to generate waveforms by letting a vector rotate about the centroid of the profile when considered to form a laminar solid. This was the same reference point chosen by Sutton and Bundalli [23]. The use of waveforms based on the centroid of the profile has been developed extensively by Beddow and co-workers [25–27]. The type of waveform generated by these investigators is described as geometric signature waveforms.

The presence of a protuberance on the outline of the profile can cause considerable increase in the amount of experimental work and analytical manipulation of the data required to obtain Fourier transform of a fineparticle signature waveform. Thus the resolution with which the value of R is measured must be greatly increased at the location of a protuberance to obtain adequate description of that protuberance. The presence of a sharp peak in a complex waveform represents the contribution of relatively high harmonic components; thus a protuberance peak or a texture-generated ripple on the waveform greatly increases the amount of effort to analyze and characterize a geometric signature waveform. Another difficulty associated with geometric signature waveforms of the type generated by the Schwartz–Shane strategy or the Beddow strategy is that the physical significance of the vector generating the waveform becomes indeterminate if it has to cross a bay in the topography of a convoluted profile. In studying such profiles such as those in carbon black agglomerate in Figure 10.10, the indeterminate nature of the rotating vector as it crosses structural features in the reentrant profile becomes a major obstacle to the extension of this type of signature waveform to the description of the morphology of the profile. Sutton has pointed out that it is possible to consider generating a type of signature waveform in which one considers the movement of a vector about some reference point external to the profile; however, this becomes difficult to interpret in terms of morphological parameters because it generates multiple-vector magnitudes for a given angle Θ.

In general, the lower frequency terms in the Fourier series of the geometric waveform of a profile corresponded to information on form (external morphology), intermediate frequency terms give information on structure (topography), and very high frequency information yields information on texture.

Erlich and Weinberg suggested that one could differentiate the basic external morphology (which they termed "form") from what they called the "structure" of a fineparticle profile by using a circularity index based on a suitable average of the first 10 harmonic amplitudes and a roughness coefficient based on a suitable average of the amplitude of the eleventh to the twentieth harmonics. Thus Figure 10.16 shows the profiles of sand grains from three different layers of an Alaskan quartz sand deposit. Erlich and Weinberg generated geometric signature waveforms for these profiles and then calculated the circularity index and the roughness coefficient from the first to the tenth and the eleventh to the twentieth harmonics, respectively. The circles are drawn for comparison purposes. Grains C, with the lowest circularity index appear more circular, and the higher roughness coefficients correspond to the "roughness" of the topography of the profile.

Beddow and co-workers have suggested the calculation of several shape descriptors from the coefficients of a Fourier series derived from the geometric signature waveform generated by a vector rotating about the centroid of a profile. Thus they have suggested what they call the *centroid aspect ratio* as a measure of external morphology. This is defined as the ratio of the maximum dimension drawn through the centroid divided by the dimension perpendicular to this line. (*Note:* The term "aspect ratio" has been widely used in fineparticle science. It is similar to the elongation defined by Heywood except that it is more usual to use the term with respect to a fiber-type fineparticle when the length is intuitively obvious and there is some symmetry about this length axis so that thickness and breadth are the same quantity.) Beddow et al. [26] have shown how this centroid aspect ratio can be derived from the Fourier coefficients. They have also suggested the use of three descriptors based on running averages of the sequence of coefficients in the Fourier series. Thus they define what they call the "lumpiness" from the average of the low-order harmonics, the "roughness" from an average of the coefficients of the intermediate harmonics, and the "texture" from the coefficients of the highest harmonics. Beddow and co-workers have shown how two-dimensional data spaces can be defined by using combinations of the centroid aspect ratio, the lumpiness, the roughness, and the texture to display the range of shapes present in a population of fineparticles. In particular, they have made extensive measurements on the two sets of fineparticles shown in Figure 10.11. At one time it was thought that a good average descriptor of the shape of a population of fineparticles could be created by combining the harmonic series for a set of individual profiles to create an average synthetic profile representative of the population. This explains the central synthetic average profile present in each

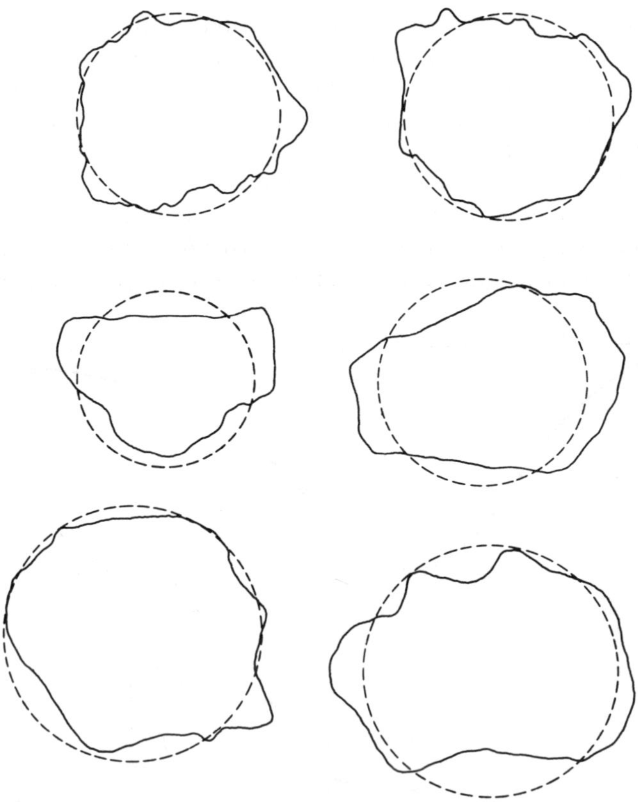

Fig. 10.16 Grain shapes of Alaskan quartz sand described by Ehrlich and Weinberg [24].

Circularity Index (Based on 1 → 10 Harmonics)	Roughness Coefficient (Based on 11 → 20 Harmonics)
A, 0.121	*A*, 0.017
B, 0.120	*B*, 0.009
C, 0.117	*C*, 0.011

portion of Figure 10.11. This does not seem to be a useful approach to the description of the average shape present in a population at least for classes of profiles similar to those shown in Figure 10.11. Beddow and co-workers have also described what they call the "symmetry" of the profile, which can again be calculated from the Fourier coefficients.

The external morphology of a profile can be studied by generating a geometric signature waveform based on the Feret diameter of a profile. To generate the Feret diameter signature waveform, referred to as the *FEDI waveform,* the profile is rotated through linearly distributed values of Θ, the angle describing the orientation of the profile with respect to a fixed direction. For convenience in data handling, it is useful to choose this direction so that the reference direction lies parallel to the maximum projected length of the profile. As the profile is rotated, the magnitude of the projected length is taken as defining the amplitude of the waveform. All values of the projected length are normalized by use of the maximum Feret diameter. The geometric signature waveform generated in this way for a carbon black profile is shown in Figure 10.17. The information contained in this geometric signature waveform is descriptive of the external

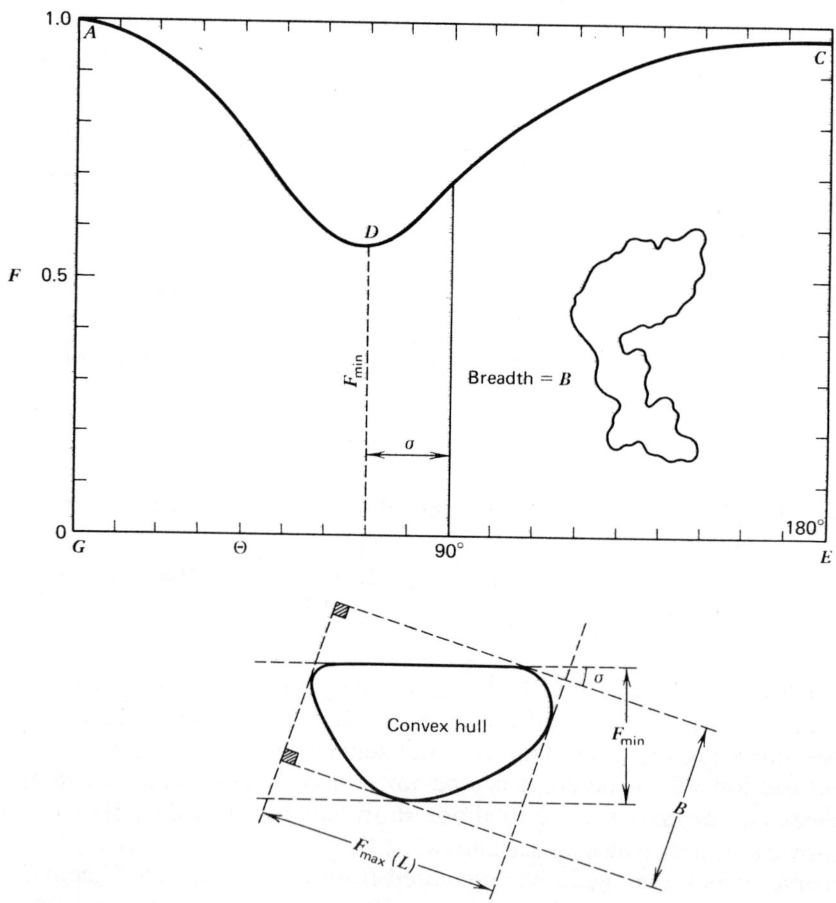

Fig. 10.17 The Feret diameter signature waveform (FEDI waveform); elongation ratio, L/B. σ the asymmetry angle is a measure of the lack of symmetry in the convex hull [18]. L, maximum projected length; B, dimension at right angles to L.

morphology of the convex hull of the profile as summarized in the lower portion of Figure 10.17. An elongation ratio for the profile can be calculated directly by taking the breadth as the width at right angles to the maximum length. An asymmetry angle descriptive of the lack of symmetry in the convex hull can be calculated from the position of the minimum Feret diameter with respect to the direction at right angles to the maximum dimension.

The FEDI waveform has two other attributes that could make it very useful in the characterization of fineparticle profiles. The first one is that the waveform is independent of the location of the point of rotation with respect to the location of the profile. This means that, for instance, the FEDI waveform could be generated for several profiles in the field of view by rotating the whole field of view and measuring the Feret diameter without first centering each profile with respect to the point of rotation. This also means that one does not need to use computer effort and time to locate a reference point before generating the waveform.

The second useful attribute of the FEDI waveform can be appreciated by considering the physical significance of the area under the FEDI waveform graph. The average value of the Feret diameter for all possible orientations is related to the perimeter of the convex hull by the relationship

$$F_{\text{av}} = \frac{\text{Perimeter of the convex hull}}{\pi} = \frac{P'_{\text{em}}}{\pi}$$

From Figure 10.17

$$\frac{F_{\text{av}}}{F_{\max}} = \frac{ADCEG}{ACEG} = \alpha \qquad (10.1)$$

$$F_{\max}\alpha = \frac{P'_{\text{em}}}{\pi}$$

Thus we can gain a dimensionless estimate of the perimeter of the profile by integrating the area defined by the FEDI waveform to calculate α for substitution in this relationship.

It can be shown that a useful descriptor of convex hull morphology, corresponding to π for a circle, can be calculated from equation 10.1 by rearranging it to produce the relationship

$$\alpha\pi = \frac{P'_{\text{em}}}{F_{\max}} = PEM = \pi_{\text{em}}$$

Thus the quantity $\alpha\pi$ is a modified measure of the ratio of the perimeter of the profile to its maximum Feret diameter and is analogous to π for a circle [31].

Kaye has pointed out that one can go from three dimensions to one for a profile in real space by using another type of geometric signature waveform that he has defined as the Cauchy signature waveform. This is generated by using the area of the projected silhouette of a fineparticle to define the amplitude of a wave. The waveform is generated by suitable rotation of the fineparticle about two independent axes. If generated in an appropriate manner, the Cauchy signature waveform would have the useful property that the area under the signature waveform is related to the area of the convex hull of the three-dimensional fineparticle. It should be noted that Bodziony has also described a measurement procedure that could be used to generate a Cauchy signature waveform. His equipment was actually used to estimate the area of the convex hull by combining estimates of the projected area antithetically; however, in the course of carrying out such measurements, he could automatically have generated a Cauchy-type waveform by interfacing his equipment directly to an automated iconometric device [32].

Signature waveforms can be used with characterization systems operating in a cybernetic mode. Advanced cybernetic systems are capable of learning from experience and making intelligent decisions. At an elementary level, an automated iconometric system could be operated cybernetically by storing in the memory of the computer a series of FEDI waveforms for standard geometric shapes. Incoming waveforms from profiles being encountered in a measurement situation could then be compared in the computer with the standard geometric shapes, and the system could decide which was the best approximation for the incoming signature waveform. Thus a printout from a computer after examining a profile could state, "the profile under test approximates to the shape of an isosceles triangle of height-to-width ratio 3:1."

Alternatively, if one were working with carbon blacks, the memory could be trained with a series of carbon blacks known to differ significantly in their external morphology. The measurement procedure would then be to match the profiles being investigated with those stored in its memory. The system could be instructed to learn from experience in that if it encountered a profile dissimilar in mathematical terms from anything stored in its memory, it could learn the shape of this new type of profile and use it as an extra category in classifying incoming signals. (For a discussion of the computation problems involved in operating a shape recognition system in the cybernetic mode, see previous works by Kaye [33–35].)

Kaye, Naylor, and others have generated characteristic signature waveforms from the diffraction patterns of fineparticle profiles. The system that they developed is known by the acronym SHADOW, from the phrase "shape analysis by diffraction originated waveform." The SHADOW system for shape characterization is a modification of a system originally developed by Redman for the automatic reading of alphanumeric characters [36, 37]. A typical set of diffraction patterns for various shapes is shown in Figure 10.18. The system used to generate signature waveforms from an optical diffraction pattern is shown in Figure 10.19. The object to be evaluated is located in the object plane of a lens. A laser beam is used to generate a diffraction pattern. One advantage of the procedure is that it is not necessary to locate the profile exactly on the axis of the lens. One only needs to isolate the profile optically from other profiles and locate it approximately within the operational area of the lens. The diffraction pattern will then be automatically produced on the axis of the lens system in the focal plane of the lens as shown in Figure 10.19. In actual practice, the profiles generating the diffraction pattern would be much smaller than the object shown in Figure 10.35, and the diffraction pattern is orders of magnitude larger than the actual dimensions of the fineparticle producing the diffraction pattern. This is an advantage from an instrument design point of view in that instead of having to make high-resolution measurements on a small object, direct physical measurements of the much larger diffraction pattern are possible. The diffraction pattern is turned into a signature waveform by interrogating it by means of the disk shown in Figure 10.19. As this disk rotates, energy from the diffraction pattern passes through the wedge-shaped portion removed from the disk and falls on the photoelectric device. The complex waveform corresponding to one complete rotation of the interrogation disk is then generated as shown in Figure 10.19. This waveform can be dumped directly into a computer and subjected to Fourier analysis to obtain a list of the coefficients of the harmonics. Alternatively, the disk can be kept static and the object rotated in the object plane of the lens. This system has the advantage that one does not have to worry about aligning the point of rotation with the optical axis of the system. An interesting aspect of the optical procedure for characterizing shape is that any speck of dirt or, for example, a fingerprint on the image of the fineparticles being characterized will not give rise to diffracted light very far from the axis of the lens system. Light scattered by speckle in the photograph and/or dirt and fingerprints will result in what optical computing scientists refer to as *dc optical signal,* which will be close to the axis of the system. This can be removed by blocking off the near-axis light signal.

The SHADOW system can be operated either in a direct mode, when it

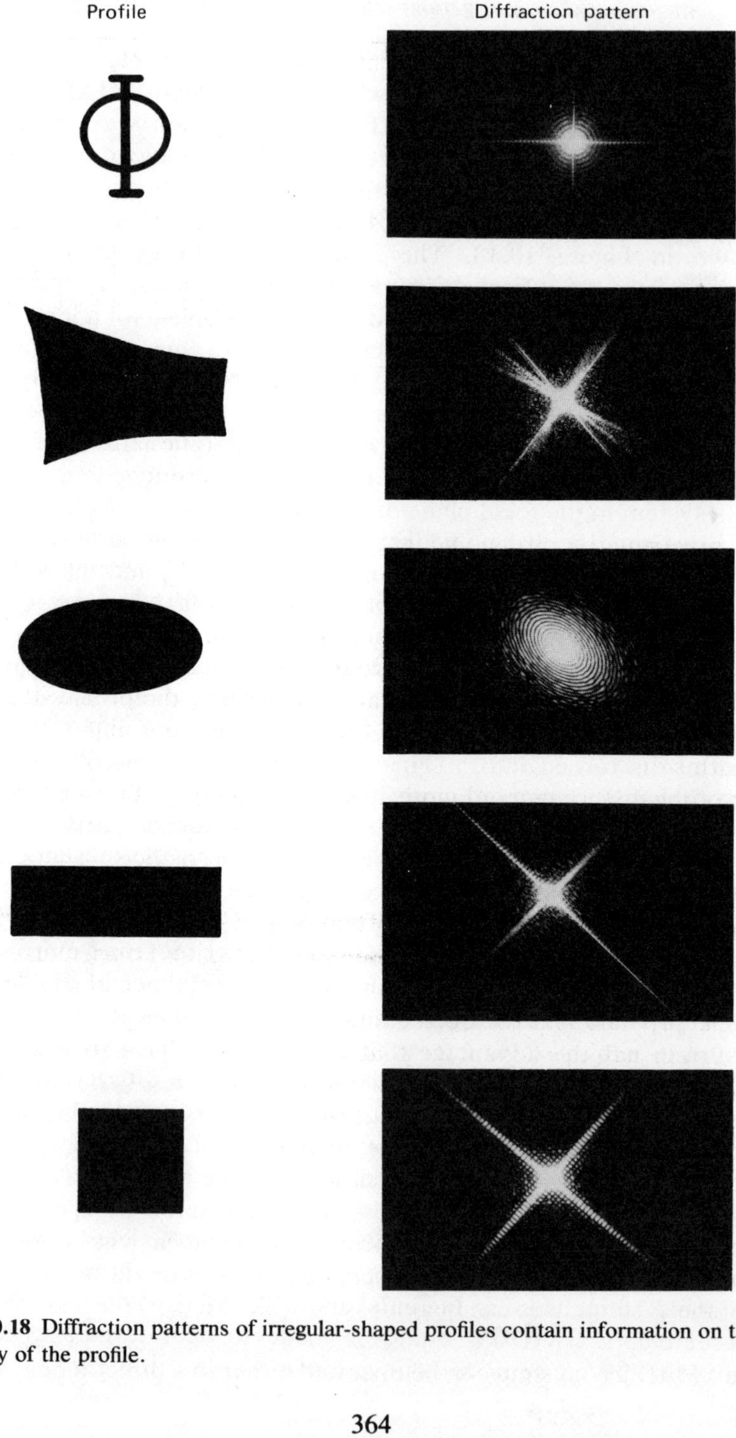

Fig. 10.18 Diffraction patterns of irregular-shaped profiles contain information on the morphology of the profile.

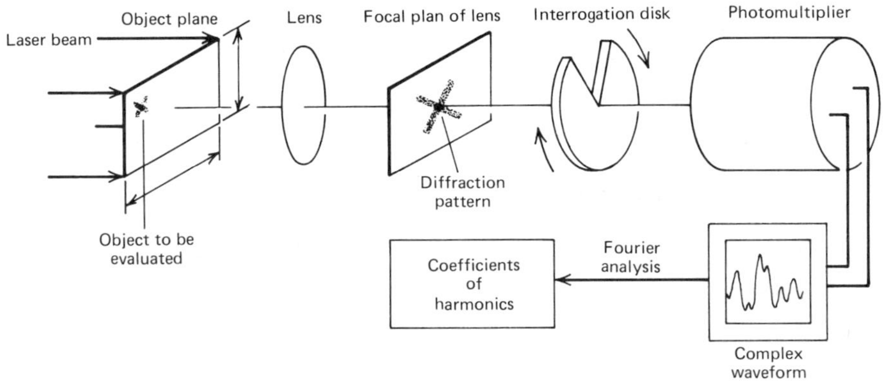

Fig. 10.19 Original system of Redman and Reid [36] for character recognition.

would print out the harmonic list descriptive of the profile, or in a cybernetic mode, whereby the shapes of the profiles being examined could be compared to a list of profile harmonics stored in the computer memory. (*Note:* If comparisons of this kind are carried out, it is not necessarily useful to proceed through the Fourier analysis of the waveform to obtain a nondimensional list describing the profile. It may be more efficient to use cross-correlation techniques to compare signature waveforms directly [33].)

One problem encountered in developing a widely applicable version of the SHADOW system is that one must have high-contrast negatives when generating the fineparticle diffraction patterns; also, the presence of texture and structure in the profile can diffract light that interacts with information generated by the basic morphology of the fineparticle [41, 42]. However, even in such a situation the interaction of texture, morphology, and structure will usually combine in any given sequence of profiles from the same population to produce unique signature waveforms that can be compared to achieve characterization of unknown profiles drawn from similar populations. Thus Murphy investigated the diffraction patterns generated by simulated carbon black profiles as illustrated in Figure 10.20. One could envisage a situation in which a whole array of technically significant shapes of carbon black profiles are used to educate a cybernetic system with regard to the categories to which fineparticles should be allocated. Such a system could also be instructed to learn from experience in that new categories of shape discovered in a series of measurements could be included in the category allocation system.

Conceptually, SHADOW could be made to operate as a cybernetic system with the capacity for discretion in varying the resolution with

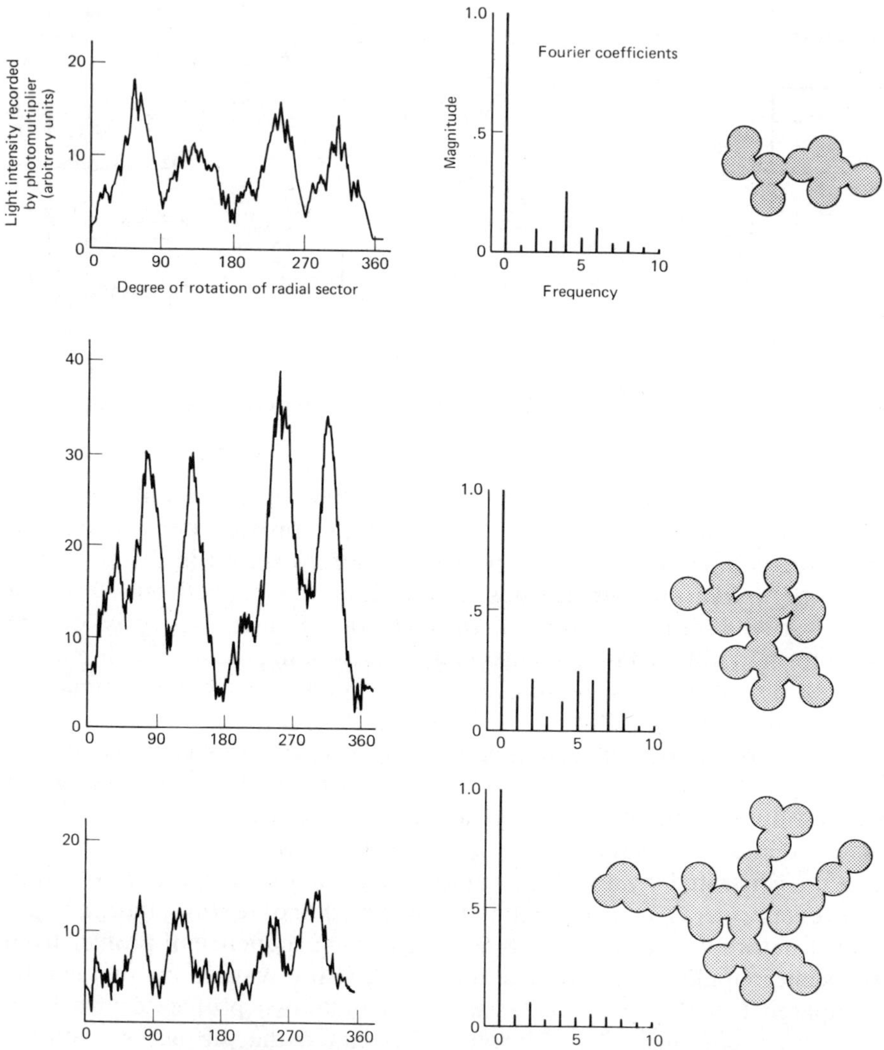

Fig. 10.20 Signature waveforms of carbon black flocs [40].

which it scrutinizes a diffraction pattern. Thus the angular span of the sector of the interrogation disk through which the energy of the diffraction pattern passes determines the fine structure of the signature waveform generated at the photocell. Thus if a relatively wide angle sector is used, only the basic features of the diffraction pattern will show up in the average signature waveform. For the basic recognition of morphology, this data compression may be advantageous in that it would reduce the high frequency spread in the signature waveform when subjected to Fourier analysis. However, for the resolution of the differences in fineparticles of similar morphology, the resolution may have to be increased. The computer could initiate any necessary higher-resolution inspection of a given profile by decreasing the width of the interrogation sector (by using two disks with sectors cut out that could be moved one over the other). Thus the system could carry out a first inspection of many shapes and only inspect with high resolution those of particular interest in a given study. Murphy [40] has briefly explored the possibility of using the SHADOW system in the description of such widely different shapes as chromosomes, viruses, and geological fossils.

10.3 FRACTAL DIMENSIONS AS DESCRIPTIONS OF FINEPARTICLE SHAPE

Mandelbrot [43, 44] has discussed the problem of describing highly convoluted or rugged boundaries that occur in nature. He has suggested that a satisfactory quantitative description of such boundaries has tended to be an elusive problem as many natural boundaries need to be studied with the aid of concepts drawn from non-Euclidean mathematics [50]. In particular, he has shown that it is useful to consider the space-filling nature of a rugged curve as being describable by a mathematical dimension that has fractional values between 1 and 2. Mandelbrot calls this dimension the *fractal dimension*. It should be appreciated that such a fractal dimension is a mathematical artifact that is useful in descriptive terminology.

In a series of publications, Kaye has shown that the concept of fractal dimension of a space-filling line can be used to describe fineparticle boundaries and fineparticle structures [45–49].

A useful way of developing the concepts associated with fractal dimensions is to consider a modified version of Archimedes' method for evaluating π [51]. Archimedes estimated the value of π from the measured perimeters of a series of inscribed and exscribed polygons of increasing number of sides. Using this procedure, he was able to place the value of π between $3\frac{1}{7}$ and $3\frac{10}{71}$. A modification of Archimedes' procedure is illustrated in Figure 10.21. This modified procedure is described as a random walk estimate of the perimeter of the profile. The simple case of the regular hexagon is given for comparative purposes in Figure 10.21. This can be regarded as having been formed by starting at point S and stepping out with a stride of magnitude $0.5D$, where D is the diameter of the circle.

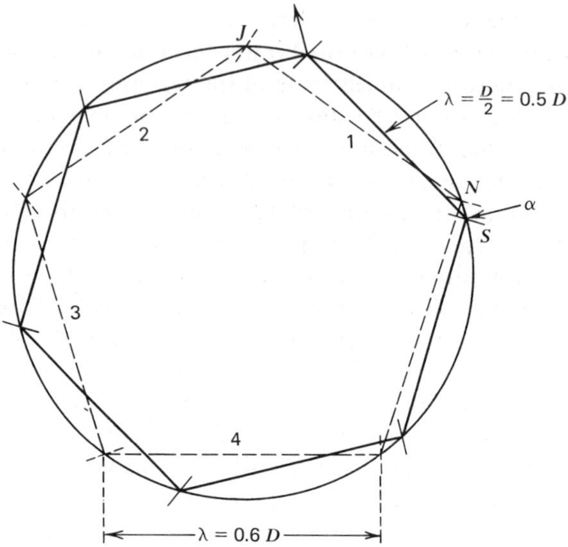

λ	π′
0.6	2.88
0.5	3.00
0.4	3.05
0.3	3.06
0.2	3.04
0.1	3.08

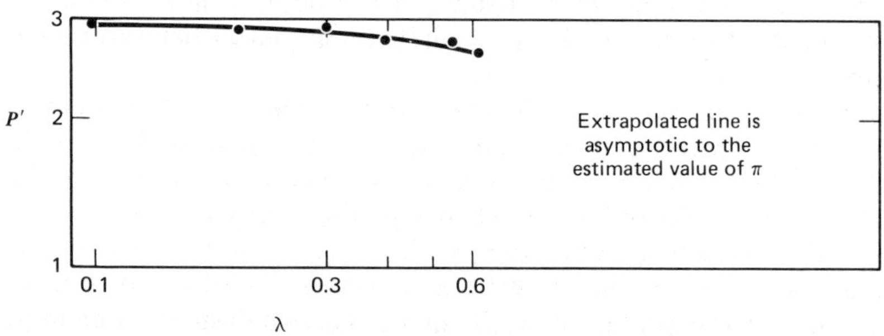

Fig. 10.21 Basic steps in a modified version of Archimedes' method for evaluation π [p', perimeter estimate (normalized) λ, stride magnitude (normalized)].

When stepping out with this stride magnitude, one arrives back at point S with a whole number of completed steps. The estimate of the perimeter now becomes the stride magnitude times the number of steps. In Figure 10.23 this perimeter estimate is denoted by P' and the stride by λ. Consider what happens if one were to step around with a stride of magnitude $0.6D$. Starting at point S, the first stride cuts the circle at J. After four complete steps, point M is reached. Completing the walk with a whole stride causes an overshoot of point S to reach point N. An unambiguous way of allowing for a fractional step in the estimate of the perimeter is to draw line MN and project orthogonal line SQ onto it. The fraction α constituting QN of line MN is taken to be the amount to be subtracted from the final step in generating an estimate of the perimeter. On the basis of this procedure, one arrives at 2.88 steps of stride $0.6D$ as an estimate of the perimeter. If the perimeter estimate is now expressed in a dimensionless form by dividing by the diameter, the perimeter estimate is 2.88λ, where $\lambda = 0.6$ of the diameter; λ is described as the step or the stride size used in estimating the perimeter of the profile by "walking" around the profile, starting at a randomly chosen point. If this procedure is repeated for a series of decreasing values of λ and the data plotted on logarithm graph paper, as shown in Figure 10.21, it can be seen that the curve tends to the value π. By $\lambda = 0.3$ of the diameter, an estimate of π correct to within approximately 2% has been obtained. Mandelbrot points out that a similar set of estimates of the perimeter for Euclidean curves such as ellipses and other regular geometric figures also tend to finite values. The situation is very different, however, for estimates of the perimeters of non-Euclidean curve boundaries. Mandelbrot shows that many naturally occurring boundaries, such as the coastline of an island, are non-Euclidean. For such boundaries, the estimates of their perimeter when plotted on a logarithmic basis against the stride used in the estimation of the perimeter does not tend toward finite limits but rather extrapolates to infinity.

To illustrate this fact, consider the carbon black profile shown in Figure 10.22. The perimeter can be estimated by stepping around the profile as illustrated. When dealing with a convoluted line, it is necessary to have a set of rules to avoid ambiguity when selecting between possible multiple intersections stepping around the profile. Thus placing one leg of a pair of compasses at point S, the arc generated for a stride of λ shown in Figure 10.22 can cut the profile at A, B, or C. To avoid such ambiguity, the following rules for generating the random walk are used:

1. The swing of the compasses is always from the last point on the perimeter inward from the outside of the perimeter.
2. The first intersection is always taken to be the definitive step.

Thus for the situation shown in Figure 10.22, A is taken to be the step from S. When these rules are applied, a step such as $B \rightarrow D$ bypasses the inlet of the perimeter as shown.

For the "walk" illustrated in Figure 10.22, after eight complete strides the compasses arrive at point Z. To complete the random walk with a fractional step, the same procedure is used as that illustrated in Figure 10.21 for the circle. Thus the step is taken from Z to the next intersection

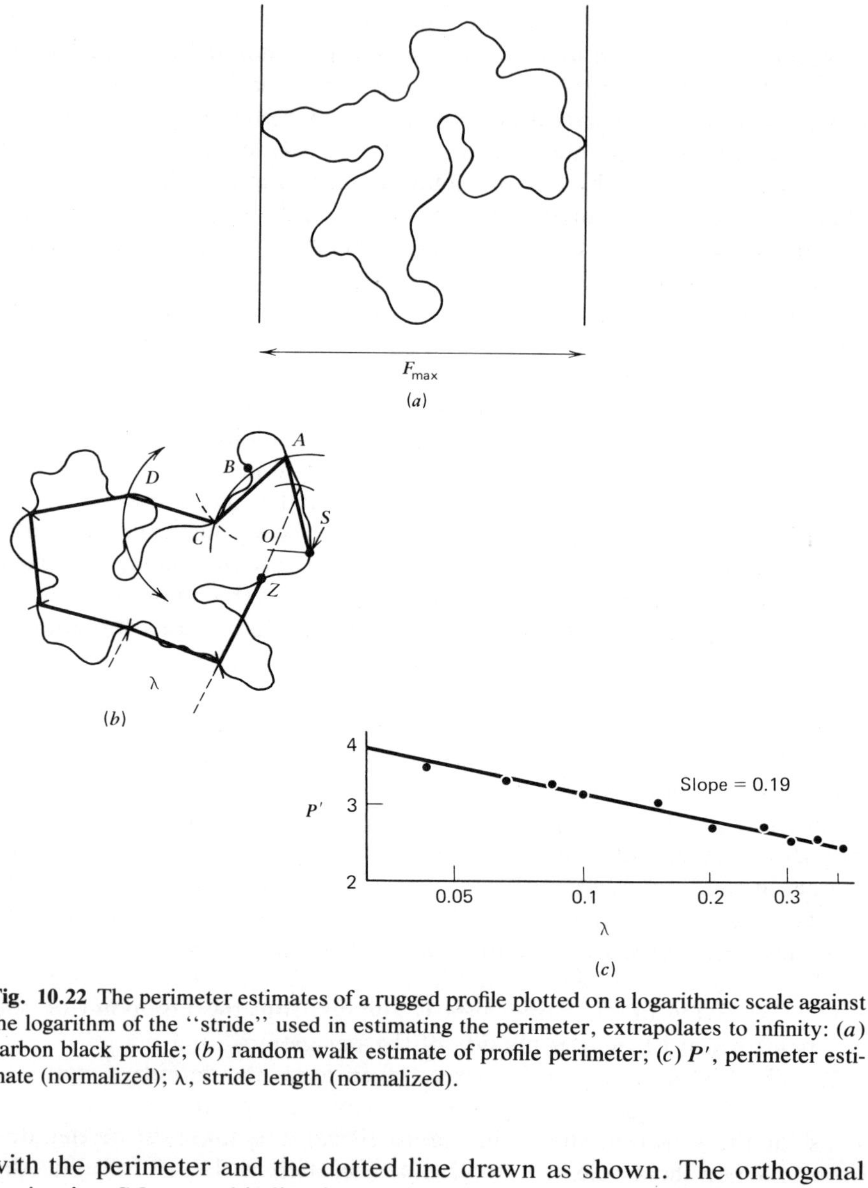

Fig. 10.22 The perimeter estimates of a rugged profile plotted on a logarithmic scale against the logarithm of the "stride" used in estimating the perimeter, extrapolates to infinity: (*a*) carbon black profile; (*b*) random walk estimate of profile perimeter; (*c*) *P′*, perimeter estimate (normalized); λ, stride length (normalized).

with the perimeter and the dotted line drawn as shown. The orthogonal projection *SO* onto this line is constructed so that *ZO* represents the fractional step completing the estimate.

The random walk depends on the starting point so that there is an inerent variability, or uncertainty, in the value of the estimated perimeter larger than the true experimental error involved in arriving at the perimeter estimate. Each of the data points of the graph shown in Figure 10.22 is an average of three estimates of the perimeter for a given stride length, with each estimate based on stepping out from a different starting point. In the case of an irregular profile the data are normalized by dividing the perimeter estimates by the maximum Feret diameter. The normalized perimeter estimates are plotted in Figure 10.22 as a function of the stride used in making the estimates, on a log–log basis. It can be seen that in this case the estimates do not tend to finite limits. If extrapolated to smaller and smaller stride lengths, the ultimate estimate of the perimeter magnitude is infinite. This is an example of the paradox that highly involuted profiles can have infinite boundaries.

Intuitively, some physical scientist might want to reject this conclusion by insisting that the meaning of the word "stride" disappears as one goes to smaller lengths. Mandelbrot has discussed the problems of allocating an intuitive meaning to infinite boundaries associated with highly convoluted profiles occurring in nature. He points out that even our traditional attribution of one, two, and three dimensions to geometric figures is not without problems and also that it is sometimes better to accept a mathematical definition rather than an intuitive picture.

Mandelbrot had been able to show that one can treat the slope of a line such as that in Figure 10.22 as defining the mathematical dimension of the curve generating the data. It can be shown that if δ is the fractal dimension of the line, the slope of the curve is given by the relationship $m = (1 - \delta)$. Kaye [45] had pointed out that this so-called fractal dimension of the line can be used to define either the *ruggedness index* or the *convolution factor* of the profile. Thus the ruggedness index is equal to the fractal dimension. Its magnitude indicates that one is dealing with a boundary since the one before the decimal point would indicate that one is dealing with a line and the portion after the decimal point would be the measure of the convolution. The convolution factor is defined as the ruggedness index minus 1. This factor has the advantage that it ranges from 0 for a straight line to 1 for the type of random walk demonstrated in a Brownian motion that will cover a surface entirely if allowed to continue for a long period of time [43].

Mandelbrot has shown that there are many curves that can be generated by recursive processes that will generate true fractal curves. When one comes to a natural boundary, one can only anticipate that a profile may be a fractal curve if the fineparticle is being generated by certain processes such as an iterative process. Thus one could anticipate that a carbon black profile might be a fractal curve since it is generated by the random accretion to a randomly structured nucleus of spherical smoke fineparticles. Unlike a mathematical curve that can be considered to have an idealized structure, there will usually be a finite limit to the fractal nature of a real boundary when the subunits forming the larger fineparticle have Euclidean boundaries.

To illustrate the potential usefulness of fractal dimensions for characterizing fineparticle boundaries, consider the simulated carbon black profile shown in Figure 10.23. Estimates of the perimeter made with various stride magnitudes are illustrated. There are two clearly discernible linear regions on the perimeter estimate–stride data plot. The first linear region at large strides corresponds to the structure of the boundary, whereas the second linear region corresponds to the strides used in exploring the texture of the surface. Measurements carried out at even greater resolution would probably yield a region that degenerates to a linear region parallel to the stride axis, thus indicating that once λ is small relative to the size of the spheres forming the subunits, one begins to

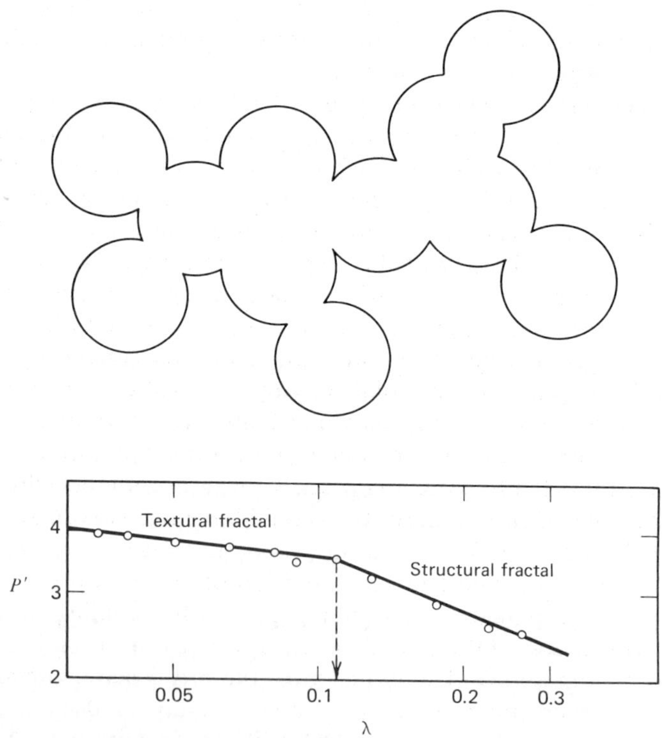

Fig. 10.23 Experiments conducted on a simulated carbon black structure have indicated that it is possible to delineate a structural fractal from a textural fractal (P' = perimeter estimate normalized; λ, stride length normalized).

approximate to a Euclidean perimeter that represents a fraction of the perimeter of the constituent circles. The data in Figure 10.23 are taken from a more extensive study (Kaye and Flook, in preparation) of the morphology of real and simulated carbon black agglomerates.

The graphical random walk procedure for generating a perimeter estimate is the most efficient procedure for carrying out that measurement. However, manual measurement of the fractal dimension by this procedure is time consuming and tedious. Fortunately, there are less efficient but unbiased procedures for measuring the fractal dimension of a curve that can be based on other image manipulation procedures. Thus Kaye has shown that the fractal dimension of a curve can be generated by the manipulation of a digital mosaic of the profile. Although such a procedure is approximately 17 times less efficient than the random walk strategy, it yields the same estimate of the fractal dimension of a convoluted profile. An alternative strategy using logic available on the Quantimet has been developed by Flook [52] at the Unilever Laboratories. Martens [53] has also indicated that several procedures for evaluating the fractal dimension can be implemented on the Omnicon.

Fractal dimensions can be used to describe the structure of agglomerates of crystals. Thus consider the crystals shown in Figure 10.24*a*, taken from a publication by McCarthy [54], who reports that the single elongated rectangular shapes can be considered to be free-standing crystals. The irregular shapes, however, are probably fused agglomerates. The description of the morphology of the agglomerated crystals is not easy by geometric waveform characterization. Since the agglomerates of crystals grow by a random joining process, it is reasonable to anticipate that the profile of an agglomerate of crystals could be describable by a fractal dimension. To explore this possibility, Kaye carried out experiments on selected clusters from the array shown in Figure 10.24. If an operator had been working with an out-of-focus picture or a poor picture or an inadequately thresholded automatic image analyzer, it is quite possible that the small crystal labeled 1*a* in Figure 10.24 could have been taken to be part of the overall cluster labeled 1. Therefore, Kaye investigated not only profile 1, but a profile constituted from the addition of 1*a* to 1 on the basis that this is the type of growth that would occur in the building up of a real cluster. The estimates of the fractal dimensions of these crystalline profiles as achieved by random walk exploration of the profiles are summarized in Figure 10.24*b*, each point represents three different random walks for a given stride length, each commencing at a different position on the profile. It can be seen that all four crystals have virtually the same fractal dimension. The use of fractal dimensions to describe rugged boundaries is a

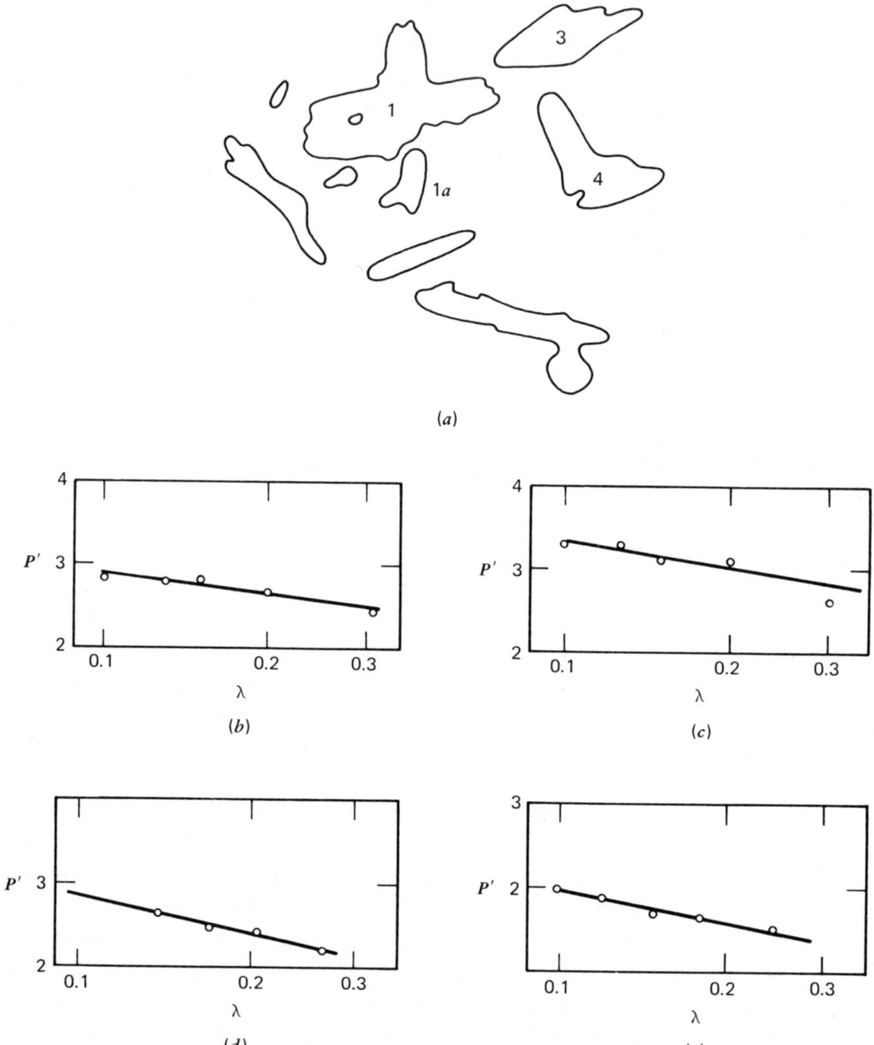

Fig. 10.24 Fractal dimensions can be used to describe the boundary of agglomerates of crystals: (*a*) crystal agglomerates described by McCarthy; (*b*) crystal 1, $\delta = 1.13$; (*c*) crystals 1 and 1*a* combined, $\delta = 1.13$; (*d*) crystal 3, $\delta = 1.17$; (*e*) crystal 4, $\delta = 1.20$. Symbols: *P*, perimeter estimate (normalized with respect to maximum Feret diameter); all fractals accurate to $\pm$ 0.03; λ, stride size normalized with respect to maximum Feret diameter; δ, fractal dimension.

rapidly expanding field of scientific endeavor; however, whether or not a specific convoluted profile has a fractal dimension has to be established empirically.

NOTES AND REFERENCES

1. R. Davies, "A Simple Feature Based Representation of Particle Shape," *Powder Technol.*, 12 (1975), 111–124.
2. H. H. Heywood, "Size and Shape Distribution of Lunar Fines Sample 12057, 72," in *Proceedings of Second Lunar Science Conference*, vol. 13, 1971, pp. 1989–2001.
3. H. Wadell, "Sphericity and Roundness of Rock Particles," *J. Geol.*, 41 (1933), 310.
4. H. Heywood, "Numerical Definitions of Particle Size and Shape," *Chem. Ind.*, 15 (1937), 149–154.
5. H. Heywood, "Particle Shape Coefficients," *J. Imp. Coll. Eng. Soc.*, 8 (1954), pp. 25–33.
6. British Standards 2955: Glossary of Terms Relating to Powders, British Standards Institution, London, 1965.
7. H. H. Hausner, "Characterization of the Powder Particle Shape," in *Proceedings of the Symposium on Particle Size Analysis*, Loughborough, England, 1967; published by the Society for Analytical Chemistry, London, England, pp. 20–27.
8. G. S. Riley, "In Search of an Easily Measurable Particle Shape Index," *J. Powder Bulk Sol. Technol.*, 1 (1977), 34–39.
9. C. Nyström, "The Use of the Ring Gap Sizer for Characterization of Particle Shape in the Sieve Range," *Powder Technol.*, 20 (1978), 83–87.
10. K. Ridgway and J. B. Scotton, "The Effect of Particle Shape on the Properties of Powders," paper presented at the Heywood Memorial Symposium, Loughborough University, England, 1973.
11. J. C. Ludwick, "Particle Shape Classification of Thickness Using Slotted Screens," *J. Sedimentary Petrol.*, 41 (1) (March 1971), 19–29.
12. M. H. Pahl, G. Schädel, and H. Rumpf, "Zusammenstellung von Teilchenform Beschreibungsmethoden," *Aufbereitungs-Technik*, (5) (1973), 7–11.
13. R. P. Loveland, Report to ASTM Committee E29.02 on Fineparticle Shape, 1975.
14. E. T. White, R. J. Batstone, J. Day, and P. G. Wright, "Modifications to Zeiss–Endter Particle Size Analyzer for Data Logging," *Powder Technol.*, 5 (1971–72), 127–130.
15. H. P. Schwartz and K. C. Shane, "Measurement of Particle Shape by Fourier Analysis," *Sedimentol.*, 13 (1969), 213–231.

16. A. I. Medalia, "Dynamic Shape Factors of Particles," *Powder Technol.*, 4 (1970–71), 117–138.

17. Those unfamiliar with the concept of the radius of gyration of a rotating mass should consult a good introductory text on physics such as D. Halliday and R. Resnick, *Fundamentals of Physics*, Wiley, New York, 1960.

18. B. H. Kaye, "The Use of Feret's Diameter Signature Waveform as a Shape Characterization Parameter in Fineparticle Science," paper given at the Symposium on Fineparticle Science held in Anaheim, California, March 1978 under the auspices of the American Chemical Society, Colloidal Science group (in press).

19. For a discussion of the use of the circularity shape factor, see L. Redman, F. A. Heckman, and J. E. Connolly, "A Practical Approach to the Reliable Quantitative Image Analysis of Carbonblack Electron Micrographs," *Proceedings of the Symposium on the Characterization of Fineparticles*, Salford University, England, September 1977; proceedings, Heyden and Sons, London, 1978.

20. B. H. Kaye, "Fractal and Shape Index Description of the Shape and Surface Structure of Metal Powders," in *Proceedings of 1980 Powder Metallurgy Conference*, Washington, D.C., June 22–27. Proceedings to be published by Metal Powder Industries Federation.

21. J. K. Beddow, M. D. Nasta, and G. C. Philip, "Characteristics of Particle Signatures," in *Proceedings of the International Conference on Powders and Bulk Handling Systems*, Chicago, May 1977. Proceedings, International Powder Institute, 222 West Adam Street, Chicago, 1977.

22. T. Church, "Problems Associated with the Use of the Ratio of Martin's Diameter to Feret's Diameter as a Profile Shape Factor," *Powder Technol.*, 2 (1968–69), 27–31.

23. H. M. Sutton and N. Bundalli, "A Suggestion for the Exact Determination of Particle Size and Shape," in *Proceedings of the Society for Analytical Chemistry*, January 1973, pp. 13–17.

24. R. Ehrlich and B. Weinberg, "An Exact Method for Characterization of Grain Shape," *J. Sedimentary Petrol.*, 40 (1) (March 1970), 205–212.

25. J. K. Beddow, A. F. Vetter, and K. Sisson, "Particle Shape Analysis," *Powder Met. Internat.*, Part I, 8 (2) (1976), 69–76; Part II, 8 (3) (1976), 107–109.

26. J. K. Beddow, G. C. Philip, and A. F. Vetter, "On Relating Some Particle Profile Characteristics to the Profile Fourier Coefficients," *Powder Technol.*, 18 (1977), 19–25.

27. J. K. Beddow and A. F. Vetter, "A Note on the Use of Classifiers in Morphological Analysis of Particulates," *J. Powder Bulk Solids Technol.*, 1 (1977), 42–44.

28. E. D. Blackham, "The Physics of the Piano," *Sci. Am.*, Vol. 213 (December 1965), 88–93.

29. T. P. Meloy, "Fast Fourier Transforms Applied to Shape Analysis of Particle Silhouettes to Obtain Morphological Data," *Powder Technol.* 17 (1977), 27–35.

30. A. G. Naylor and C. D. Wright, "Shape Analysis of Particle Profiles Using the Fourier Transform," paper presented at the Symposium on Particle Size Analysis held at the University of Salford, England, September 12–15, 1977; Proceedings, M. J. Groves, Ed., Heyden and Sons, London, 1978.

31. B. H. Kaye, *Morphology of Fineparticles*, Elsevier, Amsterdam.

32. J. Bodziony, "Cauchy's Projection Formula as a Basis for the Method of Determining Surface Area of Convex Bodies," paper presented at the Symposium of Particle Size Analysis held at the University of Salford, England, September 12–15, 1977; proceedings to be published by the Analytical Division of the Chemical Society (Great Britain).

33. B. H. Kaye and C. D. Wright, "Decision Matrix Design and Cybernetic Mode Operation of SHADOW, a System for Fineparticle Shape Characterization," *Powder Technol.* 18 (1) (1977), 27–30.

34. B. H. Kaye and C. D. Wright, "Strategy Considerations in Iconometric Characterization of Colloidal Particles," paper presented at the Symposium on Fineparticles, American Chemical Society Annual Meeting, New Orleans, Louisiana, March 21–24, 1977. Copies available from B. H. Kaye, Laurentian University.

35. B. H. Kaye, "Intelligent Machines for Controlling Fineparticle Systems," paper presented at PowTech 1977 (Harrogate, England), March 17, 1977; proceedings published by Powder Advisory Centre, London.

36. J. D. Redman, "Optical Aids to Pattern Recognition," in *House Journal of the Atomic Weapons Research Establishment*, Aldermaston, Berkshire, England, June 1967.

37. J. D. Redman and C. D. Reid, "Holography, Optical Information Processing and Fibre Optics," *J. Br. Nucl. Energy Soc.*, 8 (1967), 65–77.

38. B. H. Kaye, "Efficient Pattern Recognition in Fine Particle Science," in special issue of *Pattern Recognition*, Journal of the Pattern Recognition Society, May 1972, dealing with pattern recognition and fineparticle science.

39. B. H. Kaye and A. G. Naylor, "An Optical Information Procedure for Characterizing the Shape of Fine Particle Images," in *Pattern Recognition*, Vol. 4, Pergamon, London, 1972, pp. 195–199.

40. R. Murphy, "Optical Information Processing Procedures for Determining the Essential Features of Particle Shape," M.Sc. thesis, Laurentian University, Sudbury, Ontario, 1972.

41. A. G. Naylor, "Optical Information Procedures Applied to Powder Science and Technology," Ph.D. thesis, Reading University, England, 1973.

42. A. G. Naylor, B. H. Kaye, and G. A. Rubin, "Shape Analysis of Ceramic Profiles from Electron and Photomicrograph," paper given at the Particle Size Symposium organized by the Electrochemical Society in Boston, October 5, 1974; proceedings published by the Electrochemical Society.

43. B. P. Mandelbrot, *Fractals, Form, Chance, and Dimension*, Freeman, San Francisco, 1977.

44. For a popular discussion of the space-filling ability of fractal curves, see B. P. Mandelbrot in *New Scientist* (June 22, 1978), 808–810.

45. B. H. Kaye, "Specifications of the Ruggedness and/or Texture of a Fineparticle Profile by its Fractal Dimension," *Powder Technol.* 21 (1978), 1–16.

46. B. H. Kaye, "Characterization of the Surface Area of a Fineparticle Profile by its Fractal Dimension," paper presented at the Salford Conference on Particle Size Analysis, September 12–15, 1977; Proceedings, M. J. Groves, Ed., Heyden and Sons, London, 1978.

47. B. H. Kaye, "The Sequential Mosaic Amalgamation as a Strategy for Evaluating the Fractal Dimension of a Fineparticle Profile," (in preparation).

48. B. H. Kaye, "Fractal Dimensions and Fineparticle Science," in *Proceedings of the International Conference on Powders*, Chicago, Illinois, May 1978; published by the International Powder Institute, 222 West Adams Street, Chicago, Illinois.

49. B. H. Kaye, "Fractal Dimension—a Concept to Break a Mental Log Jam in Shape Characterization Studies of Fineparticles," article in the *Omnicon SCAN LINES,* 1 (3) (April 1978), 1–6 (a house journal of the Bausch and Lomb Corporation, Rochester, New York).

50. I. Stewart, *Concepts of Modern Mathematics*, Penguin Books, Harmondsworth, Middlesex, England; also, 41 Steelcase Road, West Markham, Ontario, Canada, 1975.

51. For an interesting review of Archimedes' technique for estimating the value of π, see D. G. Fink, *Computers and the Human Mind* (No. 43 in the Science Study series published by Anchor Books), Doubleday, Garden City, New York, 1966.

52. A. G. Flook, "The Use of Dilation Logic on the Quantimet to Achieve Fractal Dimension Characterization of Textured Surfaces," *Powder Technol.* 21 (1978), 295–298.

53. Private communication from A. Martens, Bausch and Lomb Corporation, Rochester, New York.

54. C. J. McCarthy, "Recent Advances in the Characterization of Particulates Using Automatic Evaluation of Shape Descriptors," *Microstruct. Sci.*, 4 (1976), 339–346.

AUTHOR INDEX

SUBJECT INDEX

Hershey Foods Corporation
COMMUNICATION CENTER
P.O. Box 805
Hershey, PA 17033